International Conference on Entrepreneurial Competencies in a Changing World (ECCW 2020)

SHS Web of Conferences Volume 90 (2021)

Ceske Budejovice, Czech Republic
19 November 2020

Editors:

Zdenek Caha
Michal Ruschak
Vendula Velková

ISBN: 978-1-7138-2412-1

Printed from e-media with permission by:

Curran Associates, Inc.
57 Morehouse Lane
Red Hook, NY 12571

Some format issues inherent in the e-media version may also appear in this print version.

For additional information, please contact EDP Sciences – Web of Conferences
at the address below.

EDP Sciences – Web of Conferences
17, Avenue du Hoggar
Parc d'Activité de Courtabœuf
BP 112
F-91944 Les Ulis Cedex A
France

Phone: +33 (0) 1 69 18 75 75
Fax: +33 (0) 1 69 28 84 91

contact-edps@webofconferences.org

Additional copies of this publication are available from:

Curran Associates, Inc.
57 Morehouse Lane
Red Hook, NY 12571 USA
Phone: 845-758-0400
Fax: 845-758-2633
Email: curran@proceedings.com
Web: www.proceedings.com

TABLE OF CONTENTS

https://doi.org/10.1051/shsconf/20219000001

INTERNATIONAL CONFERENCE ON ENTREPRENEURIAL COMPETENCIES IN A CHANGING WORLD (ECCW 2020)

19th November 2020
České Budějovice, Czech Republic

VŠTE

REGENSBURG

© Th e Authors, published by EDP Sciences. This is an open access article distributed under the terms of the Creative Commons Attribution License 4.0 (http://creativecommons.org/licenses/by/4.0/).

International Conference on Entrepreneurial Competencies in a Changing World (ECCW 2020)

http://www.podnikatelskekompetence.cz/

Editors: Zdeněk Caha, Michal Ruschak, Vendula Velková

About the conference ECCW 2020

Date of the conference: 19th November 2020

The conference is organized within the project Stabilization and Development of SMEs in Rural Areas (TL01000349) which is supported by the Technology Agency of the Czech Republic.

Target

The theme of the international conference is the stabilization and development of small and medium-sized enterprises in rural areas. The aim of the conference is to gather and discuss innovative ideas contributing to the development of this field at the European level with the help of scientists and practitioners. At the same time, to present specific measures applicable in the conditions of the Czech Republic.

Thematic Areas

Section 1: Stabilization and development of SMEs in rural areas

- Economic - financial management and administration
- Strategic management and decision making
- Human Resources SMEs
- Rural environment and SMEs
- Innovation
- Legislative-legal area
- Marketing as a tool for increasing the competitiveness of the company

Section 2: Support for start-up entrepreneurs and current trends strengthening the competitiveness of business entities

- Start-ups
- Business incubators and accelerators
- Corporate social responsibility
- Business ethics

About organizers

Institute of Technology and Business in České Budějovice

Website: http://www.vstecb.cz/en/

Institute of Technology and Business (VŠTE) has about 4,500 students of technically and economically-oriented study programmes, two Master´s and six Bachelor´s study programmes with a wide range of specializations in civil engineering, mechanical engineering, transport and logistics and economics. Moreover, there are about 2,000 participants in lifelong learning, who officially do not have the status of "student, but it is one of the important activities of the school. This way it is possible to study two-year programmes of the European Council of Business Administration, MBA in Financial Management and BBA in Business Valuation. VŠTE is a higher education establishment with large project activities the main goal of which is to fulfil the long-term intentions of the Institute. The Institute is currently focused on systematic and permanent improvements in quality, deepening the cooperation with the application sphere, infrastructure development, transferring research activities into the application sphere, and increasing the employability of graduates. The Institute's project activities have significantly increased over the past two years.

Faculties

- Faculty of Corporate Strategy
- Faculty of Technology
- School of Expertness and Valuation

Ostbayerische Technische Hochschule Regensburg

Website: https://www.oth-regensburg.de/en.html

Ostbayerische Technische Hochschule Regensburg (OTH Regensburg) offers its students an excellent basis for a successful career. The technical university of applied sciences has around 11,000 students. 50 practice-oriented bachelor and master degree programmes in engineering, business studies, design, architecture, health and social studies offer high quality teaching and research. Modern facilities, such as 120 state-of- the-art laboratories and an award-winning library located on an attractive campus provide the basis for successful studies. OTH Regensburg's areas of expertise are energy & mobility, information & communication, life sciences & ethics, production & systems and building & infrastructure. These areas are tied together by expertise in sensor technology and digitalization. Cooperative applied research spans fields such as renewable energy, sensor technology, mechatronics, bioengineering, computer sciences and automotive engineering.

Faculties

- Applied Natural Sciences and Cultural Studies
- Architecture
- Civil Engineering
- Business Studies
- Electrical Engineering and Information Technology
- Computer Science and Mathematics
- Mechanical Engineering
- Social and Health Care Sciences

Guaranteed by

PaedDr. Mgr. Zdeněk Caha, Ph.D., MBA, MSc.
Prof. Dr. Sean Patrick Sassmannshausen
Prof. Ing. Marek Vochozka, MBA, Ph.D.

Scientific Committee

https://doi.org/10.1051/shsconf/20219000001

Dr. Marina Shavrovskaya
Ural State University of Economics (RUS)

Mario Bogdanović, PhD, Senior research associate, College professor,
MSc in Economics & MA in Economics, MA in Psychology
(Istrian University of Applied Sciences in Pula, Croatia (HRV)

Dr. Anastasiya Pesha
Ural State University of Economics (RUS)

Johann Fabian Faltermeier, MBA
Ibi research an der Universität Regensburg GmbH (DE)

Dr. Koropets Olga
Ural Federal University (RUS)

Assoc. prof. PhDr. Ing. Ladislav Mura, PhD.
J. Selye University (SK)

Doc. Mgr. Jakub Soviar, PhD.
University of Žilina (SK)

Assoc. prof. Ing. Jiří Dušek, Ph.D.
The College of European and Regional Studies (CZ)

Prof. Ing. Alžbeta Kucharčíková, PhD.
University of Zilina (SK)

Assoc. prof. Ing. Jarmila Straková, Ph.D.
Institute of Technology and Business in České Budějovice (CZ)

Ing. Petra Pártlová, Ph.D.
Institute of Technology and Business in České Budějovice (CZ)

Ing. Marcel Kordoš, PhD.
Alexander Dubček University of Trenčín (SK)

https://doi.org/10.1051/shsconf/20219000001

Statement of Peer review

In submitting conference proceedings to *Web of Conferences*, the editors of the proceedings certify to the Publisher that

1. They adhere to its **Policy on Publishing Integrity** in order to safeguard good scientific practice in publishing.
2. All articles have been subjected to peer review administered by the proceedings editors.
3. Reviews have been conducted by expert referees, who have been requested to provide unbiased and constructive comments aimed, whenever possible, at improving the work.
4. Proceedings editors have taken all reasonable steps to ensure the quality of the materials they publish and their decision to accept or reject a paper for publication has been based only on the merits of the work and the relevance to the journal.

International Conference on Entrepreneurial Competencies in a Changing World (ECCW 2020), 19 November 2020, České Budějovice, Czech Republic

Proceedings editor(s): Zdeněk Caha, Michal Ruschak, Vendula Velková

Date and editor's signature

15. 12. 2020

17, avenue du Hoggar - PA de Courtabœuf – BP 112 - 91944 Les Ulis Cedex A (France)
Tél. : 33 (0)1 69 18 75 75 - Fax : 33(0)1 69 07 45 17 – www.edpsciences.org

The Complexity of Relationships within Business and the Importance of Common Ground

Anna Ābeltiņa[1,*], *Ketevan* Rizhamadze[2]

[1]EKA University of Applied Science, Rīga, LV-1019, Latvia
[2]Turiba University, Riga, LV-1058, Latvia

Abstract. In today's rapidly evolving digital and social media dominated world, traditional marketing approaches are becoming obsolete. Companies are trying to alter the workplace into a vibrant one that enhances the talent of the people and to manage complex relationships inside and outside the organization. The following paper illustrates the complexity of relationships that companies have to handle. It is argued that the embeddedness of relationships plays a crucial role, notably in the business-to-business (B2B) relationship and that finding common ground is pivotal. The nature of the research methodology adopted for this study is primarily quantitative. All analyses are based on secondary data obtained through books, articles, and publications. The findings of the research indicate that the B2B relationship is a complex and multi-layered subject. Furthermore, for conflict resolution, taking into consideration the cultural differences in the decision-making process is fundamental. The presented study is exploratory in nature and therefore cannot be generalized for the general public. The findings are important for managers and entrepreneurs who need to be aware of the complexity of the business-to-business context. Although the nature of the study is exploratory, the study is original because it highlights some vital aspects upon which managers can reflect and dwell on.
Keywords: B2B marketing; relationships; complexity; decision-making process; organizational structure; SMEs

1 Introduction

In the face of the rapid pace of change, an organization should mobilize knowledge and technological skills, as well as be able to generate new products, processes and services. While companies tend to be the arrangement of complex systems, structures and strategies, managing business and non-business, formal and informal relationships with other players is an overwhelming task for managers. Within the past decade, it has been empirically shown that company managers are responsible for handling complex relationship models and are liable for the responsiveness and the capability of staff members to grasp and succeed in a competitive environment. Previous research in the marketing field has consistently revealed that collaboration with customers, suppliers, research institutions and competitors is decisive for achieving success. In addition, empirical studies have shown that business relationships are correlated and interconnected with each other. Marketing relationships are a very old phenomenon and this topic is one of the most heavily researched subjects worldwide. Analysing the complexity of network relationships accounts for the fact that B2B marketing now involves networks of relationships in which parties are interactive and actively engaged in commencing and strengthening the relationship [10]. Organizations therefore function in a complex commercial environment with a different number of markets, where decisions are made within a set of different interrelationships, for which the organization is responsible to form positive relationships [20]. The interaction process between parties consists of specific episodes, in which the players have diverse goals in the exchange [28,5]. As a relationship unfolds, the parties involved in the process of exchange or in the exchange of elements thereof go through a process of adaptation, which tends to be the lasting deal that is unique to that relationship [5]. Even cautiously planned arrangements may create undesirable exchange relationships between business units. However, where planning exists, this may reflect a lack of trust and blunt the demands on the relationship. The greater danger perceived by some businessmen is therefore that one side would be performing their side of the deal and therefore losing what is called "flexibility" [25].

2 Background

By scrutinizing empirical studies in literature, it is possible to determine those variables that are fruitful predictors of relationship performance. Grayson conducted and presented the results of survey data from more than 685 direct-selling

* Corresponding author: anna.abeltina@gmail.com

agents, whereby the relationship between supervisors and supervisees was measured. The study results illustrate the role of conflict management and friendship in the work environment. Studies state that combining friendship and business in the same relationship can be beneficial, but that it can create conflict. The source of this conflict tends to be incompatible relational expectations. This article also addresses role theory and reveals that the theory is based on the premise that effective social interaction depends on a shared understanding of relationship rules—that is, the behaviours that are (in)appropriate for different people in different social situations. According to the findings, the source of conflict between business relationships and friendships is that the expectations of the former encourage an instrumental orientation whereas those of the latter encourage an exclusively intrinsic orientation [15]. In addition, Grayson argues that business relationships encourage an instrumental orientation and friendships encourage an intrinsic orientation, while Halpern expresses the idea that exchange between strangers may be characterized more as transactional exchanges, while between friends it may be seen as more relational [15,18]. Fiske and Taylor put forward the ideas of 'cognitive structures' and the 'cognition-behaviour relationship', whereby people might maintain mental models, thereby arguing that, how people behave in a particular situation depends on their definition of the relationship and on the personal goals, which they customize according to the situation [11]. In contrast, the friendship script, illustrated by Halpern, consists of a common interpretation of the voluntary nature of the relationship, of how to sustain the friendship in the future, of how to keep the 'liking' associated with the relationship, and how to achieve equality, mutual self-disclosure, and future interactions [19].

3 Methodology

The following research paper adopts a descriptive approach and is mainly focused on an extensive review and analysis of cases, articles and books on the complexity of relationships within the B2B sector. The research includes existing data that can be obtained from the internet, peer-reviewed journals, textbooks, government records and libraries. The intention was to identify some micro and macro factors from which substantial insights could be gained. The predominant research subject in this study was: What strategies do company leaders use to sustain their businesses and establish long lasting relationships? Quantitative research was deemed suitable for this research project. The overall goal of the research was to investigate a specific meaning rather than presenting the findings as absolute truth. The current research tends to be empirical, since it aimed to bring new understanding and meaning to the topic of the complexity of relationships within business and the importance of common ground. The research involved organizing, collating and analysing existing data samples for valid research conclusions. The research process started with the identification and evaluation of gaps in existing knowledge before the adoption of primary research to gather new information. As already highlighted, secondary research involved data assimilation from different sources - internet, libraries, archives, schools and organizational reports.

4 Results and Discussion

In a business-to-business setting, customer service and quality stems from products and services, the physical environment, human resources and service encounters. Some research has been conducted that pertains to appraisal of customer satisfaction to embrace both product related and service variables. These streams of research are discussed below. Relationships are about time and behind every relationship there is a history, which makes it what it is for the present and sets the guidelines for the future, no matter whether the relationship is long or short term [12,5]. A relationship, as defined by Bachnik, means a connection, whereby it is perfectly rational that a relationship is a connection to a higher-level, one beyond the relationship between oneself and another, but rather to a social order. For example, in Japan, relationships are seen not as the lowest level building blocks of society, which must be included into higher levels, but as organizing factors themselves [1]. Therefore, it makes it difficult for foreigners to break into the Japanese system as suppliers [18]. Friendship is bound by a set of alternative relationships that mark off its explicit extents and belongings and is abound within a given culture [22]. Mass marketing has given way to mass customisation, which in turn calls for companies to adapt to customers' needs quickly and to operate without geographic boundaries and to regulate multinational intra-organisational relationships [9]. Similar to many other paradigms, relationship marketing emerged as a new paradigm. Nevertheless, we can find many terms describing relationship marketing, like: "customer relationship management", "database marketing", "customer loyalty" and the most recent, "customer experience", which is the emotional embodiment of a customer to a brand [2]. Not meeting customer needs well and being indifferent to customer wants can cause reductions in sales for companies. Therefore, it becomes imperative for marketers to restructure their strategy and to follow and monitor the action of some customer segments. As a result, relationship marketing emerged [4]. The term customer relationship management (CRM) has been used since 1990s. We can distinguish 4 types of customer relationship management, as illustrated in Figure 1:

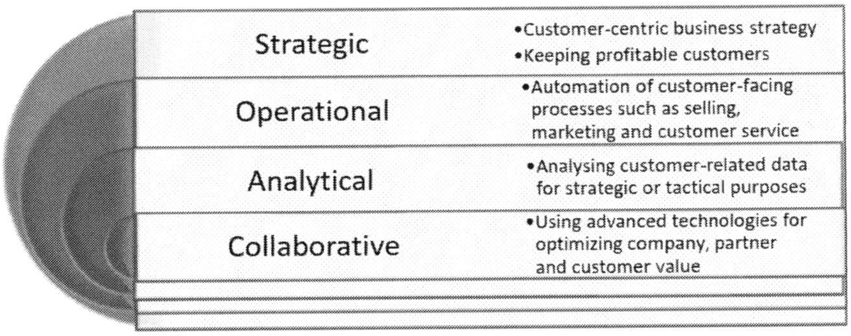

Figure 1. Types of CRM

Source: Authors based on Buttle [6].

Strategic CRM concentrates on building and establishing long lasting relationships with customers. For this purpose, the company orients its processes, technologies and people towards serving customers efficiently and effectively. Whereas, for operational CRM, the company atomizes inside procedures. For processing, interpreting, distributing, using and reporting customer-related data, analytical analyses are conducted. Examples of collaborative CRM are: category management, collaborative forecasting, joint new product development and joint market research [6].

A company has to deal with a set of relationships, that is a network of interactions. Relationships can develop as a result of the consequences of extensive collaboration and/or competition. Where the collaborative part is minor and is compensated by competition, it becomes important to divest the relationship or to work to reinforce it [16]. There is a strong correlation between customer retention and corporate profitability. The right and positive development of customer relationship marketing or customer management can be beneficial not only for companies, but for society as a whole [23]. As the main priority is retaining and sustaining long-term relationships with customers, relationship marketing therefore encompasses three levels:

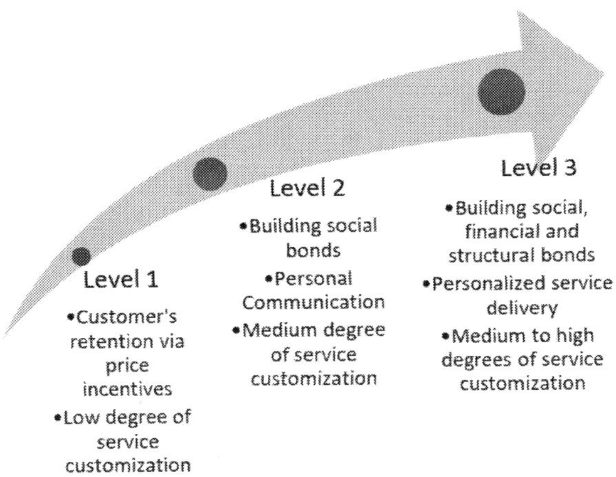

Figure 2. Three levels of relationship marketing

Source: Authors based on Berry and Parasuraman [3].

Marketing relationships should not only be seen as an exchange and interaction of networks. It should be viewed as an on-going relationship that deals with various channels of marketing and relationships. It encompasses the relationships between business marketing, marketing channels, service marketing, database marketing and direct marketing [26].

The following can be concluded on the basis of an analysis of relationship embeddedness and role theory. Montgomery suggests that role theory is a discrete alternative to rational choice theory, which might permit a more realistic demonstration of embeddedness [27]. Granovetter argues that behaviour has become much more autonomous with modernization, but is heavily embedded in social relations in premarket societies [14]. Frenzen and Davis introduced the theory of a broadened concept of utility as the basis for officially typifying embeddedness at the level of the individual purchaser [13]. The source of conflict between business relationships and friendships tends to be an instrumental or exclusively intrinsic orientation. However, commercial relationships may be attached more plainly in instrumental goals than work relationships are [29]. Embeddedness, interpersonal relationships and social bonds are crucial aspects of relationships [21]. Therefore, it is obvious to conclude that conflict is a natural and even inevitable aspect of most ongoing close relationships. So, the issue that differentiates successful from unsuccessful relationships is not whether there is conflict but how it is handled [7]. We should also take into account the fact that, adjustment of exchange relationships and dispute settlement by proceedings or the danger of it also has many costs. The gain anticipated from using this form of compulsion often fails to outweigh these costs, which are both monetary and non-monetary [25]. The Industrial Marketing and Purchasing Group (IMP Group) developed interaction approach based on the data collected from the interviews of 878 buyers and sellers from 318 firms in France, Italy, Germany, Sweden and the UK. As the results of the survey reveal, buyer-seller interaction is a dyadic interaction and the interaction at both firm and individual levels is influenced by the atmosphere, power and dependence cooperation, expectations, closeness and the environment [17]. In addition, it is concluded that it is pivotal that companies examine and assess each relationship be it national or international and to make decisions about organisational restructuring based on it [8]. Lewis and Craven examined the significance of 40 criteria in choice of a supplier and 18 criteria that would influence the replacement of an existing supplier. They found dimensions of service quality relating to product, the organization and its personnel to be the most important factors [24]. SMEs have a crucial role upon world economy and play a vital role in generating jobs and income. In this rapidly changing world, as customer needs change swiftly, SMEs are challenged to make unique selling propositions, to improve quality and to add value to products or services to accommodate those needs. For these organizations, it tends to be challenging to manage complex relationships with suppliers, customers and financial stakeholders. The capacity of SMEs to grow is rooted in abundant factors and none of them should be taken too lightly. The survey conducted by the authors related to innovation with the title: "Innovation as a pathway to business growth in SME's in Latvia" highlighted some aspects that could be linked to the present research. The research was analytical and the opinions of 110 SMEs from various manufacturing and service industries in Latvia were captured through an online survey. As the survey results revealed, 95% of respondents considered innovation to be the most critical internal factor for promoting SME growth. The second most crucial internal factor was human and technological resources. Quality improvement was selected as the most important platform for a company to drive its innovation practices, the percentage distribution was 35%. This was followed by market competition response, which was selected by 22 respondents. The results of the research show that various internal and external factors affect the development and growth of SMEs. Therefore, it is fundamental for SMEs to respond to customer needs by offering innovative services and products, as well as to draw up a strategy which takes into consideration the needs of all stakeholders and helps find common ground.

5 Research limitations

Taking into consideration that present research is exploratory in nature and it cannot be applied to general public, it has some limitations. Secondary information offers comparatively fast and inexpensive answers to many questions and generates some foundations for primary research. Nevertheless, research materials in secondary research are not exclusive to an individual or group. This means that everyone has access to the data and there is little or no information benefit obtained by those who attain the research. Consequently, coherent disadvantage was that, the data was not specifically designed to meet specific needs and were too general.

Even though the internet could afford excellent research materials, the web pages and commercial sites often has a marked bias and much of the information tends to be obsolete and not reliable. As a result, it became inevitable to conduct cross check comparison of the data from one organization with data from another source. It is important to note that this research relied on the outcomes of collective secondary research data and that it had the inherent risk of outdated research materials. In addition, it was hard to determine the authenticity of the data. Invalid data could have affected the research outcomes negatively hence, it was important to take extra care by evaluating the data before making use of it.

6 Conclusion

Organizations operating in a global environment are constantly looking to change their operations, whether it be in production or the service sectors, and to respond to worldwide challenges. From the moment a business enterprise is established, it follows a certain business model that bonds technical potential with the realization of economic value. The organization directly or indirectly develops a business model that defines the manner and means by which to create value. The quintessence of it is describing the method and process by which the services and products are developed and then delivered and supplied to customers. Subsequently, customers reimburse suppliers and service providers for the value received and these payments in turn convert into profits for organizations. Consequently, it gives the direction to an organization to scrutinize what customers desire and how those needs are met. One of the challenges companies have to overcome is to establish relationships with suppliers and clients based on trust and fruitful cooperation. It requires organizations to invest more in customer relationship management programs and to utilize all human resource potential for building a strong network of relationships. Those organizations that are capable of responding quickly and proactively to market changes are the ones who reap the benefits. The basic feature of relationships still remains communication because it is the social glue that bonds corporations together. However, communication has evolved over the years and has dramatically advanced due to novel technologies. Analysing the descriptive nature of the research, we can conclude that organizations that manage complex relationship patterns well are the ones who survive fierce competition. Also, relationship management requires common ground, a culture, a language, a set of common values and shared objectives. This article also illustrated the notion of a perfect friendship and concludes that there should be no hidden motives in the relationship and selfish and instrumental concerns should be avoided. In addition, conflict is seen as inevitable and natural to all organizations and it does exist. The success of an organization is determined by how well managers handle conflict and manage relationships in the face of those conflicts. Interpersonal relationships can facilitate problem solving and can guarantee social gratification and therefore enhance trust and assurance in a partnership and friendship. Likewise, to sum up the main findings of the research, relationships that are based on friendship are more sensitive to conflict than relationships that begin as a business relationship. Within this context, relationship embeddedness demands huge effort and time, whereby it is extremely hard to turn a business partnership into a friendship. Lastly, the study reveals that combining friendship and business in the same relationship can be advantageous, but that it can create conflict, which is often the result of a mismatch in interpersonal anticipations.

References

1. J. Bachnik, Time, Space and Person. In J. Hendry, *Japanese relationships' in Interpreting Japanese society anthropological approaches*. 2nd ed. London: Routledge, 94 (1998)
2. D. Bejou, A. Palmer, *The future of relationship marketing*. New York: Best Business Books (2006)
3. L. L. Berry, A. Parasuraman, *Marketing services: competing through quality*. New York: The Free Press, 136-139 (1991)
4. S. Biswas, *Relationship marketing: concepts, theories and cases*. Delhi: PHI Learning Private Limited, 2-4 (2014)
5. R. Brennan, L. Canning, R. McDowell, *Business-to-Business marketing*. London: SAGE publications ltd., 73 (2007)
6. F. Buttle, *Customer relationship management: concepts and technologies*. Oxford: Elsevier, 4 (2009)
7. D. Canary, W. Cupach, S. Messman, *Relationship conflict*. London: Sage publications (1995)
8. M. T. Cunningham, International Marketing and Purchasing of Industrial Goods Features of a European Research Project. *European Journal of Marketing*. **14** (5/6), 322-338 (1980)
9. B. Donaldson, T. O'Toole, *Strategic market relationships: from strategy to implementation*. Chichester: John Wiley & Sons (2007)
10. C. Fill, *Business-to-Business Marketing: Relationships, Systems and Communications*. Harlow: Financial Times Prentice Hall (2005)
11. S. T. Fiske, S. E. Taylor, *Social cognition*. 2nd ed. New York: McGraw-Hill Inc. (1991)
12. D. Ford, P. Berthon, P. Brown, *The Business Marketing Course: managing in complex networks*. Chichester: John Wiley (2002)
13. J. K. Frenzen, H. L. Davis, Purchasing Behaviour in Embedded Markets. *Journal of Consumer Research*. **17**(June) (1990)
14. M. Granovetter, Economic Action and Social Structure: The Problem of Embeddedness. *The American Journal of Sociology*. **91**(3), 481-510 (1985)
15. K. Grayson, Friendship versus business in marketing relationships. *Journal of Marketing*. **71**(October), 121-139 (2007)
16. E. Gummesson, *Total relationship marketing*. Oxford: Butterworth-Heinemann (2002)
17. H. Håkansson, *International Marketing and Purchasing of Industrial Goods: An Interaction Approach*. Chichester,: Wiley (1982)
18. J. Halpern, Elements of a script for friendship in transactions. *Journal of conflict resolution*. **41**(6), December, 835-868 (1997)
19. J. J. Halpern, The Effect of Friendship on Personal Business Transactions. *The Journal of Conflict Resolution*. **38**(4), 647-664 (1994)
20. T. Harwood, T. Garry, A. Broderick, *Relationship Marketing, Perspectives, Dimensions and Contexts*. Berkshire: McGraw-Hill Higher education (2008)
21. D. L. Haytko, Firm-to-Firm and Interpersonal Relationships: Perspectives from Advertising Agency Account Managers. *Journal of the Academy of Marketing Science*. **32**(3), 312-328 (2004)
22. D. Konstan, *Friendship in the classical world*. Cambridge: Cambridge University Press (1997)
23. P. Kotler, *Marketing insights from A to Z: 80 concepts every manager needs to know*. Hoboken, NJ: J. Wiley & Sons (2003)

24. B. R. Lewis, P. Craven, The Role of Customer Service in Buyer-Seller Relationships: Evidence from The Industrial Gases Market. In: P. W. Turnbull, D. Yorke, P. Naudé. *IMP Conference (11th): Interaction, Relationships And Networks: Past - Present – Future*. United Kingdom: Manchester (1995)
25. S. Macauley, Non-contractual relations in business: a preliminary study. *American sociological review*. **28**(1), 55-67 (1963)
26. K. Möller, H. Aino, Relationship Marketing Theory: Its Roots and Direction. *Journal of Marketing Management*. **16**(1-3), 29–54 (2000)
27. J. D. Montgomery, Toward a Role-Theoretic Conception of Embeddedness. *American Journal of Sociology*. **104**(1), 95-125 (1998)
28. P. Naudé, C. Holland, Business-to-Business Relationships. In F. Buttle, *Relationship Marketing: theory and practice*. London: Paul Chapman Publishing Ltd. (1996)
29. L. L. Price, E. J. Arnould, Commercial friendships: Service provider--client relationships in context. *Journal of Marketing*. **63**(4), 38-56 (1999)

Sustainability Management of Business Projects in Large Russian Companies

Svetlana Apenko[1,*], *Yulia* Fomina[1]

[1]Dostoevsky Omsk State University, Faculty of Economics, Mira Avenue, 55a, 644077 Omsk, Russia

Abstract. Large companies implement their business activities in the form of projects. The problem of implementing business projects based on the concept of sustainable development is becoming more urgent. The concept assumes management of business projects in such a way that allows a balance to be struck between the economic, social and environmental interests of groups of stakeholders on an institutional basis at the current time and over the long term. However, there are currently no universal and generally accepted developments in sustainability management. This study aims to develop and test a methodology for assessing and managing the sustainability of business projects that includes values, motives, and the level of maturity of sustainability. The novelty and originality of the methodology lies in the fact that it takes into account four sustainability criteria at once – economic, social, environmental and institutional. It allows the user to identify the level of sustainability, both on the basis of individual criteria and an integral indicator. It also corresponds to the current trend of studying the motives and values of implementing the concept of sustainability in projects. The proposed methodology and the test results from 36 large companies in Russia will help business project managers cope with the growing complexity of the social, environmental and economic environments.
Keywords: business projects; sustainability assessment; sustainability management

1 Introduction

Business projects related to the search for new ideas and the implementation of innovative technologies and products occupy a significant share of the activities of companies. A modern requirement for business projects, as is recognized worldwide, is sustainability, i.e. striking a balance between the economic, social and environmental effects now and over the long term.

This study focuses on the assessment of the sustainability management of business projects in large organizations. Although sustainability management is a new concept for many businesses in Russia, it has recently attracted more attention and is becoming popular. Sustainability management provides companies with the resources to cope with the growing complexity of the social, environmental and economic environments. Sustainability management is particularly relevant for large companies that are involved in large international business projects and programmes and are required to comply with sustainability requirements and report on sustainability. Large companies also use their projects to participate in the life of the regions in which they operate, so the image of business projects being sustainable becomes a significant social and economic factor.

The main and typical problem of the concept of sustainable development is its practical integration in people and processes. [1] It is important that sustainability becomes a core value and core competence of entrepreneurs, rather than a formal policy requirement. For many business projects, it is difficult to achieve a balance between economic, social, environmental and institutional principles. The problem of measuring the sustainability of business activities is also relevant.

The solution to the identified problems should be based on a well-developed theoretical basis. However, an analysis thereof shows that there are many unresolved issues.

2 Theoretical background

The sustainability management of business projects is based on the idea of sustainability. Sustainability is broadly defined as the ability of a system or process to sustain itself over a long period of time [2]. In this interpretation, sustainability is embedded in the concept of "sustainable development", proclaimed in the 1987 Brundtland Report as "our common future" and relevant at the current time [3,4]. Sustainable development can be defined as development

* Corresponding author: apenkosn@yandex.ru

[5,6]. The concept of sustainable development is aimed at enabling everyone to lead a decent life and protect the planet from damage [7, 8, 9]

Sustainability management can be defined as "organizational practices that lead to sustainable development" [10]. This practice is aimed at achieving a balance between the environmental, economic and social interests of different groups of stakeholders both in the short and long term [11, 12].

Sustainable development includes three principles: environmental sustainability, economic opportunity and social inclusion [4]. These three principles are also embedded in project management. For example, the Project Management Institute (PMI) defines sustainability as a new management paradigm that helps to cope with the complexity and dynamics of an organization. The principles of sustainable development - including economic, environmental and social; short, medium and long term; local, regional and global; as well as value-oriented ones - can be used in all project management processes, which improves the quality of project results [13].

The fourth principle of sustainable development is institutional. This principle was included by the UN Commission on Sustainable Development in its system of sustainability indicators. It consists of the commitment of a company and its projects to sustainable development through appropriate strategies and institutional framework for sustainability management.

Our research uses all four principles, although the institutional principle is currently poorly researched and belongs to the problem areas in the theoretical foundations of sustainability.

An analysis of existing research in the field of sustainability management allowed us to identify key problem areas, which then helped us formulate research questions.

The first problem is the lack of integration of a sustainable strategy into people's activities and management processes [14]. Business project managers and organizations do not always produce the expected sustainability values [15]. There is a gap between a sustainable strategy and the realization of the benefits thereof [16]. To solve this problem, it is necessary to assess and strengthen the motives for implementing a sustainability strategy in business project activities.

The second problem is the measurement of sustainability, as well as the development of universally applicable sustainability indicators [17, 18]. The methodological task, which is still under discussion and not solved, is to find indicators of project management sustainability that are reliable and accurate, but at the same time understandable for different target groups [19, 20]. Developing universal methods for assessing sustainability would simplify sustainability measurement and reporting for organizations [10]. A particular problem is the inclusion of economic, social, environmental and institutional criteria in the system of indicators [21, 22]. To solve this problem, it is necessary to develop indicators of project sustainability that meet these requirements.

3 Methodology

The identified practical problems and gaps in research allowed us to formulate the following research questions: what indicators allow us to comprehensively assess the sustainability management of business projects? What is the current level of sustainability in the business projects of large companies? What are the main motivating factors that encourage managers to implement the principles of sustainable development in management processes in Russia?

The purpose of our research is to develop a methodology for assessing and managing the sustainability of business projects, including values, motives and maturity level, as well as to test this methodology in practice.

The methodology for assessing sustainability is implemented using a new methodology developed by us. It involves assessing the level of maturity of sustainability over three modules. The first module includes questions for assessing the institutional criterion of sustainability management; the second module includes questions on the three dimensions of sustainability – economic, social and environmental; [2, 10, 11, 12, 13, 23] and the third module includes questions related to the motives for and values of sustainable development [10, 16]. The answers to the questions in each module are subsequently translated into point values. The novelty of our methodology and its difference from existing approaches is that it covers four principles of sustainability at once, thereby enabling users to evaluate the motives and values of sustainable business project management in practice.

To conduct the research, a semi-structured online survey was conducted of project managers in Russian companies. The dataset includes responses from 36 companies from twelve regions in Russia, including the cities of Moscow, Omsk, Yekaterinburg, Arkhangelsk, Chelyabinsk, Krasnoyarsk, Kazan, Kyzyl and Ulan-Ude.

The choice was made to study the business projects of large companies (more than 251 employees), which was reflected in the formulated hypothesis: large companies often implement a sustainability strategy and develop the institutional framework for sustainable business project management practices. The study confirmed this hypothesis. The organizations under study implement their projects in various sectors of the economy and include manufacturing, engineering, gas and chemical companies, as well as service companies involved in banking, IT, etc.

Eleven organizations implement their projects at the international level, eight at the national level, and seventeen at the regional level. All concern business projects related to generating economic benefits from innovative activities. The survey was conducted from November 2018 to October 2019.

4 Results and Discussion

The developed methodology enabled us to establish the level of sustainable business project management in large companies, which was determined as average. More specifically, the quantitative value of the level for all the studied companies was 23.47 points out of a maximum 40 points. To classify the results of the sustainability assessment, the following levels of sustainability management were determined:

30.1 – 40 points – high level;

20.1 – 30 points – average level;

10.1 – 20 points – below average;

0 – 10 points – low level.

The integral indicator was obtained by scoring values in two modules – module 1 institutional and module 2, which assesses the balance of economic, environmental and social sustainability criteria.

Module 1 questions related to the application of a sustainable development strategy; the time frame of the sustainable development strategy; whether the strategy focuses on the balance between environmental, economic and social interests of different groups of project stakeholders. Module 2 questions dealt with the relationship between the three principles of sustainable development - economic, social and environmental – and their role in the organization's strategy; planning the results and consequences of projects; and the economic, social and environmental goals and effects of projects.

Module 2 consisted of three submodules that were evaluated separately for economic, social and environmental sustainability criteria. The social submodule included issues related to training opportunities for project staff; gender equality; inequality in the distribution of staff income; indicators of social sustainability (human rights, poverty reduction, labour protection, support for trade unions, prevention of corruption and bribery, consumer safety, partnership between management and staff of the organization). The environmental submodule included questions about the internal assessment of the project's environmental impact at the planning stage; transport costs and cooperation with local suppliers; renewable energy; waste management; environmental sustainability indicators (water consumption, carbon dioxide emissions, other harmful gases into the atmosphere, energy consumption, materials and other resources). The economic submodule included issues related to the economic sustainability of projects; indicators for return on investment (ROI), net present value (NPV), cash flow, asset liquidity, market share, new jobs, taxes; economic efficiency of corporate social responsibility; economic efficiency of waste processing.

The maximum possible score for each module and submodule was 10 points:

7.6 - 10 points – high level;

5.1 – 7.5 points – average level;

2.6 – 5.0 points – below average;

0 – 2.5 points – low level.

The following results were obtained:

module 1 (institutional module) – 6.83 points, above-average maturity level;

module 2, submodule A (social module) – 5.21 points, average level;

module 2, submodule B (environmental module) – 4.53 points, average level;

module 2, submodule B (economic module) – 6.83 points, above average level.

The results of the study show that a significant proportion of large companies implement business projects on the basis of a developed institutional framework. First of all, they are implemented within the framework of a sustainable development strategy: 31 companies have a sustainability strategy. For the success of this strategy, management processes have been created, which include: setting and achieving a goal that assumes a balance between all sustainability criteria, managing functional areas (content, timing, risks, quality, personnel, and other project areas), taking into account sustainability criteria. Moreover, companies with a pronounced sustainability strategy have a much higher level of maturity in managing the sustainability of business projects compared to companies that do not have a sustainability strategy (Table 1). This suggests that the presence of an institutional framework is a crucial factor in achieving a high level of maturity in sustainability management.

 https://doi.org/10.1051/shsconf/20219001002

Table 1. Maturity level of sustainability management according to the presence of a sustainable development strategy (in points)

Presence of a sustainable development strategy	Institutional module	Social module	Environmental module	Economic module	Integral indicator of the level of maturity	Overall level of sustainability
All companies	683	5.31	4.53	6.81	23.47	Average
Companies with a sustainability strategy	7.71	5.92	5.13	7.29	26.05	Above average
Companies without a sustainability strategy	1.35	1.55	0.8	3.8	7.5	Low

Differences between groups are statistically significant: F-criterion = 6.225; p = 0.000.

Source: Authors.

The data in Table 1 show a higher level for the institutional and economic modules, which indicates that there is a slight imbalance in the sustainability criteria. This concerns the priority of economic interests over social and environmental indicators for the success of business projects.

A significant result was the conclusion about the impact of the type of business projects implemented on the basis of the sustainability indicators (Table 2). The overall level of sustainability of international projects was above average, whereas for national and regional levels it was only average. At the regional level, the project indicator for environmental sustainability is low. This result is due to the fact that international project managers are forced to comply with sustainability requirements and sustainability reporting, otherwise they acquire a negative reputation in international markets.

Table 2. Maturity level of sustainability management according to the type of project (in points)

Type of project	Institutional module	Social module	Environmental module	Economic module	Integral indicator of the level of maturity	Overall level of sustainability
All projects	6.83	5.31	4.53	6.81	23.47	Average
International projects	8.50	6.70	6.09	7.73	29.02	Above average
National projects	6.97	4.38	5.13	5.94	22.41	Average
Regional projects	5.68	4.85	3.24	6.62	20.38	Average

Differences between groups are statistically significant: F-criterion = 6.232; p = 0.000.

Source: Authors.

Module 3 included questions related to the motives for and values of project sustainability. Respondents chose the motives/values that could be possible reasons for implementing the sustainability management of business projects in large companies. Table 3 shows the distribution of responses. As is evident, the leading motive is related to the public opinion that the organization and its projects are considered sustainable. Other motives are more related to the need to respond to the requirements of sustainable development. The motive for internal changes in culture and thinking focused on sustainability is less marked.

Table 3. Motives for/values of sustainability management

	Motives for/Value of sustainability management	Absolutely agree	Partly agreed	Disagree
1	The public relations value of being seen as a sustainable organization	19	17	0
2	The growing cost of energy, materials and waste management	16	13	7
3	The growing body of environmental liability law and the costs of defence	17	13	6
4	The desire for a change in the culture and dominant social paradigm	11	17	8
5	The need to work with foreign partners and adhere to international standards	14	17	5
6	Digital transformation of the economy and society, the increase in open data access	17	17	2

Source: Authors.

The research presented here has therefore made a contribution to the development of the concept of sustainability management for business projects. In particular, the idea of the importance of measuring the three criteria of sustainability – social, economic and environmental was developed [2, 12, 13, 23]. The development lies in the proposed method for the quantification of these criteria using a wide range of questions and the combining of the individual estimates into a single integral indicator. In addition, the idea of the importance of taking into account the institutional criterion of sustainability was also developed [10]. The proposed methodology involves evaluating the institutional module, as well as establishing an integral indicator based on this criterion.

Beyond the above, the research also develops the idea of the importance of evaluating and developing the values and motives that stimulate companies and project managers to implement sustainable management. Thus, based on research by Cohen et al. [10, p. 22], we included questions based on motives in the methodology. This enabled us to study the basic motives, the understanding of which helps to better support the sustainability management of business projects.

5 Conclusion

One of the main results of the research presented in this contribution was the development and testing of methods for assessing the sustainability management of business projects in organizations. The advantage of the methodology is that it offers an assessment of four sustainability criteria, the three main dimensions (economic, environmental and social) plus the institutional dimension. The proposed methodology produces both an indicator for the individual criteria and an integral indicator that demonstrates the level of maturity of sustainability management.

Testing the method enabled us to obtain significant results in terms of the understanding and further development of the practice of business project sustainability management in companies. In general, the level of maturity recorded was average. However, those companies that implement a sustainability strategy and form the institutional framework for it, have a significantly higher level of maturity. The level of sustainability is higher in the institutional and economic dimensions than in the social and environmental dimensions. The environmental dimension received the lowest scores. Regional projects of companies have particularly low indicators for environmental sustainability. In other words, an imbalance was recorded in the sustainability criteria, which should be the basis for decision-making in the field of sustainability management regarding the business activities of companies.

Among the most important motives/values was the realization that implementing sustainability management in practice leads to the success of an organization in Russia. Within this context, respondents noted the value of public relations in being seen as a sustainable organization.

This study has its limitations, including the small number of participants. The thirty-six companies studied were enough to test our estimation methodology, but not enough to develop a statistical model. Future research will be based on a larger number of participants from different regions of Russia.

References

1. P. H. Dos Santos, S. M. Neves, D. O. Sant'Anna, C. H. de Oliveira, H. D. Carvalho, The analytic hierarchy process supporting decision making for sustainable development: An overview of applications. *Review. Journal of cleaner production.* **212**, 119-138 (2019)
2. B. Moldan, A. L. Dahl, Challenges to Sustainability Indicators. *Sustainability Indicators: a scientific assessment. SCOPE 67.* Washington: ISLAND PRESS, 1-24 (2007)
3. World Commission on Environment and Development (WCED). *Our Common Future.* Oxford: Oxford University Press (1987)
4. M. Del Baldo, M.-G. Baldarelli, Renewing and improving the business model toward sustainability in theory and practice. *International Journal of Corporate Social Responsibility.* **2**(3) (2017)
5. D. Opresnik, M. Seregni, M. Taisch, The Cornerstone of Sustainability Strategy in Manufacturing Companies. In: B. Grabot, B. Vallespir, S. Gomes, A. Bouras, D. Kiritsis (eds) *Advances in Production Management Systems. Innovative and Knowledge-Based Production Management in a Global-Local World.* **439**, 500-507 (2014)
6. A. Ismail, B. Johnson, Managing Organizational Paradoxes in Social Companies: Case Studies from the MENA Region. *Voluntas.* **30**, 516–534 (2019)
7. A.-L. Müller, R. Pfleger, Business transformation towards sustainability. *Business Research.* **7**, 313–350 (2014)
8. A. Marcon, J. F. de Medeiros, J. L. D. Ribeiro, Innovation and environmentally sustainable economy: Identifying the best practices developed by multinationals in Brazil. *Journal of Cleaner Production in September.* **160**, 83-97 (2017)
9. S. Cohen, W. Eimicke, A. Miller, *Sustainability Policy. Hastening the Transition to a Cleaner Economy.* Hoboken: J. Wiley & Sons (2015)
10. J. Carboni, M. González, J. Hodgkinson, *PRiSM. Project integrating Sustainable Methods. The GPM® Reference Guide to Sustainability in Project Management.* Richmond: GPM Global (2013)
11. G. Silvius, M. Kampinga, S. Paniagua, H. Mooi, Considering sustainability in project management decision making: An investigation using Q-methodology. *International Journal of Project Management.* **35**, 1133–1150 (2017)
12. M. Robertson, *Sustainability Principles and Practice.* 2nd edition. London: Routledge (2017)

https://doi.org/10.1051/shsconf/20219001002

13. R. Gareis, M. Huemann, A. Martinuzzi, *Project Management and Sustainable Development Principles*. Project Management Institute (2013)

14. V. L. M. Schropfer, J. Tah, E. Kurul, Mapping the knowledge flow in sustainable construction project teams using social network analysis. *Engineering, Construction and Architectural Management*. **24**(2), 229-259 (2017)

15. H. Kopnina, Sustainability: new strategic thinking for business. *Environment Development and Sustainability*. **19**, 27–43 (2017)

16. K. Kohl, *Becoming a Sustainable Organization. A Project and Portfolio Management Approach*. 1st ed. New York: Taylor & Francis (2016)

17. S. Karlsson, A. L. Dahl, et al. Meeting Conceptual Challenges. *Sustainability Indicators: a scientific assessment. SCOPE 67*. Edited by T. Hák, B. Moldan, A. L. Dahl, Washington: ISLAND PRESS, (2007)

18. M. Pislaru, I. V. Herghiligiu, I .B. Robu, Corporate sustainable performance assessment based on fuzzy logic. *Journal of cleaner production*. **223**, 998-1013 (2019)

19. R. B. Gibson, *Sustainability Assessment: Applications and opportunities*. London: Routledge (2016)

20. T. Bauler, I. Douglas, et al. Identifying Methodological Challenges. In: T. Hák, B. Moldan, A. L. Dahl, *Sustainability Indicators: a scientific assessment. SCOPE 67*. Washington: ISLAND PRESS, 49-64 (2007)

21. M. Gerner, Assessing and managing sustainability in international perspective: corporate sustainability across cultures – towards a strategic framework implementation approach. *International Journal of Corporate Social Responsibility*. **4**(5) (2019)

22. C. Malesios, P. K. Dey, F. B. Abdelaziz, Supply chain sustainability performance measurement of small and medium sized companies using structural equation modeling. *Annals of Operations Research*. **294**, 623-653 (2018)

23. R. Maltzman, D. Shirley, *Green Project Management*. Boca Raton, FL: CRC Press (2010)

Change Management Strategy for the Activities of Business Organizations

Irina Avdeeva[1,*], *Tatyana* Golovina[1], *Andrey* Polyanin[1]

[1]Central Russian Institute of Management, Branch of RANEPA, Department of Management and Public Administration, 302028 Orel, Russia

Abstract. Within the context of modern management, business organizations need to realize that everything around them is changing: the rules of the game, the participants, the scale of activities, and the size of markets. In other words, nothing is constant. The changes may be both local and global, whereby developments may affect different fields: executive and administrative, technical and economic, social and public, informative and analytical. Managerial decisions taken inside an organization under the aforementioned conditions require sound theoretical grounds for selection in terms of potential risks and current and future economic effects. Traditionally, long-term decisions on setting goals for an activity of a business organization and determining ways to achieve them are considered within the framework of strategic planning, as a result of which a development strategy is formed. However, under current circumstances, the place and role of change management in the strategy of business organizations requires reinterpretation. The academic originality of this contribution lies in the specification of a concept for change management with regards to the activities of business organizations based on a strategic approach and adaptation to dynamically changing market conditions, including long-term planning. As part of the study, the essence of a strategic approach to change management regarding the activities of business organizations is considered, the principles of building a change management strategy are highlighted, and an algorithm for its development and implementation is proposed.
Keywords: change management, strategy, strategic management, business organizations

1 Introduction

Within the context of the constant search for new forms and methods of work, the relevance of improving the dynamic development of the management mechanism is increasing. At the same time, a comprehensive study of the possibilities of increasing the competitiveness of business structures on the basis of the effective management of changes in their activities is becoming a promising area of academic research.

Modern business structures function and evolve under conditions of economic uncertainty and instability, which affects many processes, which is happening both globally and within the Russian economy.

This highlights the importance of logical approaches to strategic change management in order to minimize the dangers and risks generated by negative factors and to add and benefit from the effects of positive factors.

In the case of Russia, the process trends taking place within the sphere of national entrepreneurial activities (especially in SMEs) is not very positive, which is certainly having an effect on economic relations. After considerable growth in the entrepreneurial sphere in 2018, this slowed and stabilized in 2019. When assessing the overall dynamics of the number of business organizations, this decreased by 0.3%. In sectoral context, no large changes were observed, with exception to the mineral industry where the number of business organizations increased more actively. There are several reasons to explain this, including:

- global changes and the competitive recovery of the market for natural and energy resources;
- reduction of actual incomes of Russian households;
- deficit of financial assets needed for development;
- existence of considerable red-tape and executive barriers hampering the development of active business organizations and the foundation of new ones, this despite the fact that there are different federal and regional programs focused on business development and the improvement of business activity;

* Corresponding author: i-avdeeva-i@yandex.ru

- established segment structures, whereby not only individual enterprises, but also parts of entire industrial sectors remain insufficiently receptive to innovation implementation.

The goal of the research is to formulate a model of reasoning that describes the management of changes in the activities of business organizations from a strategic point of view.

2 Methodology

Within the framework of the research, a strategic review of research articles on change management was conducted. Based on the results obtained, the key principles of change management strategy for the activities of business organizations was determined, and recommendations for the implementation of a mechanism for developing a change management strategy for increasing stability and competitiveness given.

The basis of the theoretical and methodological study was academic papers by both national and international scholars. The papers touch upon management decisions regarding the development of business organizations and the enhancement of efficiency.

In general, development strategy implies how an organization implements and adheres to it in order to fulfil its main long-term goals. The authors agree with the conclusions of P.A. Argenti, who focuses on the development difficulties companies face from competitors when they only take tactical decisions aimed at achieving short-term goals, without having any strategy in place [1].

At the same time, one cannot but agree with the conclusions of Charles Lindblum, who demonstrates, on the basis of the incremental model of decision-making, that in highly uncertain, turbulent conditions, the decision-making process should be carried out through an indefinite and disordered series of small movements from "pain points" of the day, and not on the basis of predetermined goals that usually remain unattainable. Moreover, in situations of uncertainty, G. Simon shows, through his descriptive model, which takes into account personal experience and the insight of management personnel, intuitive decision-making wins.

At present, strategic management, which has quite a wide theoretical base, is a study subject for many scholars.

American authors A. Thompson and A.J. Strickland give the following model: company strategy - a combination of business organization methods aimed at achieving company goals and customer interests. Scholars point out the fact that working out a development strategy is the main task of management, thereby taking into consideration that effective management is not only based on the choice of a successful strategy and its correct implementation, but also on the unexplored impact of transformation processes at certain economic lifecycle stages as a result of the influence of the external environment [2].

According to H. Mintzberg, a strategy is a consistent, agreed and integrated structure for management decisions [3]. At the same time, scholars assign a leading role to the potential of personnel in business organizations. They believe that middle and lower-level leaders play a huge role in the implementation of changes. Teams, projects, departments and divisions managers have a unique and influential role to play. It is these leaders who must communicate the vision for change.

In literature, different approaches exist for determining the content of the stages of strategic management.

The most widespread are the models of F. David, A. Thompson and R. Lynch. David's model assumes the sequential implementation of 3 stages: strategy development, strategy implementation and strategy evaluation [2, 4]. At the same time, no attention is paid to the most important component of the strategic approach, namely the control and monitoring stage.

The problem of change management is reflected in numerous domestic and foreign works. For this study, the authors relied on the works of R. Kh. Hall, M. Hammer, D. Harrington, W. Bridges, K. Levin, T. Cummings, K. Worley, E. Cameron, P. Cotter, Rosabeth Moss Kanter, and many others. The emphasis in these studies is that many organizations have mastered the operational or structural side of change, but that they pay little attention to the human dimension of change. To obtain the desired results from a new system, initiative or a new direction, organizations need to master not only the organizational and management aspects of change, but also to understand the issues of leadership in change.

From this point of view, the research conducted by Rosabeth Moss Kanter deserves to be highlighted. She reasonably argues that no matter what strategy top managers develop, it cannot be implemented in practice if managers do not encourage subordinates to search for management decisions.

The problem within this study field is that most researchers ignore the process of change management when they consider the management mechanisms that can radically transform the activities of business organizations.

The methodological basis for the research was the theory of strategic management and the dynamic abilities of business organizations. The main methods of work are structural and logical, bibliographic and criteria analysis.

The presented concept is focused on the need to include a change management strategy in corporate strategic management. At the same time, it is necessary not only to take into account the methodology of strategic management, but also to formulate specific principles and methodological approaches that take into account the nature of changes and the magnitude of their influence on the dynamic development of business organizations. Until now, domestic and foreign scholars have not formulated the basic postulates of change management integration into the development strategy of economic systems of various levels. Currently available scientifically based methodological approaches are fragmented and focused on individual stages of the lifecycle of business organizations or specific strategic management decisions.

3 Results and Discussion

Strategic management has become an integral part of business structures and its absence not only complicates their function, but even makes it impossible.

In determining its development strategy, the administration of a business organization should clearly know the direction of travel, i.e. what should be continued and what should be implemented [5]. The answers to these questions help a company determine their strategy. The choice of business development strategy, as a development stage, is an integral part of strategic management, which enables a company to move towards its goal in the best possible way. It is the right choice of strategy that helps manifest the skills of company management, its competence and professionalism [6]. Returning to the stages of working out a strategy, and with consideration for the essence of strategic management, one can see that this activity is based on the collection, processing, analysis and exchange of information both with the internal and external environments.

Big companies are not able to react quickly to the changes in the Russian economy. Small businesses are the impetus for the general state of affairs in the Russian economy. They are the most sensitive to an increase or decrease in the level of profits within various economic sectors.

Within the context of ineffective reprocessing, large companies export fixed capital. Small companies, unlike large ones, do not have such opportunities. This contributes to the need for them to quickly revise their production and marketing activities. At the same time, improving the efficiency of their activities and their ability to overcome crisis situations is important for the country's economic recovery.

Under current market conditions, it is therefore vitally important to competently manage transformations in business organizations and increase the efficiency of their activities. It is therefore clear that there is a need to specify methodological approaches to planning a strategy for the change management of activities of business organizations [7].

The key conflicts that business organizations face on the path towards change management, as well as their characteristics, are presented in Table 1.

Table 1. Key contradictions brought about by organizational change

Conflict	Content
Between desired and needed	This is characterized by the motivational struggle between what needs to be done at the moment and what one wants to do. In most cases, the process of change dictates the need for specific actions, which often require making strong-willed executive decisions. However, there are often cases of voluntary reorganizations.
Between traditional and new	This is the conflict between traditional, often practiced and automatic actions and innovations, which require additional efforts to muster and implement them.
Between current and future benefits	This is explained by the fact that operational activities and processes that proceed according to an established scheme deliver benefits in the current period and ensure business' operations, whereas the introduction of changes promises to bring benefits in the future, but without guaranteeing them.
Between the top management and the staff	This is characterized by a three-level system of resistance factors: unwillingness to overcome difficulties, take on additional obligations; high group cohesion in a situation where the goals of the group and the organization do not coincide; lack of a clear strategy for the development of the organization; imperfect system of organizational communications.

Source: Authors.

Due to the fact that the transformation process of business organizations under current operating conditions is of a constant nature, organizations therefore face difficulties and controversies because of change [8]. Although certain management decisions are viewed as an adequate response to such challenges in the micro- and macro-environment, they in turn generate even greater changes [9].

Successful change management inside business structures means the adoption and implementation of managerial decisions, which is a sufficient condition for a desired outcome and the achievement of goals with respect to resources of the company.

The analysis of the approaches to the development of change management strategy allowed us to come to the conclusion that it is necessary to establish scenario- and project-based approaches and to ensure conformity with the company-wide business development strategy [11].

The scientific explanation of the methodological developments under such conditions is predetermined by the theoretical basis, which, as a rule, is a set of principles that must be taken into account (see Figure 1).

In order to properly understand the strategic plan of a company in terms of initiating and implementing measures for the adaptation of business structures in a dynamic and unpredictable environment, it is necessary to talk about change management strategy within the context of having a functional strategy, one that is subordinate to the general corporate strategic goals and principles [12], while having its own set agenda for a determined time period [13]. The aforementioned strategy should therefore conform with (or be developed within the set boundaries of) the general corporate strategy [14] in order to:

1) promote or ensure the achievement of the goals adopted under the general corporate strategy;
2) establish the basis for further development of competitiveness inside business organizations;
3) maintain and strengthen current positions of business organizations (market, technological, financial, etc.).

At the same time, the process of keeping and strengthening the current position of the company should not be considered separately from strategic planning because current or expected issues must be dealt with using the appropriate tools for the development, composition and implementation of both the general corporate and functional strategies (see Figure 2).

The role of a change management strategy is subject to the strategic goals of business organizations and therefore changes accordingly [15, 16, 17], it may range from a complete profile change to the maintenance of current activities and sustainability.

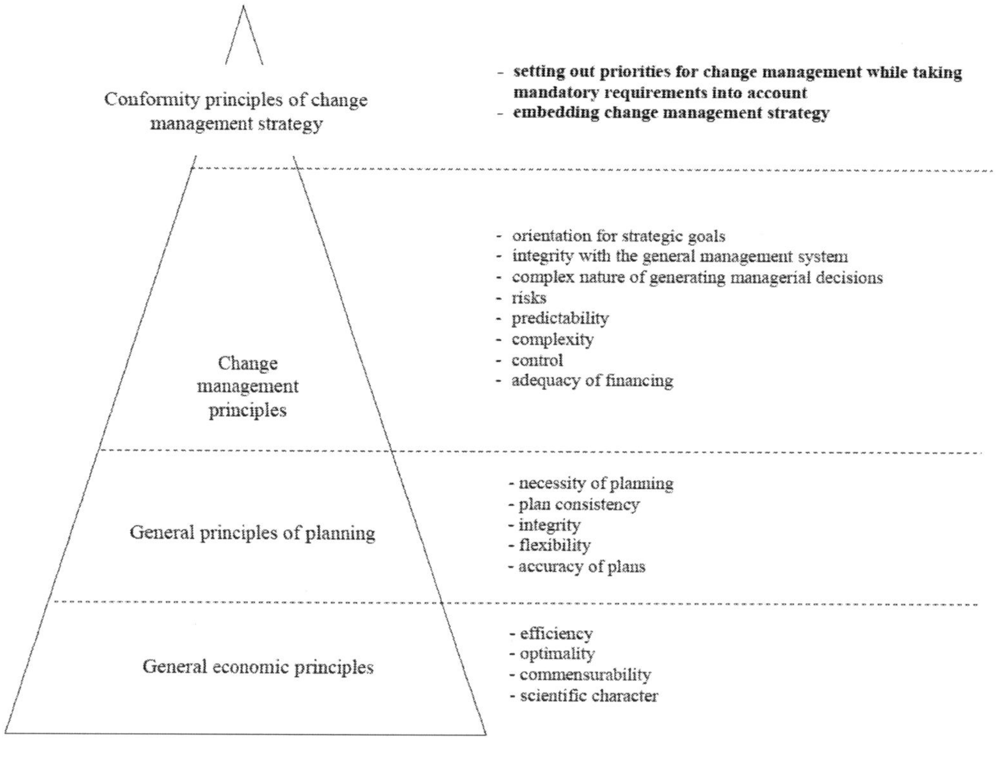

Figure 1. Recommended principles for the development and implementation of a change management strategy in business organizations

Source: compiled by the authors.

4 Conclusion

The uncertainty and instability of the external environment has a significant impact on the quality of the management activities of business organizations. This is what is driving the increased interest in this field, both in terms of theory and entrepreneurial activities in practice.

The above referenced framework actively employs the advantages of a strategic approach to change management because it:

- presents an opportunity to combine and sustain subsystems relative to various levels of complexity;
- conveniently demonstrates existing advantages and disadvantages;
- ensures self-management and implementation of information and communication technologies with the perspective to minimize human participation in the analysis and elaboration of the best possible solution based on the elements of artificial intelligence.

The theoretical significance of the study lies in the improvement of business management based on the use of a strategic approach, one that is appropriate for a period of economic instability. The proposed change management tools make it possible to effectively solve management problems in order to fulfil the economic interests of business organizations.

The practical importance of the study is driven by the use of the original algorithm for change management strategy development, which enables improvements in the sustainability and competitiveness of business organizations.

Future research will be linked to increasing the efficiency of change management through the active use of information systems, including artificial intelligence and neural networks.

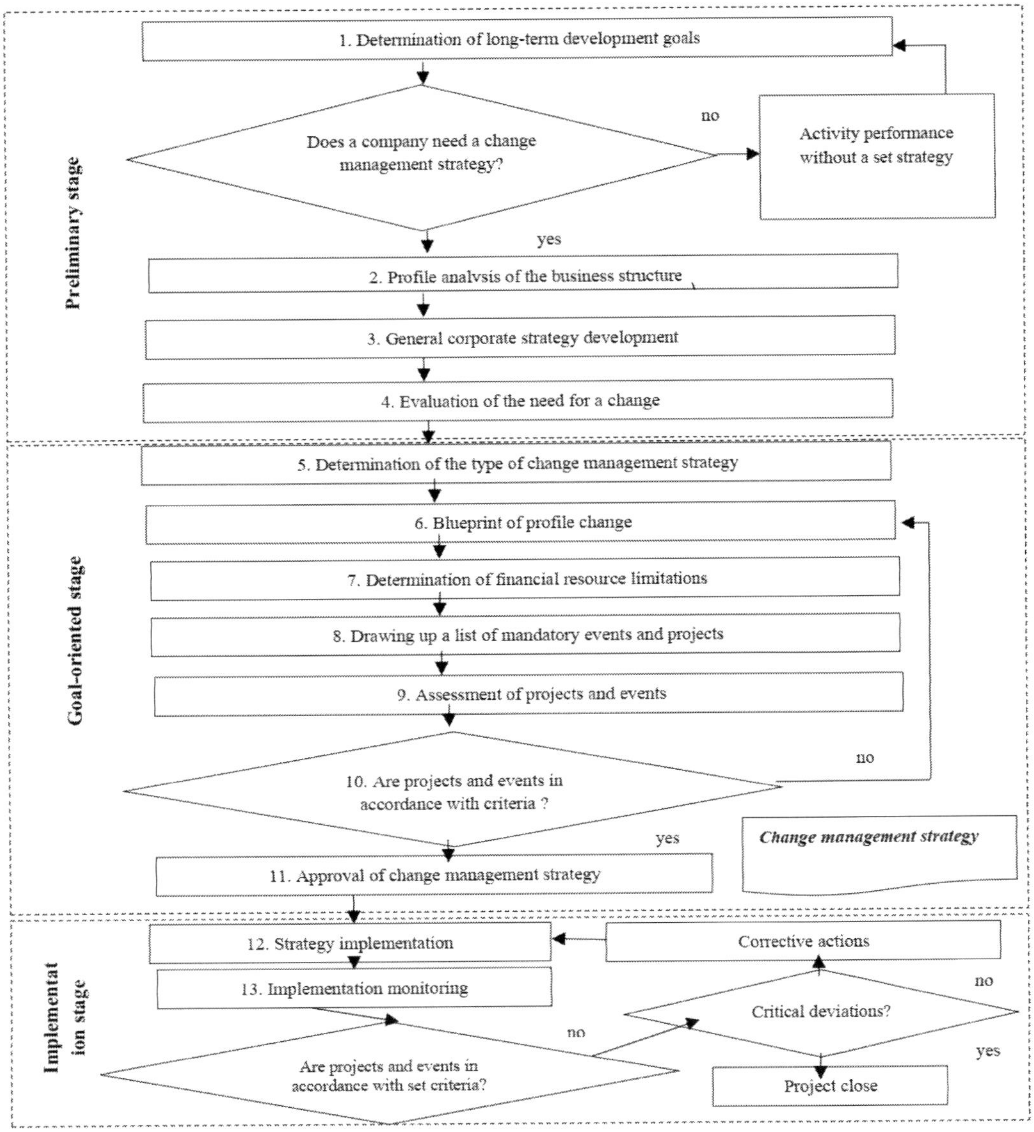

Figure 2. Algorithm for change management strategy development and implementation in business organizations

Source: compiled by the authors.

References

1. P. A. Argenti, R. A. Howell, K. A. Beck, The strategic communication imperative. *MIT Sloan Management Review*. **46**(3), 82–90 (2005)
2. A. Thompson, A. J. Strickland, *Crafting and Implementing Strategy: Text and Readings*. New York: Irwin Professional Publishing (1997)
3. H. Mintzberg, J. A. Waters, Of strategies, deliberate and emergent. *Strategic Management Journal*. **6**(3), 257–272 (1985)
4. A. Awang, A. Kassim, A. M. Noor, N. Shukor, A. Z. Shaari, S. Amran, S. A. Khalid, Strategic Entrepreneurship Model for Economic Transformation: Malaysian Evidence. *Asian Social Science*. **11**(7), 19–34 (2015)
5. M. A. Hitt, R. D. Ireland, R. E. Hoskisson, *Strategic management: Competitiveness and globalization*. Stamford: Cengage Southwestern Publishing Co. (2004)
6. S. Elbanna, B. A. Katheeri, M. Colak, The harder firms practice strategic management, the better they are. *Strategic Change*. **29**(5), 561–569 (2020)
7. F. Hofmann, M. Jaeger-Erben, Organizational transition management of circular business model innovations. *Business Strategy and the Environment*. **29**(6), 2770–2788 (2020)
8. H. Linkweiler, Change management and post-acquisition considerations for smoothly implementing new management strategies. *Journal of Airport Management*. **13**(1), 31–38 (2019)
9. D. Binci, S. Belisari, A. Appolloni, BPM and change management: An ambidextrous perspective. *Business Process Management Journal*. **26**(1), 1–23 (2019)
10. F. Deutscher, F. B. Zapkau, C. Schwens, M. Baum, R. Kabst, Strategic orientations and performance: A configurational perspective. *Journal of Business Research*. **69**(2), 849–861 (2016)
11. V. M. Repnikova, O. N. Bykova, P. V. Stroev, D. E. Morkovkin, S. N. Kurbakova, Improvement of approaches to the strategic development management of enterprise. *Entrepreneurship and Sustainability Issues*. **8**(1), 301–312 (2020)
12. I. Vandangeon-Derumez, A. Djedidi, E. Szendy, An experiential approach to learning about change management. *Journal of Management Development*. **38**(9), 708–718 (2019)
13. N. Yunita, N. Nasir, M. Adam, N. Darsono, A structural relationship of change management and its supply change strategy for improving officer and organizational performance. *International Journal of Supply Chain Management*. **9**(4), 1210–1221 (2020)
14. M. C. Díaz-Fernández, M. R. González-Rodríguez, B. Simonetti, The Moderating Role of Top Management Team Diversity in Strategic Change in a Multicultural Context. *European Management Review*. **6**(4), 957–973 (2019)
15. R.-G. C. Dirksen, The strategy of communication in organizational development 2.0. *Gruppe. Interaktion. Organisation. Zeitschrift fur Angewandte Organisationspsychologie*. **51**(3), 345–352 (2020)
16. G. Cattani, J. F. Porac, H. Thomas, Categories and competition. *Strategic Management Journal*. **38**(1), 64–92 (2017)
17. M. Y. Veselovsky, M. A. Izmailova, E. N. Lobacheva, P. P. Pilipenko, G. A. Rybina, Strategic management of innovation development: Insights into a role of economic policy. *Entrepreneurship and Sustainability Issues*. 7(2), 1296–1307 (2019)

Hiring Graduates for Marketing – Advantages, Disadvantages and Competency Requirements

Radim Dušek[1,*], *Monika* Hrubá[1], *Karolína* Brodská[1], *Petra* Skálová[1]

[1]Institute of Technology and Business, Faculty of Corporate Strategy, Department of Tourism and Marketing, Nemanická 436/7, 370 10 České Budějovice, Czech Republic

Abstract. The selection of suitable employees is becoming more and more demanding because of the increasing number of job applicants that want to get into a marketing career. The aim of this contribution is to find out which marketing competencies a competent graduate of economics should possess to be attractive for marketing related jobs in SMEs in the Region of South Bohemia, as well as to identify the positives and negatives of hiring such graduates. Based on data obtained from 93 SMEs with their own marketing department, the desired competent graduate for marketing related jobs can be described as a communicative and creative individual with the ability to analyse information. The main advantages of hiring graduates include creativity, the ability to work with modern technologies and/or enthusiasm for learning. On the other hand, a lack of experience and/or the longer training time are the main disadvantages of hiring fresh graduates for marketing-related jobs.
Keywords: competency models; marketing competencies; graduates; HR; SME

1 Introduction

In developed countries, major efforts have been made by universities to prepare graduates of business or economic degrees, and who want to be marketers, to meet actual labour market requirements. Now, more than ever, if candidates for marketing related jobs, including university economics graduates, want to start a successful career in, for example the marketing communication field, they should possess key competencies, as determined by employers. These requirements are no longer based only on theoretical knowledge, but especially on higher level abilities, ones which provide a framework for the behaviour and skills that are expected of competent marketers in various types of marketing specializations [1, 2, 3]. Key competencies can be defined as a generic complex of knowledge, values and attitudes, but primarily life skills, such as problem solving, communication, collecting and analysing information, ability to work independently, leadership, creativity, ability to cooperate effectively, and the desire to participate in lifelong learning [4, 5]. These competencies allow individuals to act appropriately and efficiently within various professional and personal life situations. They represent and incorporate a wide range of long-term qualifications and requirements for a flexible response to changing labour market demands, as well as ensure maximum performance in the defined tasks [6].

The selection of suitable employees is becoming more and more demanding because of the increasing number of job applicants that want to get into a marketing career. Nowadays, due to the requirements for matching the needs of companies with individual employee abilities, companies are increasingly using a competency-based approach to selecting the most suitable candidate. A competency-based approach involves the specification of the desired framework of the individuals' key competencies for specific positions by means of a competency model as a basis for the selection of suitable applicants for a given position [7-9].

A competency model defines a combination of priority skills, abilities and knowledge, that is, individual competencies necessary to perform the tasks related to a specific position within an organization, for example in the marketing department. Competency models can be divided into two basic types: a) generic - for universal application throughout a company; and b) specific – wherein competencies that are a prerequisite for the required performance of a given employee in a specific position within a company are defined [10, 11].

To find component and capable potential employees for marketing related jobs, the recommended competency model should consist of the following particular groups of desired key competencies, each of which should be tested on job applicants during the hiring process [12-17]:

* Corresponding author: dusek@mail.vstecb.cz

1. Competencies related to problem solving:
   a. *Creative thinking*
   b. *Negotiating skills*
   c. *Public speaking*
   d. *Analysing and assessing information*
2. Social (interpersonal) competencies:
   a. *Ability to communicate*
   b. *Ability to work in a team*
   c. *Ability to lead and motivate people*
3. Self-management:
   a. *Ability to work independently*
   b. *Ability to work under pressure*
   c. *Ability to manage a budget*

Most of these competencies are behavioural, so if students have certain basic individual capabilities, their competencies can be developed throughout higher education [1]. With regards to the development of marketing competencies in universities and the preparation of students for their future successful marketing careers, there are plenty of methods that have been proven to help. These include case studies [18], business simulation games [19, 20] or project-based learning [21, 22].

The latter is considered to be one of the most suitable pedagogical tools for the development of students' marketing competencies. This approach is characterized by the creation of student research teams that work on long-term projects (usually one or two semesters) connected to real world business practice with the purpose of solving the actual business tasks and problems [23, 24].

2 Methodology

The aim of this contribution is to find out which marketing competencies a competent graduate of economics should possess in order to be attractive for marketing related jobs in small and medium-sized enterprises (SMEs) in the Region of South Bohemia. A partial goal is to identify the positives and negatives associated with employing graduates in marketing departments.

Following an analysis of secondary resources [12-17], the research focused on a set of eight marketing competencies, namely:

1. Creative thinking;
2. Communication;
3. Working independently;
4. Effective team collaboration;
5. Negotiating skills;
6. Analysing and assessing information;
7. Public speaking;
8. Ability to work under pressure.

The following criteria were defined for the selection of suitable respondents from among SMEs:

a) companies with 25 - 249 employees;

b) companies with a marketing department;

c) companies in the Region of South Bohemia.

The minimum number of 25 employees was set due to the assumed existence of marketing departments in such SMEs. The data was collected in 2018 – 2019 on the basis of a questionnaire survey conducted among 93 companies that met the aforementioned requirements. Data was primarily obtained from those persons responsible for recruitment – HR managers or directors in some cases. Using a 5-point scale (from 1 to 5, where 1 = very significant, 2 = significant, 3 = important, 4 = not significant, 5 = unimportant), the respondents evaluated the importance of specific marketing competencies. Subsequently, in the open-ended question, the respondents also defined their idea of the optimal behaviour of a competent graduate for each specific marketing competency. In the second part of the questionnaire, respondents were asked to identify, from lists of 15, the advantages and disadvantages of employing graduates. They could choose a maximum of 4 options from each group. For the statistical analyses, STATISTICA 9 software was used. As the obtained data do not follow normal (Gaussian) distribution, only a non-parametric test was used for the analysis thereof (Friedman ANOVA).

3 Results and Discussion

3.1 Preferred marketing competencies

Using the Friedman ANOVA to determine the significance of a specific marketing competence, the following hypotheses were tested at the significance level of $\alpha = 0.05$:

H_0: The average significance of the individual marketing competencies determined by the surveyed companies is identical.

H_A: non H_0

Results of Friedman ANOVA:

ANOVA = 237.5801; df = 8; N = 93; p-value = 0.000

Taking into consideration the obtained p-value, the zero hypothesis is therefore rejected. This means that in the process of finding a competent graduate for a marketing department, there are identifiable differences in the companies′ perception of the marketing competencies required. The average significance of every competency of the analysed group of marketing competencies based on the SMEs points of view are presented in Table 1. The closer the individual means are to 1, the greater the importance the specific competency has to employers.

Table 1. Importance of marketing competencies (means, n = 93)

Marketing competencies	Arithmetic mean	Standard deviation
Creative thinking	1.27	0.43
Assessing and analysing information	1.34	0.54
Working independently	1.39	0.57
Communication	1.48	0.61
Effective team collaboration	1.67	0.69
Ability to work under pressure	1.93	0.77
Public speaking	2.56	0.73
Negotiating skills	2.83	0.87

Source: Author.

Based on the results, it is possible to define the basic profile of a desired competent graduate for marketing related jobs as follows: a communicative individual with the ability to provide original and unexpected solutions to problems, both independently and as a member of a team. When solving problems, the graduate should be able to obtain necessary information, purposefully and efficiently, from various resources, find a core message, and distinguish between relevant and irrelevant information. Also of importance is the ability to work towards a common predefined goal in cooperation with co-workers and comply with deadlines. On the other hand, the competent graduate does not have to be a good public speaker and/or have the ability to negotiate.

3.2 Advantages and disadvantages of hiring graduates

Employing graduates of universities of economics has many advantages, which is why SMEs hire them, in particular for entry level marketing jobs. The most important advantages for the surveyed SMEs are presented in Figure 1. Notably in the field of marketing, hired graduates see everything from a different angle, come up with different and creative ways and ideas, or point out mistakes that were ignored in the past. They are often skilled in working with modern technologies and social networks, which opens up new ways of marketing, especially for small companies. Moreover, graduates are very enthusiastic about learning new things and usually pay more attention to their tasks because they are at the start of their careers.

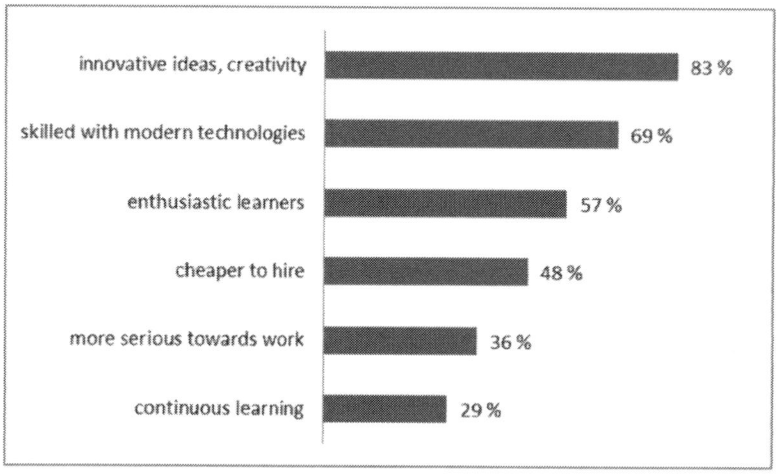

Figure 1. Advantages of hiring graduates (n = 93)

Source: Author.

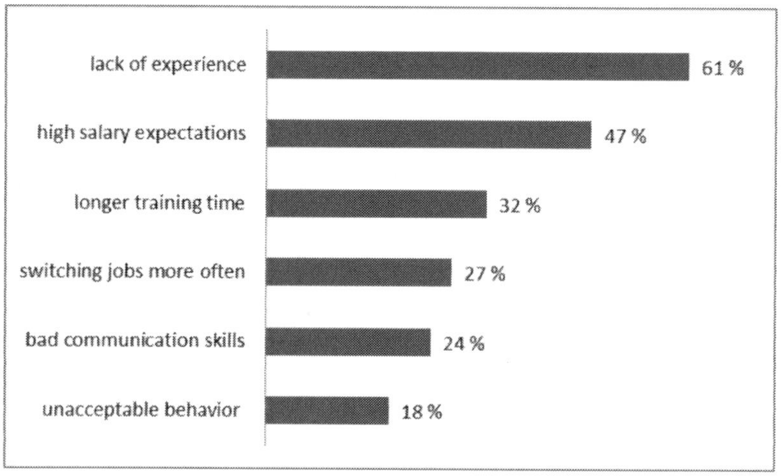

Figure 2. Disadvantages of hiring graduates (n = 93)

Source: Author.

It is clear that there are also disadvantages for the surveyed SMEs in hiring graduates for marketing departments (see Figure 2). Sometimes graduates of universities of economics have no work experience in the marketing industry, so it takes a longer time to develop their skills. As a result, the management will have to understand that the hired graduate will not immediately generate positive change within the organization. That is why every newly hired graduate has to be trained properly, but this period may be longer than in the case of more experienced employees. Unfortunately, one of the main problems of hiring fresh graduates are their high salary expectations. Sometimes they cannot understand that they have just joined a company and therefore cannot expect to be paid like an experienced marketing communications specialist. At the same time, young people are very often searching for the right career and salary is an important reason for switching jobs, every year if needs be.

4 Conclusion

The aim of this contribution was to find out which marketing competencies a competent graduate of economics should possess in order to be attractive for marketing related jobs in SMEs in the Region of South Bohemia, as well as to identify the positives and negatives of hiring such graduates. Based on data from 93 SMEs with their own marketing department, the appropriate candidate can be described as a communicative individual with the ability to provide original and unexpected solutions to problems, both independently and as a member of a team. When solving problems, the graduate should be able to obtain necessary information, purposefully and efficiently, from various resources, find a core message, and distinguish between important and unimportant information. Also of importance is the ability to work towards a common predefined goal in cooperation with co-workers and to comply with deadlines. On the other hand, the competent graduate does not have to be a good public speaker and/or have the ability to negotiate.

It is clear that for the surveyed companies, hiring graduates for their marketing departments has its advantages and disadvantages. The main advantages include creativity, ability to work with modern technologies and/or enthusiasm for learning. On the other hand, the lack of experience, high salary expectations and longer training time are the main disadvantages of hiring fresh graduates for marketing-related jobs.

References

1. M. Hay, Business schools: a new sense of purpose. *Journal of Management Development*. **27**(4), 371–378 (2008)
2. V. Vojtko, R. Dušek, Filling the gap between theory and practice by business simulation games. *Conference Proceedings of the 20th Annual International Conference Enterprise and Competitive Environment*. Mendel University in Brno, 908–918 (2017)
3. M. Tomlinson, 'The degree is not enough': students' perceptions of the role of higher education credentials for graduate work and employability. *British journal of sociology of education*. **29**(1), 49–61 (2008)
4. M. Armstrong, *Armstong's Handbook of Human Resource Management Practice*. Philadelphia: Kogan Page (2011)
5. A. Wiek, L. Withycombe, C. L. Redman, Key competencies in sustainability: a reference framework for academic program development. *Sustainability Science*. **6**(2), 203–218 (2011)
6. T. Weiss, S. Kolberg, *Coaching Competencies and Corporate Leadership*. Florida: CRC Press (2003)
7. D. D. Dubois, W. J. Rothwell, *Competency-Based Human Resource Management*. California: Davies-Black Publishing (2004)
8. C. J. Gonsalvez, F. L Calvert, Competency-based Models of Supervision: Principles and Applications, Promises and Challenges. *Australian Psychologist*. **49**(4), 200–208 (2014)
9. M. A. Campion, A. A. Fink, B. J. Ruggeberg, L. Carr, G. M. Phillips, R. B. Odman, Doing competencies well: best practices in competency modeling. *Personnel Psychology*. **64**(1), 225–262 (2011)
10. S. Sanghi, *The handbook of competency mapping: understanding, designing and implementing competency models in organizations*. New Delhi: Response Books (2004)
11. D. Liebenow, S. Uedelhoven, M. Stricker, J. Nachtwei, Quality in Competency Management: An Explorative Study to Develop a Benchmark for Competency Modeling and Competency Models. *Zeitschrift fur Arbeits-und Organisationspsychologie*. **64**(2), 120–148 (2020)
12. G. W. Stevens, A Critical Review of the Science and Practice of Competency Modeling. *Human resource development review*. **12**(1), 86–107 (2013)
13. J. Hadrich, New Key Competencies for Authorities, Companies and their Managers Part 2: Obstacles and Success Factors of modern Leadership. *Deutsche Lebensmittel-Rundschau*. **116**(5), 186-196 (2020)
14. J. M. Loveland, S. A. Thompson, J. W. Lounsbury, D. Dantas, Is diffusion of marketing competence necessary for a market orientation? A comparative investigation of marketing managers and their defining traits. *Marketing Intelligence & Planning*. **33**(3), 469–484 (2015)
15. DQ. Xin, X. Zhao, The Construction of College Students Self-marketing Competency Model. Recent advance in statistics application and related areas, vols I and II. *Proceedings from the 2nd Conference of the International Institute of Applied Statistics Studies*, Qingdao, China (2009)
16. M. Rieckmann, Future-oriented higher education: Which key competencies should be fostered through university teaching and learning? *Futures*. **44**(2), 127–135 (2012)
17. S. Nwankwo, A. Gbadamos. *Entrepreneurship Marketing: Principles and Practice of SME Marketing*. Oxon: Rouletge (2020)
18. E. Anderson, B. Schiano, *Teaching with cases: a practical guide*. Harvard Business School Publ, Boston, Mass (2014)
19. S. R. Tiwari, L. Nafees, O. Krishnan, Simulation as a pedagogical tool: Measurement of impact on perceived effective learning. *The International Journal of Management Education*. **12**(3), 260–270 (2014)
20. M. Pratt, S. Hahn, Effects of Simulation on Student Satisfaction With a Capstone Course. *Journal of Hospitality & Tourism Education*. **27**(1), 39–46 (2015)

21. W.N. Bender, *Project-based learning: differentiating instruction for the 21st centrury*. Thousand Oaks: Corwin (2012)
22. R. Maltese, *Project Based Learning: 25 Projects for 21st Century Learning*. Indianapolis: Dog Ear Publishing (2012)
23. M. Almulla, The Effectiveness of the Project-Based Learning (PBL) Approach as a Way to Engage Students in Learning. *Sage Open*, **10**, 3 (2020)
24. J. Atkinson, *Developing teams through project-based learning*. Hampshire: Gower Publishing (2001)

Self-determination Theory as Mediator in the Nexus of Gamification and Customer Purchasing Behaviour

Lubica Gajanova[1,*], *Mladen* Radišić[2]

[1]Department of Economics, The Faculty of Operation and Economics of Transport and Communications, University of Žilina, Univerzitna 1, 010 26, Zilina, Slovakia
[2]Department of Industrial Engineering and Engineering Management, Faculty of Technical Sciences, University of Novi Sad, Trg Dositeja Obradovića 6, 210 00 Novi Sad, Serbia

Abstract. Gamification in marketing can be understood as the use of game mechanisms to motivate customers and encourage them to behave in a specific way and interact in some activities or communities in order to increase their satisfaction, loyalty and commitment, thereby contributing to brand building. There are many foreign works confirming the benefits of gamification. However, there is an absence of any known studies explicitly conducted into gamification within a marketing context in the Slovak Republic. This paper therefore attempts to address this kind of research gap and answer the question of whether gamification is a more effective form of promotion for improving the purchasing behaviour of customers in Slovakia. The results of the research confirm that gamification as part of sales promotion has a positive effect on the purchasing behaviour, search intentions and attitudes of Slovak customers. The results also indicate the positive mediation role of motivation based theory related to gamification and customer purchasing behaviour.
Keywords: gamification; motivation; self-determination theory; purchasing behaviour

1 Introduction

The term gamification has been known in research since 2002, but increased attention has been paid to it since 2010 [1]. Between 2011 and 2013, the concept of gamification quickly came to the forefront of information technology trends due to its emplacement in the Gartner Hype Cycle. However, Gartner´s 2014 Hype Cycle refuted this assumption and shifted gamification down the curve to the Trough of Disillusionment - to hybrid clouds and augmented reality [2]. The reasons why gamification can be considered a future trend are: the growth of the gaming industry [3], the expansion of the use of smartphones and wearable technologies [4], and the increasing application of the concept of quantified self [5]. Gamification complements all these areas perfectly.

The positive effect of gamification in marketing has been confirmed by many scientists and practitioners [6-12]. This can be explained by the fact that companies are able to achieve a competitive advantage using innovation and knowledge about users in a progressive form of marketing communication [13-17]. However, there is an absence of any known studies explicitly conducted into gamification within a marketing context in the Slovak Republic. The research is also justified when taking into account the singularity and parochialism of Slovak peoples´ perception of brand, as described by Kliestikova and Janoskova [18]. They say that Slovak customers are, from the psychological point of view, still conservative, with strong tendencies towards their historical socio-political heritage, this despite accelerating globalization and related processes and phenomena, which they are not aware of. This postulate has been proven in practice when foreign and domestic brands have applied either their original branding patterns or patterns copied from effective foreign practice for brand building and which have led to the opposite effect – an undesirable decline in brand value. This paper therefore attempts to address this kind of research gap and answer the question of whether sales promotions with gamification elements are a more effective form of promotion in terms of improving the purchasing behaviour, search intentions and attitudes of Slovak customers.

2 Literature Review

To date, there is no generally accepted definition of gamification [19]. According to Shauchenka [20] the most common definition comes from Deterding, Dixon and Khaled [6]: gamification is the use of game design elements in non-game contexts. The Oxford Dictionary defines gamification as the application of typical elements of game playing (e.g. point scoring; competition with others; rules of play) to other areas of activity, typically as an online marketing

* Corresponding author: lubica.gajanova@fpedas.uniza.sk

technique that encourages engagement with a product or service. In their publication on the implementation of game elements in web and mobile applications, Zichermann and Cunningham [21] define gamification as the process of game-thinking and game mechanics to engage users and solve problems. It follows from their definition that the main goal of gamification is to achieve interaction with the customer, and on this basis to achieve the desired goal [22, 23]. According to Graessley et al. and Princes, gamification can be applied to any problem that can be solved by influencing human motivation and behaviour [24, 25]. Chaffey [26] defines gamification as the process of applying game principles and mechanics to ordinary activity in order to actively engage customers through rewards, engagement or amusement. The goal of the successful application of gamification elements should be to increase customer loyalty. Burke [19] argues that gamification focuses on motivating players to meet their goals with the intention of meeting the specific goals of the organization that implements gamification. Gamification in marketing is integrated into an external form of gamification, which aims to target customers through playful ways in the external environment of the company. However, the goal is not to force customers to participate in these games. Rather, companies are trying to adjust well-established processes to make them more fun and playful for their customers. Gaming mechanisms are then a way for customers to become addicted to the company and be more loyal to its products or brand [27]. In general, gamification is considered to be a system that can force customers to enjoy participating in business services, resulting in increased customer loyalty or increased customer retention [28].

According Koch et al. [29], 25% of gamification comes down to technology, the remaining 75%, psychology. The success of gamification is obviously due to psychological reasons [30, 31]. The aim of gamification is to motivate users to do things that they would not have been able to do without the use of game elements. It's about influencing existing behaviour and creating a positive experience for the player to motivate them to repeat the activity. Motivation is the key element with which gamification works. To better understand it, it is necessary to explain the basic psychological theory dealing with the phenomenon of motivation in the context of gamification.

Self-Determination Theory (SDT) is based on the traditional theory of intrinsic and extrinsic motivation. The structure of internal motivation describes a person's natural propensity for adaptation, acquisition, spontaneous interest and exploration, which is so important to cognitive and social development and which is a vital source of pleasure and vitality in life. In other words, if we are motivated internally, we do the activity not because of the separable effect or outcome but for our own satisfaction or interest and for the pleasure of the activity itself [32, 33, 34]. External motivation then refers to an activity that is performed for the purpose of a certain result that is separable from the activity, for example reward. When comparing people who are motivated internally to those who are more controlled from the outside, the first group show greater interest, excitement and confidence, which in turn translates into increased performance, stamina and creativity [35]. Internal motivation also has a longer-term effect [36]. If we combine this theory with gamification, it can be said that while its aim is to increase internal motivation, gamification itself is actually an application of external motivators [37]. Gamification combines the above-mentioned types of motivation. On the other hand, it motivates from the outside by means of play elements and subsequent rewards. It tries to evoke feelings or needs that motivate us internally. According to self-determination theory, the dominant form of motivation for playful behaviour can be assigned to intrinsic motivation [38]. Since gamification makes use of playful elements, this could lead to intrinsically motivated behaviour. In the context of gamified sales promotions, this means that gamification, through the playful elements, can stimulate an intrinsic confrontation of the customer's involvement in game elements, and therefore in the promotional offer. From a company's point of view, the intrinsic motives caused by gamification could therefore have a positive effect on variables such as customer relationship, promotion perception, customer satisfaction, loyalty and purchasing behaviour.

3 Methodology

The research question is whether sales promotions with gamification elements are a more effective form of promotion for improving the purchasing behaviour, search intentions and attitudes of Slovak customers. The survey was conducted in March 2020 by means of the CAWI method. A total of 399 respondents participated in the study.

A basic set of surveyed respondents was formed of Slovak citizens older than 15 years. That age limit was set as one of the essential prerequisites for autonomous decision-shopping, in accordance with valid Slovak law [39]. The size of the representative sample was determined by the following equation:

$$n \geq [t\alpha\text{^}2\times p \times(1\text{-}p)] /d\text{^}2 \qquad (1)$$

Where: n – the minimum number of respondents; tα – the critical value of the normal distribution at significance level $\alpha = 0.05$; p – the likely sample proportion, expressed as a decimal; d – the confidence interval, expressed as a decimal [40]. The confidence level was set at 95%. The critical value of the normal distribution at confidence level $\alpha = 0.05$ was 1.96. This is based on the fact, that % of the area of the normal distribution is within 1.96 standard deviations of the mean. For those cases where the likely sample proportion was not known, p was set at 50%. The confidence interval was set at 5%. With these parameters, the minimum sample size is 384. So the number of respondents in the survey can be considered as sufficient [41]. To answer the research question, the hypothesis was set as follows. There is a statistical dependence between the preferred type of promotion and its impact on purchasing behaviour, search intentions and the attitudes of customers.

https://doi.org/10.1051/shsconf/20219001005

The hypothesis is based on the basic competence of gamification, namely that the use of game mechanisms and game elements can direct human behaviour in a desired direction [6, 42]. Gamification motivates people to show a desired behaviour [12]. For example, Murray and Bellman [43] report a positive effect of gamification on purchasing behaviour, and Lounis, Neratzouli, and Pramatari [12] have shown that gamification can increase the willingness to pay for sustainable products. Siemens et al. [44] also concluded that gamification has a positive effect on consumer attitudes. Based on this, gamification is expected to have a motivating strengthening effect, which leads to improved promotional efficiency by guiding purchasing behaviour in a desired direction. In particular, gamification is expected to increase shopping intentions, reduce search intentions and have a positive impact on consumer attitudes (consisting of satisfaction and loyalty to the retailer and satisfaction with the shopping experience). Given that gamification is mostly a psychological aspect, the hypothesis was analysed not only in terms of total direct effect between the variables examined but also with the use of mediators to specify the relationship between independent and dependent variables through the inclusion of explanatory variables (motivation based theory).

Respondents' answers to the following questions were used as input variables: (1) Which form of promotion do you prefer?; and (2) to what extent does the preferred form of promotion influence your shopping behaviour? In order to properly understand the questions and thus ensure relevant answers, explanatory notes were also included in the questionnaire. In the first question, concerning the determination of the preference of the form of promotion, the respondents had the possibility to choose between classical sales promotions and gamified sales promotions. Classical sale promotion was described as a non-monetary sale promotion because it is comparable to gamified sales promotion in terms of intrinsic motivation. In addition, a study by Kevin & Clancy shows that only 15-35% of consumers are price sensitive [45]. Respondents could express the degree of influence of the preferred form of promotion on their shopping behaviour using a 5-point Likert scale. Respondents were asked to rate the effects of the promotion with regards to increasing their purchasing intentions, reducing their search intentions and the positive impact on their attitudes (consisting of satisfaction and loyalty to the retailer and satisfaction with the shopping experience).

When confirming the existence of a statistical dependence between the two examined variables, it is necessary to deal with the internal structure of the relationships. To achieve this, the data was more comprehensively analysed using an index, an expression that measures each value based on its relative significance to the other values in the contingency table. The index indicates the relative significance of a particular value in a pivot table cell (nij) relative to other cells and the total number of responses (n). The index calculates the weight that an item contributes to the overall total. The index compares each value to its row total, its column total, and the overall total, using a weighted average [46].

Given that gamification is mostly a psychological aspect, the hypothesis was analysed not only in terms of total direct effect between the examined variables, but also with the use of mediators to specify the relationship between independent and dependent variables through the inclusion of explanatory variables (motivation based theory) [47]. Mediators are an interesting research method to investigate the expected indirect effect of gamification on purchasing behaviour, and therefore on promotional effectiveness. For this, Intrinsic Motivation Inventory (IMI) was used. This is a multidimensional measurement device intended to assess participants' subjective experience related to a target activity. The instrument assesses participants' interest/enjoyment, perceived competence, effort, felt pressure and tension, perceived choice and usefulness while performing a given activity, therefore yielding six subscale scores [48]. Each aspect was described and respondents were asked to answer using a 7 point Likert scale, where 1 was "not at all true" and 7 was "very true". Aspects were described as follows:

- I enjoyed doing this activity very much because it was fun to do, it was interesting and quite enjoyable (Interest/Enjoyment).
- I think I am pretty good at this activity compared to other ones as well, so I felt competent, satisfied with my performance and pretty skilled at this activity (Perceived Competence).
- I put a lot of effort into this, I tried very hard on this activity because it was important to me to do well at this task (Effort/Importance).
- I did not feel nervous, anxious and tense at all while doing this, on the contrary I was very relaxed (Pressure/Tension).
- I believe I had some choice about doing this activity and I did it because I wanted to (Perceived Choice).
- I believe this activity could be useful, important, beneficial and of some value to me (Value/Usefulness).

Figure 1 illustrates the investigation of multiple mediation within the context of this paper.

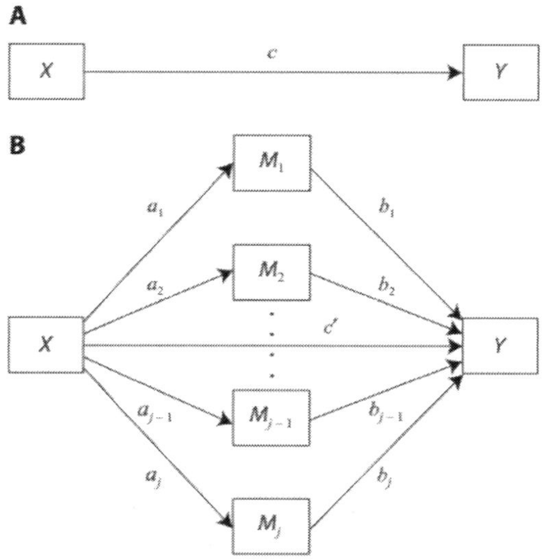

Figure 1. Illustration of multiple mediation with j mediators. (A) X affects Y. (B) X is hypothesized to exert indirect effects on Y through M1, M2, . . .Mj.

Source: Preacher and Hayes [49].

IBM SPSS statistical software was used to test the mediation effect by PROCESS – a tool developed for conditional process modelling in SPSS. In all the analyses presented below, the significance of the indirect effect was obtained by computing 95% confidence intervals using 5,000 bootstrap samples.

4 Results

When testing the mediating path, it is possible to find out the direct and indirect effect of the independent variable to the dependent variable. The values for the direct effect of the preferred type of promotion on purchasing behaviour (in terms of shopping, search intentions and the attitudes of customers) are shown in Table 1.

Table 1. Direct effect of promotion type on purchasing behaviour

Direct effect of X on Y						
Effect	se	t	P	LLCI	ULCI	c'_ps
0.3984	0.0899	4.4314	0.0000	0.2217	0.5752	0.3063

Source: Authors' calculation – SPSS.

Based on the results ($p \leq 0.05$) it is clear that there is a statistically significant relationship between the dependent and the independent variable. After confirming the dependence, we examined the internal relationships between the variables using the Index. The calculations are shown in Table 2.

Table 2. Indexes of responses

Type of promotion	Strongly agree	Agree	Neutral	Disagree	Strongly Disagree
Gamified sales promotion	1.279828	1.366815	0.668924	0.716432	0.467031
Classical sales promotion	0.607229	0.485133	1.464703	1.398021	1.748083

Source: Authors based on Dalgleish [39].

Put simply, the index expresses the degree of protrusion or backwardness in a given sample. That is, how much the index for a particular cell loses to or exceeds 1. We can therefore say that gamified sales promotions have a clear positive effect on purchasing behaviour, while the preference for classical sales promotions has a negative one. It is therefore established that there is a statistically significant impact of the type of promotion on purchasing behaviour that may be mediated. The results of the structural model analysis in Table 3 and 4 suggest that the type of promotion has a statistically significant impact on purchasing behaviour, search intentions and the attitudes of customers, even when a six path mediator is included.

Table 3. Indirect effect of promotion type on mediators

Mediators	Impact of X on mediators					
	coeff	se	t	P	LLCI	ULCI
Interest/Enjoyment	0.3596	0.1554	3.5927	0.0004	0.2528	0.8639
Perceived Competence	0.3660	0.1639	3.6590	0.0003	0.2775	0.9221
Effort/Importance	0.2947	0.1634	2.9292	0.0036	0.1574	0.7998
Pressure/Tension	0.3349	0.1600	3.3393	0.0009	0.2197	0.8487
Perceived Choice	0.3697	0.1609	3.6974	0.0002	0.2786	0.9114
Value/Usefulness	0.3381	0.1657	3.3718	0.0008	0.2330	0.8847

Source: Authors' calculation – SPSS.

Table 4. Indirect effect of mediators on purchasing behaviour

Mediators	Impact of mediators on Y and mediating impact of X on Y					
	coeff	se	t	P	LLCI	ULCI
Interest/Enjoyment	0.3184	0.0388	9.2641	0.0000	0.2101	0.3233
Perceived Competence	0.1969	0.0275	5.6851	0.0000	0.1022	0.2103
Effort/Importance	0.2607	0.0275	7.5938	0.0000	0.1548	0.2629
Pressure/Tension	0.0938	0.0274	2.7889	0.0055	0.0226	0.1304
Perceived Choice	0.1205	0.0288	3.3827	0.0008	0.0408	0.1540
Value/Usefulness	0.2288	0.0273	6.5852	0.0000	0.1263	0.2338
Type of promotion	0.3063	0.0899	4.4314	0.0000	0.2217	0.5752

Source: Authors' calculation – SPSS.

All estimated paths for the indirect effect were statistically significant ($p \leq 0.05$), so the mediators of intrinsic motivation Interest/Enjoyment, Perceived Competence, Effort/Importance, Pressure/Tension, Perceived Choice, Value/Usefulness completely mediates the path between the type of promotion and purchasing behaviour (Figure 2).

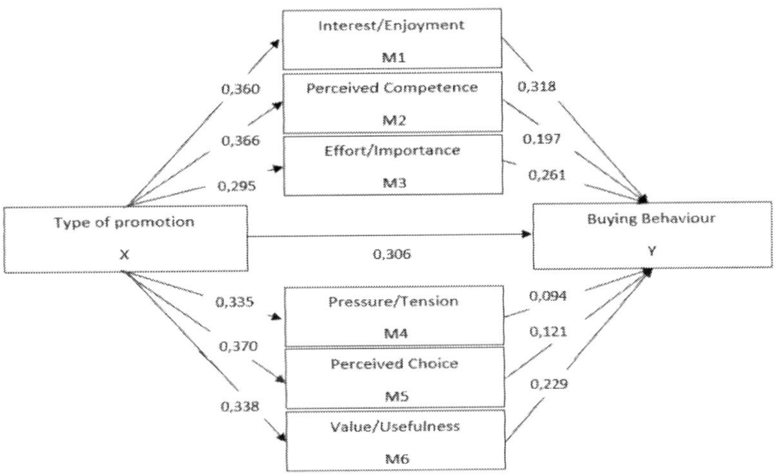

Figure 2. Multiple mediation model

Source: Authors.

5 Discussion

Based on the confirmed effect of the promotion type on the mediators analysed, it is possible to conclude that gamified sales promotions generate intrinsic motivation among consumers. Intrinsic motivation caused by gamified sales promotion is desirable for companies, because internally motivated customers buy not only because of price promotion, which is a manifestation of extrinsic motivation. The choice of a product by internally motivated customers is based on positive feelings such as satisfaction or affinity. Intrinsic motivation is also linked to greater stability [35], and therefore it can be assumed that truly motivated customers will continue to buy from the same retailer, company or brand. In support of this, Rigby [50] states that the meeting key psychological needs [51] in relation to games maintains the engagement of participants. In addition, intrinsic motivation leads to greater enjoyment [38], which could be expressed in a business context by satisfaction with the purchase and / or seller; because, according to Rheinberg, Vollmeyer and Engeser [52], intrinsic motivation has a positive effect on consumer behaviour. Based on these findings and in line with the results of the survey, we can say that the internal motivation caused by gamified sales promotion has a positive effect on consumers' shopping behaviour, and therefore leads to improved promotional effectiveness. Intrinsic motivation therefore acts as a mediator between gamification and purchasing behaviour. Previous research has shown that intrinsic motivation usually motivates people to play games [38]. Therefore, we assume that the application of game mechanisms and elements of gamification evoke internal motives. With a suitable game design that is based on self-determination theory and, above all, satisfies key internal psychological needs, gamified sales promotions are expected to evoke intrinsic motivation.

6 Conclusion

Consumer trends and the tough competitive situation in retail make it clear that retailers need to be creative in order to successfully initiate purchasing impulses through promotions. This situation indicates the need for innovative promotional concepts such as gamification. This is based on the assumption that, according to foreign secondary data, gamification has the potential to bring an additional advantage, and therefore achieve a strengthening effect, which in turn increases the effectiveness of promotions. This paper focused on confirming this assumption for Slovak customers, too. Within this context, this paper attempts to answer the question of whether sales promotions with gamification elements are a more effective form of promotion for improving the purchasing behaviour, search intentions and the attitudes of Slovak customers. The results clearly show that gamified sales promotion has a statistically significant positive impact on purchasing behaviour, both in general terms and in accordance with the specific relationships between variables according the use of mediators of intrinsic motivational character.

The limitation of the research may be the fact that it is not possible to draw clear conclusions from the perspective of consumers. The actual point of view may differ from reality. Sometimes it happens that a respondent deliberately conceals or does not admit facts for various reasons, even if the questionnaire is anonymous, which results in an idealized and partially distorted response [53]. To remove limitations of this kind, it is advisable to repeat the questionnaire survey on another sample of respondents. The questionnaire survey was conducted only in the presence of a geographically limited population. From a certain point of view, this may represent another limitation to this paper. This can be removed if the research results are subsequently verified and compared within an international environment. If these barriers were removed, more accurate results could be achieved.

Acknowledgement

This contribution is a partial output of the project APVV-15-0505: Integrated Model of Management Support for Building and Managing Brand Value in the Specific Conditions of the Slovak Republic.

References

1. F. Groh, Gamification: State of the Art – Definition and Utilization. In M. Asaj, et al., *Proceedings of the 4th Seminar on Research Trends in Media Informatics.* 39–46 (2012)
2. J. Simoes, Using Gamification to Improve Participation in a Social Learning Environment. In I. Buchemet al., *Proceedings of The PLE Conference 2013: Learning and Diversity in the Cities of the Future.* 167–184 (2013)
3. I. Koksal, *Video Gaming Industry & Its Revenue Shift* [online]. Available at: https://www.forbes.com/sites/ilkerkoksal/2019/11/08/video-gaming-industry--its-revenue-shift/#46336bdd663e (2019)
4. A. Holst, *Wearable technology - Statistics & Facts* [online]. Available at: https://www.statista.com/topics/1556/wearable-technology/ (2020)
5. R. Rawassizadeh, E. Momeni, C. Dobbins, P., Mirza-Babaei, R. Rahnamoun, Lesson Learned from Collecting Quantified Self Information via Mobile and Wearable Device. *Journal of Sensor and Actuator Networks*. **4**(4), 315–335 (2015)
6. S. Deterding, D. Dixon, R. N. Khaled, L. Nacke, From Game Design Elements To Gamefulness: Defining "Gamification". In A. Lugmayr et al., *Proceedings of the 15th International Academic MindTrek Conference: Envisioning Future Media Environment.* 9–15 (2011)
7. E. Plessis, *Jak zákazník vnímá značku: Nahlédněte s pomocí neurovědy do hlav spotřebitelů* [*How the customer perceives the brand: With the help of neuroscience, look into the heads of consumers*]. Brno: Computer Press (2011)

8. I. Kuo, *Gigya's Social Gamification Boosts Content Discovery by 68%* [online]. Available at: https://www.gamification.co/2013/03/28/gigya-boosts-content-discovery/ (2013)
9. D. Zadákov, *Trendy na sociálních sítích v roce 2019. Co by vám nemělo uniknout? [Trends on social networks in 2019. What should you not miss?]* [online]. Available at: https://www.evisions.cz/blog-2018-11-12-trendy-na-socialnich-sitich-v-roce-2019-co-by-vam-nemelo-uniknout/ (2018)
10. Z. Zéman, A. Bogdan, Marketing strategy and building brand value on the Hungarian market. *Ekonomicko-manazerske spectrum*. **13**(2), 1–9 (2019)
11. J. Hamari, Transforming Homo Economicus into Homo Ludens: A Field Experiment on Gamification in a Utilitarian Peer-To-Peer Trading Service. *Electronic Commerce Research and Applications*. **12**(4), 236–245 (2013)
12. S. Lounis, X. Neratzouli, K. Pramatari, Can Gamification Increase Consumer Engagement? A Qualitative Approach on a Green Case. In: C. Douligeris et al., *Proceedings of Conference on e-Business, e-Services and e-Society*. 200–212 (2013)
13. W. Sroka, J. Cygler, B. Gajdzik, The Transfer of Knowledge in Intra-Organizational Networks: A Case Study Analysis. *Organizacija*, **47**(1), 24–34 (2014)
14. A. Domańska, Cooperation between knowledge-based institutions and business: empirical studies and network theories. *Forum Scientiae Oeconomia*. **6**(4), 81–94 (2018)
15. L. Lizbetinova, P. Starchon, S. Lorincova, D. Weberova, P. Prusa, Application of Cluster Analysis in Marketing Communications in Small and Medium-Sized Enterprises: An Empirical Study in the Slovak Republic. *Sustainability*. **11**(8), 2302 (2019)
16. L. Lizbetinova, P. Starchon, D. Weberova, E., Nedeliakova, M. Jurikova, The Approach of SMEs to Using the Customer Databases and CRM: Empirical Study in the Slovak Republic. *Sustainability*. **12**(1), 227 (2020)
17. H. Palus, H. Matova, A. Krizanova, J. Parobek, A survey of awareness of forest certification schemes labels on wood and paper products. *Acta facultatis xylologiae Zvolen*. **56**(1), 129–138 (2014)
18. J. Kliestikova, K. Janoskova, Branding with understanding: how national profile of consumer influences brand value perception. *Marketing and management of innovations*. (3), 149–157 (2017)
19. B. Burke, *Gamify: How gamification motivates people to do extraordinary things*. Brookline, MA: Biblomotion. (2014)
20. N. Shauchenka, Gamification. In: A. Ternes, I. Towers, *Internationale Trends in der Markenkommunikation*. Wiesbaden: Springer Fachmedien, 33–50 (2014)
21. G. Zichermann, C. Cunningham, *Gamification by design: Implementing game mechanics in web and mobile apps*. Sebastopol: O'Reilly Media (2011)
22. E. Alani, S. Kamarudin, L. Alrubaiee, R. Tavakoli, A model of the relationship between strategic orientation and product innovation under the mediating effect of customer knowledge management. *Journal of International Studies*. **12**(3), 232–242 (2019)
23. J. Kliestikova, A. Krizanova, T. Corejova, P. Kral, E. Spuchlakova, Subsidies to Increase Remote Pollution? *Science and Engineering Ethics*. **24**(2), 755–767 (2018)
24. S. Graessley, J. Horak, M. Kovacova, K. Valaskova, M. Poliak, Consumer Attitudes and Behaviors in the Technology-Driven Sharing Economy: Motivations for Participating in Collaborative Consumption. *Journal of Self-Governance and Management Economics*. **7**(1), 25–30 (2019)
25. E. Princes, Boosting impulse buying behaviour in marketing management: customer satisfaction perspective. *Polish Journal of Management Studies*. **20**(2), 403-413 (2019)
26. D. Chaffey, F. Ellis-Chadwick, *Digital Marketing. Strategy, implementation and practice*. Pearson Education Limited (2016)
27. A. Krizanova, J. Majerova, T. Kliestik, P. Majercak, Theoretical Aspects of Brand Building in Seafood Industry. *Naše More*. **60**(5–6), 105–112 (2013)
28. A. Asquer, Not Just Videogames: Gamification and its Potential Application to Public Services [online]. *Digital Public Administration and E-Government in Developing Nations: Policy and Practice*. 146–165 Available at: https://papers.ssrn.com/sol3/papers.cfm?abstract_id=2429345 (2014)
29. M. Koch, F. Ott, S. Oertelt, *Gamification von Business Software – Steigerung von Motivation und Partizipation [Gamification of business software - increasing motivation and participation]*. Neubiberg: Universität der Bundeswehr München (2013)
30. M. Heidt, *Gamification: Grundlagen der Implementierung spieltypischer Elemente Wirtschaftskontext* [Gamification: Basics of the implementation of game-typical elements Economic context]. Saarbrücken: AV Akademikerverlag (2012)

31. M. Krstić, N. Pavlović, Behavioral Economics: New Dimension in Understanding the Real Economic Behavior. In Akkucuk, U., *Handbook of Research on Sustainable Supply Chain Management for the Global Economy*. Hershey: IGI Global, 281–298 (2020)

32. D. K. Richards, M. R. Matthew, A. C. Field, Profiles of motivations for responsible drinking among college students: A self-determination theory perspective. *Addictive Behaviors*. **111**, 106550 (2020)

33. H. Lee, Intrinsic and extrinsic motivations affecting impulse-buying tendency in mobile shopping. *Social Behavior and Personality*. **46**(4), 683–694 (2018)

34. N. Miura, H. C. Tanabe, A. T. Sasaki, Neural evidence fort the intrinsic value of action as motivation for behavior. *Neuroscience*. **352**, 190–203 (2017)

35. R. M. Ryan, E. L. Deci, Self-determination theory and the facilitation of intrinsic motivation, social development, and well-being. *American Psychologist*. **55**(1), 68–78 (2000)

36. D. Dicheva, C. Dichev, G. Agre, G. Angelova, Gamification in Education: A Systematic Mapping Study. *Journal of Educational Technology & Society*. **18**(3), 75–88 (2015)

37. G. Richter, D. R. Raban, S. Rafaeli, Studying Gamification: The Effect of Rewards and Incentives on Motivation. In: T. Reiners, L. C. Wood, *Gamification in Education and Business*. Cham: Springer International Publishing, 21–46 (2015)

38. R. M. Ryan, C. S. Rigby, A. Przybylski, The Motivational Pull of Video Games: A Self-Determination Theory Approach. *Motivation and Emotion*. **30**(4), 344–360 (2006)

39. J. Majerova, Analysis of Slovak Consumer's Perception of the Green Marketing Activities. In C. Bektas, *Proceedings of the World Conference on Business, Economics and Management (WCBEM)*. 553–560 (2015)

40. D. Moravcikova, A. Krizanova, J. Kliestikova, Green Marketing as the Source of the Competitive Advantage of the Business. *Sustainability*. **9**(12), 2218 (2017)

41. K. Valaskova, J. Kliestikova, A. Krizanova, Consumer Perception of Private Label Products: An Empirical Research. *Journal of Competitiveness*. **10**(3), 149–163 (2018)

42. S. Deterding, Gamification: Designing For Motivation. *Interactions*. **19**(4), 14–17 (2012)

43. K. Murray, S. Bellman, Playing Games Efficiently: Hedonic Learning and Loyalty. In: G. J. Fitzsimons, V. Moritz, M. N. Duluth. *Advances in Consumer Research*. MN: Association for Consumer Research, 247–248 (2007)

44. J. C. Siemens, S. Smith, D. Fisher, A. Thyroff, G. Killian, Level Up! The Role of Progress Feedback Type for Encouraging Intrinsic Motivation and Positive Brand Attitudes in Public versus Private Gaming Contexts. *Journal of Interactive Marketing*. **32**, 1–12 (2015)

45. P. Kotler, K. L. Keller, *Marketing Management*. Hoboken: Pearson Education (2009)

46. D. Dalgleish, *Beginning Pivot Tables in Excel 2007*. New York: Apress (2007)

47. D. P. Mackinnon, *Introduction to Statistical Mediation Analysis*. Abingdon: Routlege (2008)

48. CSDT, *Intrinsic Motivation Inventory* [online]. Available at: https://selfdeterminationtheory.org/intrinsic-motivation-inventory/

49. K. J. Preacher, A. F. Hayes, Asymptotic and resampling strategies for assessing and comparing indirect effects in multiple mediator models. *Behavior Research Methods*. **40**(3), 879–891 (2008)

50. C. S. Rigby, Intrinsic and Extrinsic Player Motivation: Implications for Design and Player Retention. [online]. *Celebration*, FL: Scott Rigby Immersyve Inc. Available at: https://www.gdcvault.com/play/1015568/Intrinsic-and-Extrinsic-Player-Motivation (2012)

51. E. L. Deci, R. M. Ryan, *Handbook of self-determination research*. Rochester, NY: The University of Rochester Press (2004)

52. F. Rheinber, R. Vollmeyer, S. Engeser, Die Erfassung des Flow-Erlebens. In: J. Stiensmeier-Pelster, F. Rheinberg. *Diagnostik von Motivation und Selstkonzept* [*Diagnostics of motivation and self-concept*]. 261–280 (2003)

53. J. Hendl, J. Remr, *Metody výzkumu a evaluace* [*Research and evaluation methods*]. Praha: Portál (2017)

Inclusive Growth and International Labour Migration

Sergey Chepel[1], *Evgenia* Tukhtarova[2,3], *Natalia* Neklyudova[2,*]

[1]Institute for Forecasting and Macroeconomic Research, Tashkent, Uzbekistan
[2]Institute of Economics, Ural Branch of the Russian Academy of Sciences, Russia
[3]Ural State Economic University, Russia

Abstract. This contribution presents the results of an econometric analysis of the interconnections between migration, economic diversification, the quality of state institutions and other qualitative characteristics of economic growth that have developed in developing countries of the world over the past two decades. The study highlighted the most important factors that contribute to the creation of conditions and prerequisites for the inclusive development of the economy. The analysis showed that among other factors, the most important are the quality of state institutions, technological development, which creates the conditions for investing in human capital. The study also made it possible to determine thresholds for these factors as guidelines for the transition of the CIS countries to the inclusive development of their economies and to limit disordered labour migration. Another important result of the study is the fact that large-scale and poorly controlled migration is acting like a brake on the sustainable development of the whole post-Soviet space. To overcome this process, it is necessary to create large-scale joint integration, knowledge-based and technological projects that will not only reduce the scale of migration to Russia, but also contribute to increasing human capital in the CIS countries and to the creation of knowledge-based jobs in their labour markets.

Key words: labour migration; inclusive growth; income inequality; technological development; state institutions; human capital

1 Introduction

The problem of economic growth is a major topic of modern scientific discussion. As a fundamental point, it is not so much a return to high economic growth, but rather the formation of a new growth model that would be able to involve all available resources in the process. Moreover, this process would ensure the development of both state institutions and technological structures, as well as the comprehensive development and accumulation of human capital. This approach reflects the understanding that the wellbeing of society is derived not only from the growth of real GDP and the material income of the population, but is a multidimensional concept, including the accumulation of physical and human capital. It is also important to take into account the fact that the model should be based not only on one parameter of "sustainability" of economic growth, in modern realities this is no longer enough, but should be inclusive, that is, have a positive impact on the wellbeing of the general population.

The stagnation of the world economy in the context of fiscal constraints associated with the crises of globalization poses risks to sustained growth and employment. More households face the risk of poverty and poor access to education and medicine, which ultimately exacerbates inequalities between countries. As is known, high levels of inequality can lead to falling aggregate demand, loss of social stability and, ultimately, uncontrolled migration flows [33].

The growing inequality is caused not only by globalization, but also by uneven technological shifts across countries. For example, the combined impact of institutional and technological changes led to increased economic and social inequality in European countries as early as in the 1980s [1].

In this regard, it is becoming clear that the current model of development based on stimulating consumer demand and expanding production is no longer the basis for sustained economic growth. The continuation of this approach will have many undesirable social, economic and environmental consequences. To overcome modern challenges, a rethinking of theoretical and methodological approaches is necessary in order to develop a new solution to current problems. We believe that the concept of sustainable development (sustainable development is a concept based on the analysis of the nation's

* Corresponding author: nnp81@mail.ru

assets, one of whose elements is human capital) needs to be refreshed and recognize a deeper consideration of the social dimension in linking the qualitative characteristics of human capital and economic growth.

The modern concept of inclusive economic growth, where the quality and role of human resources play a primary role, can offer options for overcoming the current crisis. This contribution presents the results of research into and solutions for the transition of the CIS countries towards inclusive growth.

In this contribution, "inclusive growth" refers to an approach based on the effective use of all available resources, especially the enhancement of the role of human capital and its innovative activities.

2 Theoretical Review

The scientific community has now developed *three main paradigms* for sustained economic growth:

1) *the model of investment growth*, which is based on a high level of capital investment in the development and implementation of technologies created in national economies [2];
2) *the model of innovative growth* - here growth, unlike the previous model, is embedded in "advanced innovations", the development of new technologies and their distribution outside national borders [3,4];
3) *the inclusive growth model* based on the effective use of human capital and innovative activities [5-7].

The first two models, based on the technological development of countries, ceased to provide growth in labour productivity. Although technological factors of production continue to play a role in labour productivity, human resources and the quality of institutions in a country are becoming increasingly important [8]. So, according to foreign researchers, a country's growth can be attributed to inclusivity if the following condition is met: if simultaneously with the increase in a country's GDP there is an increase in the human development index (HDI) [9]. In this case, the role of the state in achieving the balance between investment policies and the development of the national innovation system is increasing.

Labour migration, which is an integral part of the labour market, can both increase the inclusiveness of a country's economic growth and contribute to its decline in world ratings for this indicator.

The effect depends on the immigrants' qualifications and skills. The influx of foreign unskilled workers can lead to an increase in the supply curve for labour and an increase in investment in physical capital. On the other hand, labour migration can affect not only the level of physical capital, but also the sectoral distribution. Thus, the arrival of highly skilled immigrants in a country causes the growth of knowledge-intensive industries and contributes to sustainable, inclusive economic growth [10], whereas the influx of low-skilled immigrants helps to attract capital to labour-intensive medium and low-tech sectors of the economy, which can stagnate it due to the intensive operation of infrastructure without modernizing it. In other words, capital is replaced by work and the development of a country is being carried out at the expense of extensive sources.

Uncontrolled illegal and disordered migration unambiguously has negative consequences. Diaz, Saenz and Kwan showed that the influx of Mexican illegal migrants in Arizona over four years (2006-2009) led to an increase in unemployment amid stagnant economic growth in this state [11]. The negative link between illegal migration and economic growth is due to the fundamental basis of social inequality and the decline in a country's inclusive potential. This is due to the need to direct budget flows not to socio-economic development, for example, healthcare or education, but to ensure law and order and strengthen law enforcement agencies in the country.

The impact of labour migration on the budget is reflected not only in the tax revenues from migrant labour, but also in the need for their integration and social benefits. Moreover, immigration affects transfer payments, increasing the burden on the financial sphere due to the difference in interest rates between the local population and foreigners, as a result of which there is an inadequate redistribution of the country's financial resources, and as a result - a decrease in the quality (inclusiveness) of growth. According to the survey by Kerr and Kerr [12], based on retrospective studies, on average the state budget has a slight positive net financial effect for host countries. At the same time, the large influx of unskilled labour seriously limits the country's high-quality (inclusive) development. This is confirmed by other studies. For example, Mayr also estimated a small positive overall financial effect of migration processes on Austria's budget [13].

If the influx of skilled personnel has a positive impact on economic growth in the country of arrival, the sending state remains in a difficult position. Numerous studies on the outflow of human capital indicate a negative impact on the overall level of intellectual development of sending countries and on the qualitative indicators of economic growth. In particular, the study by Beine, Docquier and Rapoport [14] showed that large-scale outflows of skilled labour even in the medium term have a negative impact on the development of certain sectors of the economy in the sending country. At the same time the long-term impact of the outflow of skilled personnel is reflected in changes in the quality and quantity of human capital in the country of migrants' origin. As a result, this launches a negative multiplier effect, reducing the scientific and economic (that is inclusive) potential of the country.

The heterogeneous effect of labour migration on economic growth is evidenced in other studies [15-17]. For example, Bildirici et al. established a positive link between migration and economic growth, which can be fair for importing countries, while for sending countries this link can be reversed [15]. This result was obtained for 77 countries between 1990 and 2001. In our opinion, this may be explained by the fact that in a situation of redundancy of human resources, the released places in the labour market reduce the burden on it. At the same time, state institutions can distribute the remaining human resources in the labour market more efficiently and on an even basis. For developed countries, this burden is increasing, leading to an imbalance between the labour market and productive forces. Due to the presence of strong state institutions in the country that are able to effectively manage the availability of resource potential, prerequisites for inclusive growth are created. While in other countries, state institutions that cannot cope with the imbalance in the country, leads to a decrease in the qualitative components of economic growth.

In countries with strong institutions, the inflow of migrant workers is manageable and even selective. In addition to the fact that developed countries allocate quotas for attracting the number of migrants, state institutions also calculate the number of necessary specialists in various sectors. This ensures both equilibrium in the labour market and increased quality growth in these sectors.

A number of other studies also show the important role of existing state institutions in a country, through which countries can benefit from migration processes and have a positive effect on poverty reduction and income growth in the country of origin. Thus, studies conducted at different times by foreign authors for individual countries recorded such a relationship through the following simplified mechanism of state regulation of migration processes:

"labour migrants → remittances → consumption growth → production growth → economic growth" [16, 18-20].

Consequently, the relationship between poverty and migration is a rather complex economic process, which is important to manage and for which strong, high-quality state institutions are required.

In the absence of such institutions, migration and its impact on economic growth can be reversed. So, there are research results of various authors with the opposite effect of remittances for the country. For example, Catrinescu et al. [21] noted that remittances can increase *long-term economic growth* only in countries where there is public confidence in the management decisions of state institutions. In countries with low-quality public institutions, usually with low incomes, the positive effect is observed only in the medium term. For example, this conclusion was made by Goschin [22] and Ziesemer [23] when investigating the impact of migration on economic growth in countries with an average income of less than $1,200 per capita. Other researchers came to the same conclusion, explaining this effect by the fact that, despite the positive impact of remittances on economic growth and the reduction of poverty in the country of the migrant exodus, they are sources of external shocks in the event of crises in the world economy and, as a result, lead to the instability of economic development [24].

Another explanation of the feedback between migration and economic growth is that a sharp increase or decrease in remittances can have a strong negative effect of weakening a country's trade balance and reducing its competitiveness in foreign markets [25, 26].

Numerous recent studies show that in developing countries, a large proportion of remittances of migrants sent home are not always invested in the productive needs of the country. For example, in poor countries, remittances are often used to repay household loans and debts, bypassing the mechanism for transforming money into investment [21, 27-29].

The study by Anderson and Kroeger on the example of migrants from Kyrgyzstan also testifies to the fact. During the study, the authors recorded the positive impact of migration on the country's economic growth only in 2001-2008 [30]. However, even for the period from 2009 to 2011 the scientists noted that this potential was exhausted due to the deterioration of the country's human capital. Thus, the high rate of migration of skilled personnel led to the depletion of the human capital of Kyrgyzstan and worsened the country's economic development indicators in subsequent years. This can be explained by the fact that remittances were not invested in the human capital of young generations or in the country's economy. Since most of these transfers were directed to household personal consumption, this predetermined the decline in the country's economic potential. In this case, the country falls into the "poverty trap".

The phenomenon of the "poverty trap" was revealed by Lucas [31]. On the example of a study of various countries, Lucas showed that only a country that was able to direct investment effectively not only in physical but, first of all, in human capital can make a significant breakthrough in its development. The researcher has amply demonstrated, through successful countries, how investment in human capital creates a long-term link between technology and sustained economic growth.

Thus, empirical research shows the importance of migration processes in explaining the dynamics of economic growth. At the same time, the most important constraints to inclusive growth are weak state institutions, the poor quality of human capital and imbalances between productive forces and the availability of labour in a country. All that leads to unsustainable economic growth.

Inequality in the development of countries is largely explained by the degree of access to education. As numerous studies show, the level of education is a very important prerequisite for the formation of conditions for the inclusive

 https://doi.org/10.1051/shsconf/20219001006

development of a country, as it creates the basis for the development of technology, which in turn is the basis for an inclusive leap into the future [32].

In 2016, Russia was the third largest recipient of migrants in the world, after the United States and Germany. The processes taking place in the world are also reflected in Russia. This is large-scale, poorly controlled migration, with poor quality state institutions, and economic development based on a raw-material orientation. While developed countries are currently transiting to the sixth digital technology architecture, Russia (in the presence of high human potential) lacks other prerequisites for the inclusive leap into the future. This is caused by the poor quality of its institutions and the lack of a balanced investment policy for the development and accumulation of physical and migration potential, as well as the creation of sustainable, adequately paid jobs in new sectors of the economy. A balanced and well-thought out migration policy with the participating countries is necessary for the transition towards inclusive sources of development. To this end, we propose investigating the impact of migration on inclusive development, based on the experience of successful countries.

3 Methodology

To examine the relationship between the qualitative characteristics of economic growth and the rate of migration in the current stage of economic development, an econometric study based on World Bank data for 76 developing countries was conducted. Countries with unstable political regimes, low incomes per capita (according to the per capita GDP criterion), as well as countries with a population of less than 5 million people were excluded from the category of developing countries.

The following indexes were chosen as indicators of economic development and inclusiveness: GDP per capita according to PPP at constant prices in 2011, the inclusive development index, the Gini coefficient and the human development index. World Bank data were used to estimate migration: the share of net migration in percentage (%) of the local population, as well as UN data: net migration on average for five-year periods and the number of migrants in percentage (%) of the local population. In addition to these indicators, the most important factors of inclusive development were also included: the quality of state institutions, indicators of economic diversification and scientific and technological development, the investment climate and labour market indicators (see Table 1). The statistical database developed in such a way made it possible to assess the link between migration and inclusive development, as well as to identify factors that play a critical role in sustained economic growth.

Table 1. Classifier of variables

Variable	Explanation
a. Level of economic development and inclusiveness (quality) of growth	
gdp pc ppp	GDP per capita for PPPs in USD
idi	Inclusive Development Index (ranging from 1 to 10)
gini	Gini coefficient
hdi	Human development index
b. Migration	
nmigr	Net migration (% of the population using WB statistics - average estimates over five years)
nm95-00 (nm00-05,….)	Net migration according to the UN estimates for 1995-2000, 2001-2005, 2006-2010, 2011-2015 (in thousands people)
nm00-15pop	Ratio of net migration (by ***nm00-15*** indicator) to population
nmbmigr	Number of migrants (% of the population, from the UN International Migration Report 2015)
c. Employed in labour markets	
empl	Employment rate in % of the population
emp ind	Employed in industry (% of total employment)
d. Quality of state institutions	
rol	Rule of Law Index
cpi	Corruption perceptions index (from 0 to 100, average for 2012-2015)
goveff	Government efficiency
e. Economic diversification and scientific and technological development	
manuf	Manufacturing (% of GDP)
htexport	High-tech exports (% of manufacturing export)
intus	Internet users (per 100 people)
patapp	Patents (residents per 1 million people)
r&dexp	Expenditures for R&D (% of GDP)
publ	Number of publications in scientific journals per 1 million people
labter	Number of employees with higher education (%)

Variable	Explanation
yearter	Number of years of university education
	f. Investment activity and resource endowment
fdi	Net inflows of foreign direct investment (% of GDP)
inv	Gross fixed investment (% of GDP)
res	Resource endowment of a country (1- resource-rich, 0 - other countries)

Note: specific values of variables are given in the annual average for the period of 2000-2017.

Source: Authors.

In order to select the indicators of migration flows that reflect economic processes more effectively, statistical tests were conducted on their adequacy and reliability. The quantitative analysis and testing yielded the following results:

1) Number of migrants (in % of the population, from the UN International Migration Report 2015, **nmbmigr**) - an indicator reflecting the number of registered migrants living in the territory of the country, showed that for the absolute majority of countries (50 countries out of 73 that have relevant statistics for this indicator, see figure with histogram for **nmbmigr**) its value is less than 5%. This means that most developing countries are labour service providers. The advantage of this indicator, unlike the other two, is that its value is positive, which facilitates the interpretation of the results of regression analysis. The leading countries according to this indicator are UAE (81%), Kuwait (70%), Oman (41%), Jordan (36%), Saudi Arabia (29%), etc. Among the CIS countries, Kazakhstan (20%), Belarus (11%) and Russia (8%), i.e. the countries with the highest level of development (USD 17-24,000 per capita) in the post-Soviet space, differed due to the high level of migration in recent years, which was one of the incentives for the influx of migrant labour. On the other hand, the number of migrants in Central Asian countries was less than 5% (Tajikistan - 3%, Uzbekistan - 4%, Kyrgyzstan - 5%). This reflects their position as labour-exporting countries, formed by, among other things, the low level of development of their economies (USD 3-6 per capita) and therefore the low incomes of the population.

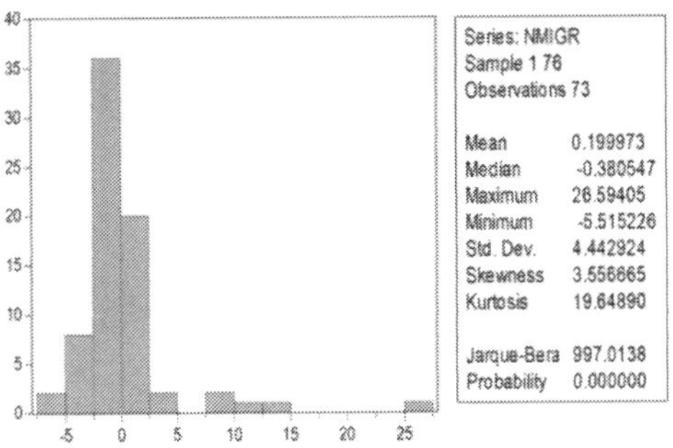

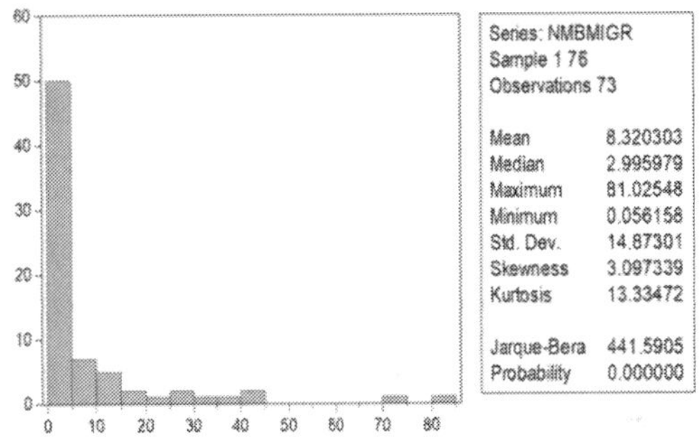

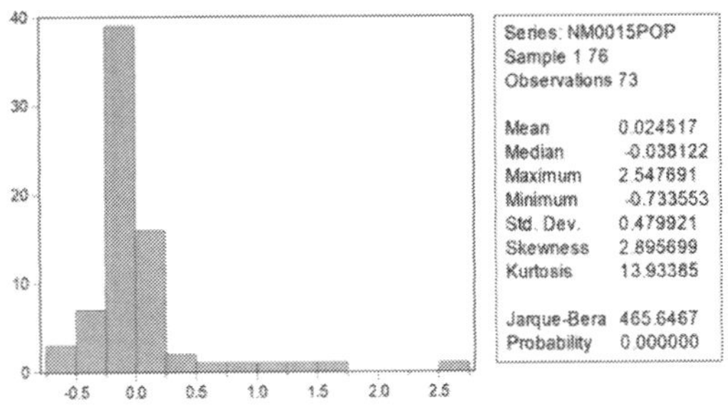

Migration indicators:
nmigr - net migration (% of the population using WB statistics - average estimates over five years);
nmbmigr - the number of migrants in the country (% of the population, from the UN international migration report 2015);
nm95-00 (nm00-05,....) - net migration according to the UN estimates for 1995-2000, 2001-2005, 2006-2010, 2011-2015. (in thousand persons **nm00-10pop** per capita)

Figure 1. Distribution of studied states – countries with migrant inflows and outflows

Source: Authors.

2) The indicator of *net migration* (**nmigr**, in % of the population according to World Bank statistics) varies in our sample of countries in the range from 5.5% to 26.6%. The advantage of this indicator is that it makes it possible to separate countries with an inflow of migrants that exceeds their outflow (**nmigr** > 0, countries that are consumers of foreign labour services) from countries that provide labour force (**nmigr** < 0). At the same time, the number of countries of the second category (48) significantly exceeds the number of the first group of countries (25). It should be noted that the leading countries with a maximum inflow of migrants are almost the same as for the previous indicator (**nmbmigr**): Kuwait (11.1%), Malaysia (2.0%), Oman (13.0%), Saudi Arabia (3.4%), etc. Among the CIS countries, net labour inflows in

2000-2015 occurred in Russia (2.2%), Kazakhstan and Belarus (0.1%). At the other extreme, the leading labour exporting countries includes states that were part of the former USSR: Moldova (-5.5% of the population, average estimate for 2000-2015), Georgia (-4.8%), Armenia (-3.7%), Lithuania (-3.6%), Latvia (-2.8%); as well as Albania (-5.5%) and El Salvador (-4.8%). For Central Asian countries, the values of this indicator were the following: Kyrgyzstan -3.4%, Uzbekistan -1.7%, Tajikistan -1.2%, with an average median estimate for developing labour exporting countries of -1.3%.

3) The indicator of *net migration according to UN* **nm00-10pop** is similar in value to the World Bank **nmigr** indicator and differs only in the procedure for obtaining averaged estimates, as well as a higher average level of net migration (10.5% versus 1.3% for **nmigr**), which makes it preferable to the World Bank indicator.

The primary findings were as follows: although all indicators are based on existing statistical reporting (migrant registration) and significantly underestimate the actual absolute extent of migration, they all provide an objective picture of the differences across any of these indicators for the vast majority of developing countries in the world; the vast majority of them were migrant suppliers; only a small number of developing countries have a significant number of registered migrants (the first 5 countries account for half of all migrants). There is a strong statistically significant correlation between these three indicators (pairwise correlation coefficients from +0.85 to +0.95). All this proves the validity of the hypothesis discussed above and the possibility of using these indicators in further analysis.

Econometric methods were used to examine the relationship between these indicators. At the same time, the peculiarity of the matrix structure according to the indicators included in the analysis was taken into account. This consists primarily of matrices (indicators) of varying degrees for some sample countries. An example is omissions in the dynamics of indicators such as, among other things, spending on education, science and technology, the Gini coefficient, and the number of employees with higher education.

This makes it very difficult to use classical panel analysis methods (RE, FE, etc.). The only alternative in this case is the use of the cross-section method for averaged estimates of indicators (see Table 1) throughout the reporting period (2000-2015) and for all selected developing countries.

Both the traditional form of multifactor regression equations and the form with threshold values were used, which made it possible to substantiate the role and significance of state institutions in solving the problem of streamlining migration flows and improving the quality of economic growth for countries located in the post-Soviet space.

4 Results

In order to investigate the link between migration and inclusive development, two hypotheses were formulated and tested.

Hypothesis 1. *The low-cost labour force created by disordered large-scale labour migration is a deterrent for developing countries, preventing the diversification of their economies and the growth of scientific and technological potential (the analogue of the resource curse hypothesis).*

Based on the information opportunities provided by the resources of the World Bank, the United Nations and other international organizations (see variable classifier, Table 1), the indicator of the level of development of the manufacturing industry was used to test the hypothesis. This indicator generally correlates well with the breadth of the range of products produced (level of diversification) and is an indicator of the share of high-tech products in a country's exports. The control variables were indicators of investment activity, the quality of state institutions, number of employees, and the quality of human capital.

The result of searching for a large number of combinations of control variables with migration factors (**nmigr, nmbmigr, nm00-15**) showed (see Table 2) that the indicator of economic diversification **manuf** is statistically significantly associated with the migration factor **nmbmir** (number of migrants in % of the population). In all the obtained regressions (Nos. 1-3 in Table 2), the corresponding coefficient has a stable negative value for various combinations of control variables (share of employees in manufacturing industry, the quality of state institutions, resource endowment, etc.). This means that for countries with high levels of labour migrants (Kuwait, UAE and other OPEC countries, Russia, Kazakhstan, etc.) an increase in this factor reduces the level of diversification of the economy and enhances its commodity character, vulnerability to external shocks and instability of world markets. At the same time, an additional increase in net migration by 10 percentage points is associated with a decrease in the share of manufacturing industry (level of economic diversification) by an amount of 1.1 to 1.5 percentage points.

A similar equation was obtained for the indicator of technological development (share of high-tech products in exports (**htexport**), equation 4). The number of migrants (**nmbmig**) shows a negative impact on this indicator, while the effectiveness of the government and the number of employed in the economy (in % of the population) increases the level of technological development. These results are the arguments in favour of the hypothesis under consideration.

 https://doi.org/10.1051/shsconf/20219001006

It is important to note another result of econometric analysis. In almost all the regressions obtained, the control variables included factors characterizing the qualitative component of economic development - government efficiency (**goveff**) and the quality of human capital (**hdi**). In terms of content, this may mean not only that these factors are important for the development of finished-product industries, but also that these indicators can influence the quality of migration flows, their orderliness and ultimately their reduction.

Table 2. Econometric assessment of migration impact on economic diversification and technological development

Explaining variable (factors)	Dependent variable - **manuf** (manufacturing)			Dependent variable **htexport**
	№ 1	**№ 2**	**№ 3**	**№ 4**
Control variables Employed in industry **(emp_ind)**	0.036***	0.27***	0.22**	
Government efficiency **(goveff)**		2.46**		9.61***
Resource endowment of a country **(resdummy)**		-2.15	-1.75	
Corruption perceptions index in resource-rich countries **(res*cpi)**				-0.39
Human development index **(hdi)**			0.13**	
Human development index in resource-rich countries **(res*hdi)**				0.18
Employed in economy (**empl**)				0.34***
Number of migrants (**nmbmigr**)	-0.11***	-0.15***	-0.14***	-0.17*
Constant	8.78	11.4	4.50	-6.8
R^2	0.24	0.32	0.31	0.25
Number of countries	73	73	72	71

Note: *** – significant at 1%, **– 5%, * – 10%.

Source: Authors.

The conclusions above find additional confirmation when moving from the traditional form of multifactorial regression equations to regressions with threshold values. Most often, the use of such regressions is aimed at determining the threshold value of the quality of state institutions, the excess of which creates conditions for a positive and statistically significant effect of one of the control variables on the dependent variable, while improving the quality of the entire equation. The form of the regression equation in this case is the following:

$$div = c + a_1*migr+a_2*CV_1+a_3*CV_2 +...+b*(th - QI)* CV_3 \quad (1)$$

where ***div*** is an indicator of the level of economic diversification (or its technological development);

QI - quality indicator of state institutions (corruption perceptions index - ***cpi***, government efficiency - ***goveff***, etc.);

$CV_{1,2...}$- control variables (***inv*** - investment, ***res*** - conditional variable for countries rich in natural resources, ***empl*** - employed in economy, etc.);

c - constant, $a_{1,2...}$ ***b, th*** - econometrically estimated regression coefficients, including the threshold value for the quality of state institutions (***th***).

If the results of the regression parameter assessment show the negativity and significance of coefficient ***b***, as well as a positive and significant estimate of parameter ***th***, then the growth of factor CV_3 will lead to an increase in the diversification of the economy (***dif***) only when the state institution quality indicator (***QI***) for a particular country exceeds the threshold value (***th***) obtained from the evaluation of the equation.

Based on the results of testing this form of relationships, two equations were obtained. The first shows the conditions for growth of diversification according to the indicator of the share of manufacturing industry in the structure of the economy and has the following form:

$$manuf= 10.37-0.13*nmbmigr + 0.32*imp\_ind - 0.013*fdi*(52.0 - cpi) \quad (2)$$

P-val (0.0017) (0.0003) (0.019) (0.0001)

$R^2 = 0.30$

where the values of confidence probability (p-val) are given under each coefficient of factors included in the model. Since they are less than 0.1, all these coefficients, and therefore the corresponding relationships, are statistically significant.

In this equation, the foreign direct investment indicator (***fdi***) is the new control variable, while the corruption perceptions index (***cpi***) is an indicator of the quality of state institutions. The threshold value obtained (***th*** = 52.0%) can be interpreted as follows: the inflow of foreign direct investment contributes to the diversification of the economy only if the value of this index exceeds 52% (in this case, the expression in brackets becomes negative and when multiplied by the negative value of the coefficient ***b*** = - 0.013 causes a positive increase in ***manuf*** with ***fdi*** growth).

It is noteworthy that the CIS countries do not meet this requirement and have estimates of this indicator significantly below the required threshold (for example, for Russia 28%, Uzbekistan 18, Tajikistan 23, Kazakhstan 28, Kyrgyzstan 26, Azerbaijan 28, Belarus 31%). This largely explains the lack of progress in diversifying the economies of these states, even in the face of the significant inflows of foreign direct investment that have taken place in these countries over the past 15 years. In the obtained equation, as in the above, the migration factor ***nmigr*** has a negative impact on the level of diversification of the economy.

The second equation obtained gives a similar result for another indicator - the share of high-tech products in the export structure ***htexport*** (see equation below). The migration factor ***nmbmigr*** is also negatively associated with a dependent variable in it. All this argues for the hypothesis we are considering that large-scale and irregular labour migration has a negative impact on economic diversification and technological development in our sample of developing countries.

$$htexport = -9.75 - 0.16*nmbmigr + 0.35*empl + 4.24*goveff - 0.033*fdi*(30.1\ \ cpi) \quad (3)$$

P-val (0.1) (0.008) (0.14) (0.04) (0.04)

$R^2 = 0.30$

To create the necessary conditions for inclusive growth in developing countries, including states in the post-Soviet space, a second hypothesis was formulated and tested.

Hypothesis 2. *The main factors in reducing disordered large-scale labour migration for developing countries are active investment policies, improved quality of government institutions and inclusive economic growth.*

To test this hypothesis and apply the experience of these countries for the post-Soviet space, a subset of countries with negative net migration (49 countries) was allocated from the total number of developing countries (76 states).

Dependent variables in this case, unlike the equations obtained and described above, are migration indicators for labour exporting countries (***nmigr, nm0015pop***), the values of which, for the convenience of interpreting the results, are taken in terms of absolute values, i.e. with a plus sign. Indicators of investment activity (***inv, fdi***), quality of state institutions (***cpi, rol, goveff***) and of the inclusiveness of economic growth (***gini, idi, emp_ind, manuf***) were used as factors for reducing disordered migration. The level of economic development (***gdp_pc_ppp***) and scientific and technological development (***r&dexp, patapp, etc.***) were used as control variables.

A large number of variants of these variables showed that the highest values of the statistical significance of interrelationships are achieved when using the ***nm0015pop*** indicator (UN indicator) as a dependent variable. Another condition identified in the course of the analysis is the use of regression with a threshold value on the quality indicator of state institutions, which gives better results than traditional multifactor regressions.

The obtained regression with the best statistical characteristics has the following form:

$$nm0015pop = -0.34 - 0.14*r\&dexp - 0.008*emp\_ind - 0.001*fdi*(25.2 - cpi) \quad (4)$$

P-val (0.09) (0.05) (0.01) (0.02)

$R^2 = 0.31$

All factors included in the regression are statistically significant (since p-val < 0.1), and negative signs of coefficient values indicate a decrease in migration when the factor grows.

The result makes it possible to draw a number of important conclusions for measures to limit disordered migration. Firstly, the growth of investment activity (according to the foreign direct investment ***fdi*** indicator) does limit the scale of irregular migration, but only if one of the most important indicators of the quality of state institutions - corruption perceptions index ***cpi*** - exceeds the econometrically obtained threshold of 25.2 (on a 100-point scale). This result is well within the earlier conclusion on indicators of diversification and technological development (***manuf, htexport***), although the previously obtained thresholds were slightly higher (30 and 52).

As shown above, the average annual estimates of this indicator for the CIS countries (labour exporters) that we analysed range from 18 (Uzbekistan) to 26 (Kyrgyzstan), i.e. on the border or even below this threshold. Consequently, unless Central Asian governments take decisive measures to combat corruption, all efforts to stimulate investment to create new jobs and thereby limit migration to Russia and other countries with significantly higher labour costs will not yield the expected results, since the investment resources attracted will be used extremely inefficiently.

As for the hypothesis of the importance of inclusive economic growth to limit labour outflows, it was not fully confirmed. Using the ***idi*** (Inclusive Development Index, introduced by Asian Development Bank, is an annual assessment of 103 countries' economic performance that measures how countries perform on eleven dimensions of economic progress in addition to GDP. It has 3 pillars; growth and development; inclusion and; intergenerational equity – sustainable stewardship of natural and financial resources) indicator as an inclusion factor shows its strong statistical insignificance (p-val > 0.9). A similar result is characteristic of the Gini index.

Slightly better statistical estimates are obtained using combinations of these factors with the conditional variable ***res*** (country endowment of natural resources), i.e. in the form of ***idi * res***. In this case, p-val decreases to 0.3 but does not reach the required threshold value (0.1). This result can be interpreted as saying that inclusive development plays a special role for countries with rich natural resources.

Significance appears only when the variable of the proportion of employed in industry (***emp_ind***) is used as an inclusion factor. While not strictly an indicator of inclusiveness, this variable nevertheless reflects the process of improving the quality of economic growth, since in developing countries, as the experience of China, Turkey and Malaysia shows, the contribution of economic growth to the expansion of sustainable and adequately paid employment is primarily associated with the creation of new jobs in non- extractive industries.

The analysis also showed that the most stable factor in all their combinations included in the regression was the factor of expenditure on R&D (***r&dexp)***. This underlines the importance of organizing joint integration projects between Russia and Central Asian countries in the field of research and development to limit irregular migration, which will create a significant number of new jobs in the field of medium- and high-tech industries that have a significant multiplier effect on the entire economy in terms of generating new employment.

5 Conclusion

The tested logical interrelationships and hypotheses made it possible to form guidelines for inclusivity for the post-Soviet space. The analysis showed that the most important condition for inclusive development is the quality of state institutions. In this case, state institutions in the countries of the post-Soviet space need to create conditions for the effective management of migration processes. In our opinion, to increase the inclusiveness of economic growth in labour exporting countries, such as CIS countries, it is necessary to prioritise the creation of conditions for accelerating the development of manufacturing industry.

As world experience shows, the development of manufacturing industry depends heavily on the quality of education in a country. In this case, investment in human capital has an important role to play. The gap in income between CIS countries is also explained by the difference in quality indicators of human potential. Therefore, in order to equalize income between countries, it is necessary to create unified educational standards and interstate centres of competence, the directions of which correspond to the selected priorities of technological development in the countries of the post-Soviet space.

Sharing technological experience is one of the most important conditions for increasing inclusiveness in the world. Therefore, it is necessary to move to a unified technological policy focused on the joint creation of unified technological platforms in key areas of economic activity for the CIS countries (resource conservation, transport, medicine, new materials, information technology, banking systems, etc.). The implementation of a unified technology policy will allow the CIS countries to reduce costs in the transition to the production of new generations of equipment and consumer products, as well as reduce the degree of scientific and technological dependence on western countries and China.

To implement a unified technological policy, it is necessary to create joint integration interstate projects, which, in addition to the technological breakthrough, can significantly mitigate the problem of employment in countries that send migrant workers.

Trade in technology provides the basis for the harmonization of customs regulations, tax policies, accounting, as well as the simplification of procedures for resolving economic disputes and the creation of independent (transnational) judicial bodies to resolve emerging issues.

The successful implementation of the proposed measures requires the effective and coherent work of state institutions in the CIS countries. In our opinion, it is necessary to create a supranational structure that will regulate and direct this process to increase the inclusiveness of the entire post-Soviet space, thereby involving state institutions, business and society, which will ultimately allow all countries to become active participants in the development of the world economy.

References

1. T. V. Merkulova, Economic growth and inequality: Institutional aspect and modeling of the relationship. *Mir Rossii. Sotsiologiya. Etnologiya.* **19**(2), 59–77 (2010)
2. R. Harrod, An Essay in Dynamic Theory. *The Economic Journal.* **49**(193), 14–33, (1939)
3. H. Etzkowitz, L. Leydesdorff, The dynamics of innovation: from National Systems and "Mode 2" to a Triple Helix of university–industry–government relations. *Research Policy.* **29**(2), 109-123 (2000)
4. P. Aghion, S. Durlauf, *Handbook of economic growth*. Amsterdam: Elsevier (2005)
5. T. McKinley, Inclusive Growth Criteria and Indicators: An Inclusive Growth Index for Diagnosis of Country Progress. *ADB Sustainable Development Working Paper Series*. No.14. – Mandaluyong City, Philippines, Asian Development Bank (2010)
6. A. R. Ramos, R. Ranieri, J. Lammes, *Mapping Inclusive Growth*. Working Paper, No.105, International Policy Centre for Inclusive Growth (2013)
7. R. Anand, S. Mishra, and S. J. Peiris, *Inclusive Growth: Measurement and Determinants*. [online]. Available at: https://www.imf.org/external/pubs/ft/wp/2013/wp13135.pdf (2013)
8. L. V. Smorgunov, The State, Cooperation and Inclusive Economic Growth. *Vlast'.* **11**, 22–30 (2017)
9. A. Roy, Inclusive Growth through MFIs' Outreach in Assam. *Asia Pacific Business Review.* **7**(3), 47–62 (2011)
10. S. V. Chepel, A. K. Bondarenko, Is external labour migration factor of economic growth. Econometric analysis and conclusions for the CIS. *Journal of the New Economic Association.* **4**(28), 142–166 (2015)
11. P. Diaz, D. Saenz, V. Kwan, Economic Dynamics and Changes in Attitudes Toward Undocumented Mexican Immigrants in Arizona. *Analyses Of Social Issues And Public Policy.* **11**(1), 300–313 (2011)
12. S. P. Kerr, W. Kerr, Economic Impacts of Immigration: A Survey. *Harvard Business School Entrepreneurial Management Working Paper.* (2008)
13. K. Mayr, The Fiscal Impact of Immigrants in Austria – A Generational Accounting Analysis. *Empirica.* **32**(2), 181–216 (2005)
14. M. Beine, F. Docquier, H. Rapoport, Brain drain and economic growth: theory and evidence. *Journal of Development Economics.* **64**(1), 275–289 (2001)
15. M. Bildirici, S. Sunal, E. A. Alp, M. Orcan, Determinants of Human Capital Theory, Growth and Brain Drain: An Econometric Analysis for 77 Countries. *Applied Econometrics and International Development.* **5**(2), 109–140 (2005)
16. H. Chen, I. Fang, Migration, social security, and economic growth. *Economic Modelling.* **32**, 386–399 (2013)
17. C. Di Maria, E. Lazarova, Migration, Human Capital Formation, and Growth: An Empirical Investigation. *World Development.* **40**(5), 938-955 (2012)
18. S. Diaz-Briquets, J. Perez-Lopez, Refugee Remittances: Conceptual Issues and the Cuban and Nicaraguan Experiences. *International Migration Review.* **31**(2), 411–437 (1997)
19. C. Woodruff, R. Zenteno, Remittances and Microenterprises in Mexico. *Latin American Economics.* (2001)
20. M. Sana, D. Massey, Household Composition, Family Migration, and Community Context: Migrant Remittances in Four Countries. *Social Science Quarterly.* **86**(2), 509–528 (2005)
21. N. Catrinescu, M. Leon-Ledesma, M. Piracha, B. Quillin, Remittances, Institutions, and Economic Growth. *World Development.* **37**(1), 81–92 (2009)
22. Z. Goschin, Remittances as an Economic Development Factor. Empirical Evidence from the CEE Countries. *Procedia Economics and Finance.* **10**, 54-60 (2014)
23. T. H. W. Ziesemer, Worker remittances, migration, accumulation and growth in poor developing countries: Survey and analysis of direct and indirect effects. *Economic Modelling.* **29**(2), 103–118 (2012)
24. K. S. Imai, R. Gaiha, A. Ali, N. Kaicker, Remittances, Growth and Poverty: New Evidence from Asian Countries. *Journal of Policy Modeling.* **36**(3), 524–538 (2014)

25. P. Guha, Macroeconomic effects of international remittances: The case of developing economies. *Economic Modelling*. **33**, 292–305 (2013)

26. E. Nyamongo, R. Misati, L. Kipyegon, L. Ndirangu, Remittances, financial development and economic growth in Africa. *Journal of Economics And Business*. **64**(3), 240–260 (2012)

27. D. Mosse, S. Gupta, M. Mehta, V. Shah, J. Rees, K. Team, Brokered livelihoods: Debt, Labour Migration and Development in Tribal Western India. *Journal of Development Studies*. **38**(5), 59–88 (2002)

28. O. Stark, J. Taylor, S. Yitzhaki, Migration, remittances and inequality. *Journal of Development Economics*. **28**(3), 309-322 (1988)

29. A. Barguellil, M. H. Zaiem, M. Zmami, Remittances, Education and Economic Growth: A Panel Data Analysis. *Journal of Business Studies Quarterly*. **4**(3), 129–139 (2013)

30. A. Kroeger, K. Anderson, *Remittances and Children's Capabilities: New Evidence from Kyrgystan, 2005-2008* [online]. Available at: https://www.diw.de/documents/publikationen/73/diw_01.c.388327.de/dp1170.pdf (2011)

31. R. Lucas, *Lectures on economic growth*. Cambridge, Mass.: Harvard University Press (2002)

32. S. V. Chepel, N. M. Ibragimova, Economic Growth and Accessibility of Higher Education: Hypotheses, Channels, Interrelations and Its Importance For Developing Countries. *Human capital and vocational education*. **4**, 4–16 (2016)

33. World Economic Forum, *Global Risks Report 2016*. [online]. Available at: http://www3.weforum.org/docs/GRR/WEF_GRR16.pdf (2016)

Analysis of Change in the Corporate Sphere of the Czech Republic

Tsolmon Jambal[1,*], *Jaroslav* Stuchlý[1]

[1]Institute of Technology and Business, Faculty of Corporate Strategy, Okružní 517/10, 370 01 České Budějovice, Czech Republic

Abstract. This contribution presents research into the changes taking place in companies in the Czech Republic. The aim was to determine whether companies in the Czech Republic are undergoing fundamental changes, what influences are forcing them to change, and whether managers prefer radical changes to incremental improvements. The research was carried out in the form of a questionnaire survey, which consisted of four sections and 60 questions, conducted by teachers and students of VSTE in České Budějovice in 2019. The article analyses the characteristics of the selected statements and examines the interdependencies between the variables, including the dependence on the identified identification variables.
Keywords: change, radical change, incremental improvement, urgency of change, influence of surroundings

1 Introduction

There are many publications in scientific literature that deal with change management. However, although the issue of change management dates back to the middle of the 20th century, it only recently began to appear in literature published in the Czech Republic. In the Czech Republic, for example, Častorál [1], Kubíčková and Rais [2], Machan [3], Veber [4], Řehoř [5] Bauer and Hauriová [6] and others deal with the issue of change management. In the Czech Republic, the theoretical literature on change management and organizational change has only come to the fore among experts and the public in the last decade. Veber et al. [4] define change management as a complex of management activities that lead to a shift in an individual, team, department, entire organization from the current state to a state required in the future. Řehoř [5] considers change management as a methodology, process, and professional discipline that should facilitate and accelerate the course of change by grouping resources at a given time so that it is possible to effectively implement the chosen strategy and speed up the return on investment.

Worldwide, we find interesting contributions by Drucker [7], Kotter and Cohen [8], Heffernan [9], Hodges [10], Burnes [11], Dawson and Andriopoulos [12], McGrath and Bates [13], Armenakis and Bedeian [14], Masood and Kern [15], Kim and Choi [16], Ma Jiang and Liu [17] French, Sherwood and Bradford [18], Schmuckler [19], Frankenhoff and Grager [20], Darmon [21] and many others.

Armenakis and Bedeian [14] examine theoretical and empirical organizational change through four research areas. The research areas concern content, context, process, and criterion. Content issues focus on the essence of current organizational changes. Context issues focus on the forces or conditions that exist in the external and internal environment of the organization. Process issues concern the actions taken during the implementation of the intended change. Criterion issues concern the results normally assessed within the context of organizational change efforts. The authors also draw attention to the monitoring of affective and behavioural responses to change and the dynamics of current requirements in the workplace.

Kotter and Cohen [8] defined eight stages of successful large-scale change: increase urgency; build the guiding team; get the vision right; communicate for buy-in; empower action; create short-term wins; don't let up; and make change stick. Although the individual steps look simple, none is worth skipping. Kotter writes about evoking a sense of urgency. If a company has the right people, then under certain circumstances there may be a sense of urgency for change almost as a team. Kotter also points out the core of the problem, which is how to change people's behaviour. Behavioural change happens in highly successful situations mostly by speaking to people's feelings and emotions.

Several Bachelor's and Master's theses have focused on this issue, most often reflecting on a situation in a selected company. However, research focused on the Czech Republic is still lacking. This contribution was created within the newly included subject of change management in the field of study Business Economics at the Department of

* Corresponding author: jambal@mail.vstecb.cz

Management of the Institute of Technology and Business in Ceske Budejovice. The aim of this contribution is to examine the content and contextual issues of organizational change in companies in the Czech Republic.

2 Methodology

The research was carried out in cooperation with students of the Institute of Technology and Business in České Budějovice from 28 April 2019 to 28 May 2019 with the intention of increasing the quality of subjects in the field of Business Economics, involving students in research, and obtaining data usable for the creative activities of teachers and students. Students obtained data for the whole of the Czech Republic via an online questionnaire survey:

https://docs.google.com/forms/d/e/1FAIpQLSffbC5y2QswGtNWSEbTzAu8pA8xBVyxWq27ebBvSZgCq5FyA/viewform?usp=sf_link

In total, results were obtained from 433 respondents or managers of companies. The questionnaire consisted of 4 parts: a) general change management in the company; b) aspects of change in relation to corporate culture; c) characterization of changes within the company; d) identification data about the company. For this article, only the data from the first and last parts of the questionnaire are used.

To describe the general management of change from different perspectives, 18 statements were put forward, which the respondents evaluated on a five-point scale. The purpose of doing this was to objectively characterize the statements and statistically evaluate them using averages and distributions of their absolute and relative frequencies. In addition, it enabled the introduction and evaluation of a quantity (so-called "changeability" variable) for describing the general tendency to implement business changes. For "changeability", a variable was taken that is equal to the average evaluation of all 18 statements. To evaluate the changeability variable, its numerical characteristics, graphical course (boxplot diagram and histogram), and analysis of its dependence on the identification variables obtained in part d) were used.

Basic methods of descriptive and inductive statistics, including statistical tests, conditional numerical characteristics and analysis of variance, were used to obtain the results. Statistical analyses were performed in Excel and R [22, 23].

Testing statistical hypotheses is one of the basic methods of statistical induction and one of the simplest methods of quantitative decision theory. When solving a test problem, we look for answers to questions such as: Does smoking cause cancer? A statistical hypothesis is a statement about the parameters of a basic set or about the parameters of multiple sets. We call the statistical hypothesis test the procedure by which, on the basis of the results obtained from a random sample, we verify whether the statistical hypothesis about the population can be considered correct or not. The procedure of statistical testing is as follows: formulation of hypotheses; selection of test criterion and its distribution; choice of significance level; the definition of the critical field; calculation of test criterion value from measured values; and formulation of test conclusion and its factual interpretation.

In order to test the null hypothesis against the alternative, we must choose the test statistics and the critical value. We oppose the null hypothesis H_0: $\Theta = \Theta_0$ and the alternative hypothesis H_1: $\Theta \neq \Theta_0$ (or H_a), where Θ_0 is the hypothetical value of the parameter [23].

3 Results and Discussion

For the following claims, a score of consent from 1 to 5 points was given. 1 means strongly disagree; 2 rather disagree; 3 neutral positions; 4 tend to agree and 5 strongly agree.

Table 1. General statements on the state of change in the company

	Claim	Average score
C01	Our organization is going through a number of serious and often painful changes	2.87
C02	The sense of urgency of the need for change is greater than 10 years ago	3.60
C03	The sense of urgency of the need for change is greater than 5 years ago	3.52
C04	The sense of urgency of the need for change is greater than a year ago	3.15
C05	We have to deal with the effects of strong macroeconomic influences	3.26
C06	We have to deal with strong local competitive influences	3.41
C07	We have to deal with strong international competitive influences	3.15
C08	We have to deal with strong customer influences	3.66
C09	We have to deal with strong supply influences	3.21

 https://doi.org/10.1051/shsconf/20219001007

	Claim	Average score
C10	We have to deal with the effects of globalization in the industry	3.23
C11	The pressure to reduce costs is increasing	4.06
C12	The pressure to increase the quality of products and services is increasing	4.13
C13	The pressure to increase productivity is increasing.	4.10
C14	The pressure to reduce waste of production resources is stronger than before.	3.77
C15	The pressure to reduce personnel costs is stronger than before.	3.42
C16	The pressure to look for new business development opportunities is greater than before.	3.66
C17	Management prioritizes radical changes.	2.94
C18	Management prioritizes incremental improvements.	3.46

Source: Authors.

Claim CO1 examines whether companies are undergoing critical changes. This average rating is the lowest (2.87) of all. 198 respondents (45.73%) answered strongly or rather disagree, 76 respondents (17.55%) had a neutral attitude and 159 respondents (36.72%) answered rather or strongly agree.

Claims CO2 - CO4 examines the urgency of changes over time (1 - 10 years). The value of the urgent need for change for one year is around 3, which is a neutral position. This means that the urgent need for change compared to the previous year is less than five or ten years ago.

Table 2. Urgency of the need for change over time in absolute and percentage frequency

	1	2	3	4	5	Median
The urgency of the need for change is greater than 10 years ago	18 (4.16%)	76 (17.55%)	78 (18.01%)	151 (34.87%)	110 (25.40%)	4
The urgency of the need for change is greater than 5 years ago	16 (3.69%)	66 (15.24%)	91 (21.02%)	197 (45.49%)	63 (14.55%)	4
The urgency of the need for change is greater than a year ago	24 (5.54%)	101 (23.32%)	126 (29.10%)	148 (34.18%)	34 (7.85%)	3

Source: Authors.

Claims C05 - C16 concerned environmental influences and claims C17 – C18 management's preference for fundamental versus gradual change. The managements' of the companies prefer (61.43%) gradual change, with only 30.71% of respondents indicating they prefer radical change. If the average values of the individual influences are ranked, the following results are obtained: increasing the quality of products and services (4.13); productivity growth (4.10); reducing costs (4.06); reducing waste of production resources (3.77); finding new business development opportunities and customer influence (3.66); reduction of personnel costs (3.42); the action of local competitors (3.41); macroeconomic influences (3.26); globalization in the field (3.23); supplier influence (3.21); and international competitive influences (3.15).

The following graph shows the evaluation of the individual change management questions on a scale of 1 to 5.

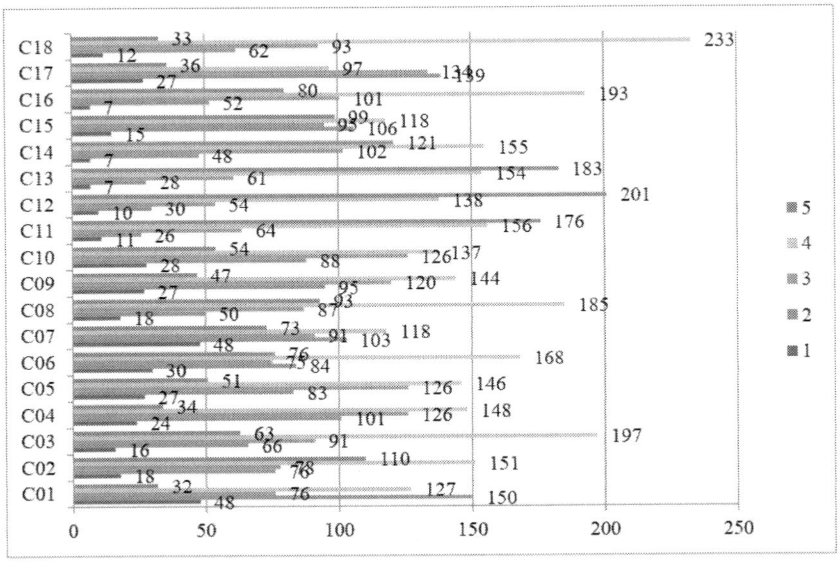

Figure 1. Evaluation of individual statements about change management

Source: Authors.

By averaging the point evaluations for the individual evaluation statements about the management of changes in the companies, a variable was generated, which we have chosen to call the "changeability" variable.

3.1 Numerical characteristics of the changeability variable

The numerical characteristics of the changeability variable are given in the following tables.

Table 3. Numerical characteristics of the changeability variable

Characteristics	Changeability
Average	3.480113
Median	3.555556
Standard deviation	0.5465154
Interquartile range IQR	0.6666667
Coefficient of Variation	0.1570396
Minimum	1.333333
Maximum	4.777778

Source: Authors.

The Shapiro-Wilk test was subsequently applied to determine whether the variability of the changeability variable has a normal distribution. The results were: W = 0.9744; p-value = 6.624 * 10-7. This means that the data are not normally distributed at the 5% level of significance. This is also clear from the boxplot diagram and histogram in Figure 2.

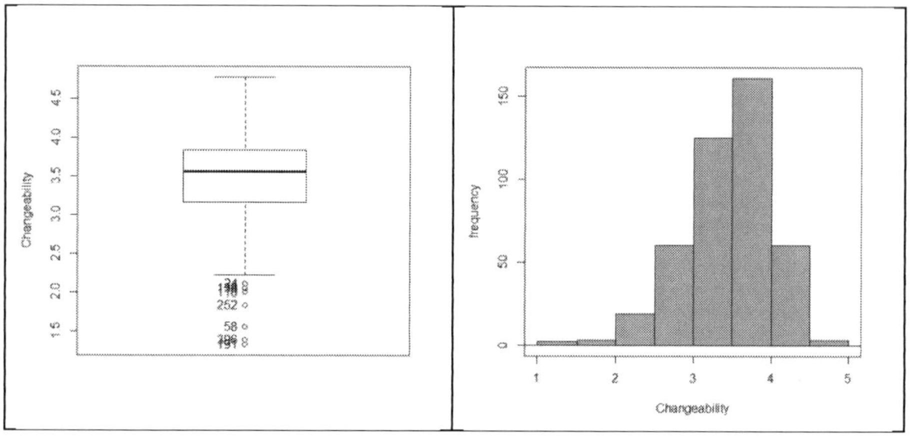

Figure 2. Boxplot and histogram of the changeability variable

Source: Authors.

It is therefore possible to use an asymptotic one-sample test for the population average of the changeability variable. At the 5% level of significance, the null hypothesis was tested to see whether the population average of changeability is equal to 3 (centre of all possible scores). The resulting p-value was $2.2 * 10^{-16}$, which means that the null hypothesis is rejected. The sample average of changeability is 3.48 points and with 95% confidence is located between 3.428 points and 3.532 points.

3.2 Dependence of changeability variable on identification variables

Analysis of variance can be used to verify the dependence of the changeability variable on the form of the enterprise, industry, number of employees, and place of business. The results from the ANOVA tables are summarized in Table 4, where F is the value of the test statistic, Df is the degree of freedom, P is the p-value of the F-test.

Table 4. Analysis of the dependence of the changeability variable on identification variables

	Changeability		
Factor	**F value**	**Df**	**P**
Form of enterprise	2.970	7	0.00512
Industry	2.604	12	0.00235
Number of employees	12.060	3	$1.36*10^{-7}$
Place of business	2.196	13	0.00912

Source: Authors.

At the 5% level of significance, the statistically significant dependence on the variables form of enterprise, industry, number of employees, and place of business was demonstrated. This is also confirmed by the tables of characteristics, where the contingent mean values vary considerably depending on the number of employees.

Tables 5 to 8 show the numerical characteristics of the changeability variable for variability according to form of enterprise, industry, the number of employees, and place of business.

Table 5. Numerical characteristics of the changeability variable by form of enterprise

Form of enterprise	Mean	Standard deviation	data n
Joint-stock company	3.604762	0.4955206	140
Cooperative	3.477778	0.4886924	10
Other	3.444444	0.5241101	7
Another trading company	3.477778	0.3586915	15
Self-employed	3.318182	0.5121056	22
Contributory and budgetary organization	3.500000	0.4297794	14
Limited liability company	3.462281	0.5717998	190
State enterprise	3.180952	0.633366	35

Source: Authors.

Table 6. Numerical characteristics of the changeability variable by industry

Industry	Mean	Standard deviation	data n
Tourism	3.401961	0.5930180	17
Transport and logistics	3.440898	0.5751851	47
Finance	3.138889	1.2177950	2
Other	3.488889	0.5420583	5
Marketing	2.666667	0.5499719	2
Trade	3.515432	0.5337173	90
Food production	3.629630	0.6789001	3
Industry	3.621176	0.4751761	138
Services	3.359903	0.6302419	69
State administration and self-government	3.406907	0.4822028	37
Building and construction	2.953704	0.4613676	6
Education	3.58889	0.5085686	5
Agriculture	3.143519	0.3815750	12

Source: Authors.

Table 7. Numerical characteristics of the changeability variable by number of employees

Number of employees	Mean	Standard deviation	data n
0-9	3.048246	0.7137025	38
10-49	3.449954	0.5241790	121
50-249	3.484634	0.5050497	141
250 and more	3.626149	0.4882231	133

Source: Authors.

Table 8. Numerical characteristics of the changeability variable by location of business

Place of business	Mean	Standard deviation	data n
Prague	3.376296	0.51935939	75
South Bohemian Region	3.452540	0.55489262	199
South-Moravian Region	3.702381	0.31016886	14
Karlovy Vary Region	3.500000	0.83296288	6
Vysocina Region	3.563973	0.39163741	33
Karlovy Vary Region	3.722222	0.07856742	2
Liberec Region	2.388889	1.49278098	2
Moravian-Silesian Region	2.638889	0.43212081	2
Olomouc Region	3.338889	0.43212081	2
Pardubice Region	3.261905	0.39246279	7
Pilsen Region	3.561254	0.55815108	39
Central Bohemian Region	3.676329	0.52578302	46
Usti Region	3.388889	NA	1
Zlín Region	3.433333	0.78015510	5

Source: Authors

4 Conclusion

From the evaluation of the individual statements in part a) we can see that: the first statement "Our organization is going through a number of serious and often painful changes" is the least evaluated; the urgency of the need for change compared to the previous year is less than five or ten years ago; the management of the company prefers gradual changes; from the effect of individual influences, the increase in the quality of products (services) is rated the highest, and international competitive influences are rated the lowest. These results are also confirmed by the graph of the frequency distribution in terms of the predominance of positive or negative evaluations.

The changeability variable, describing the tendency to undertake change, ranges from 1.3 to 3.8 points, is homogeneous and has a slightly asymmetric distribution with a significant shift of the mean to a more positive value, with a mean distribution of 3.48 points. Its values depend significantly on all the identification variables used, mostly on the number of employees of the company and the industry in which it operates, and less so on the form of the enterprise and place of business.

References

1. Z. Častorál, *Strategic management of change and knowledge*. Prague: Jan Amos Comenius University (2010)
2. L. Kubickova, K. Rais, *Change management in companies and other organizations*. Prague: Grada (2012)

3. R. Machan, *Change management*. 2nd ed. Prague: University of Economics and Management (2013)
4. J. Veber, *Innovation management*. Prague: Management Press (2016)
5. P. Řehoř, *Change management*. Ceske Budejovice: University of South Bohemia in Ceske Budejovice (2016)
6. M. Bauer, I. Haburiová, *Leadership using kaizen and lean: fairy tales for tired managers*. Brno: BizBooks (2015)
7. P. F. Drucker, *Management in a time of great change*. Prague: Management Press (1998)
8. J. P. Kotler, D.S. Cohen, *The Heart of Change: Real Stories of How People Change Their Organizations*. Prague: Management Press. (2003)
9. M. Heffernan, *Beyondmeasure: the big impact of small changes*. London: Ted Books, Simon & Schuster (2015)
10. J. Hodges, *Managing and leading people through organizational change: the theory and practice of sustaining change through people*. Philadelphia: Kogan Page (2016)
11. B. Burnes, *Managing Change*. 7th ed. New York: Pearson (2017)
12. P. Dawson, C. Andriopoulos, *Managing Change, Creativity and Innovation*. 3rd ed. London: Sage Publications Ltd. (2017)
13. J. McGrath, B. Bates, *89 most important management theories for practice*. Prague: Management Press (2015)
14. A. Armenakis, A. Bedeian, Organizational change: A review of theory and research in the 1990s. *Journal of Management*. **25**(3), 293–315 (1999)
15. T. Masood, M. Kern, J. Clarkson, Characteristics of changeable systems across value chains. *International Journal of Production Research*. (2020)
16. J. Kim, S. O. Choi, The intensity of organizational change and the perception of organizational innovativeness; with discussion on open innovation. *Journal of Open Innovation: Technology, Market, and Complexity*. **6**(3) (2020)
17. S. Ma, Z. Jiang, W. Liu, A design change analysis model as a change impact analysis basis for semantic design change management. *Proceedings of the Institution of Mechanical Engineers, Part C: Journal of Mechanical Engineering Science*. **231**(13), 2384–2397 (2017)
18. J. R. French, J. J. Sherwood, D. L. Bradford, Change in Self-identity in a Management Training Conference. *The Journal of Applied Behavioral Science*. **2**(2), 210–218 (1966)
19. E. Schmuckler, How to move in Management – Executives guide to changing Jobs. *Personal psychology*. **20**(3), 362–363 (1967)
20. W. P. Frankenhoff, Ch. Granger, Strategic Management – New managerial concept for an era of rapid change. *Long range planning*. **3**(3), 7–13 (1971)
21. R. Y. Darmon, *Leading the sales force: A dynamic management process*. Cambridge: Cambridge University Press (2007)
22. J. Stuchlý, J. Jambal, Customer evaluation of loyalty programs in selected companies of the South Bohemian region. *SHS Web of Conferences – Innovative Economic Symposium 2018 (IES2018): Milestones and Trends of World Economy*. (2019)
23. J. Stuchly, *Statistical data analysis: study support for full-time and combined study: bachelor's study program*. Ceske Budejovice: VSTE (2015)

Comparison of Value Added within EU in Terms of Corporate Investment in Research and Development

Yaroslava Kostiuk[1,*], *Kristina* Korená[1]

[1]Institute of Technology and Business, Faculty of Corporate Strategy, Okruzni 517/10, 37001 Ceske Budejovice, Czech Republic

Abstract. Globalization has accelerated the growth of innovation in many countries outside the European Union. In order to remain competitive, companies and individual EU economies therefore support the development of innovations through investments in research and development.
This contribution is focused on the specification of value added in terms of investments in research and development in the construction and manufacturing sectors. Using statistics of the European Commission for the period 2013 – 2017, a dataset of 269,892 companies in the EU28 countries was generated.
The purpose of this contribution is to determine to what extent investment in research and development participates in the creation of value added. A research question was formulated to determine the ideal ratio between value added and investment.
Methods: To achieve the objectives set, regression analysis of neural networks was applied using Statistica software.
The results of the research indicate that the optimal values in the construction sector were achieved by Great Britain (~EUR 11.3 billion.) and France (~EUR 16.1 billion), and in the manufacturing sector by France (~EUR 11.42 billion), Italy (~EUR 11.41 billion) and Great Britain (~EUR 10.6 billion).
The authors consider the method of regression analysis and neural networks to be appropriate for examining the optimal ratio between investment in R&D and value added.
Keywords: value added; investment; research and development

1 Introduction

Investment in research and innovations has been an important part of the EU´s coordinated response to the pandemic outbreak in 2019/2020, which is necessary for sustaining an inclusive recovery within Europe. These incentives for research and development are relevant because they increase the resistance of the manufacturing sector, competitiveness of economies, prepare individual countries for the future, and play a key role in achieving the goals of the European Union [1]. This contribution deals with the influence of corporate investment in research and development on the value added of the EU countries.

For the EU countries, innovations are an important component of sustainable international competitiveness. It is for this reason that the European Commission creates a multi-annual budget and prepares an investment plan to determine which direction all the member states will go in in terms of investment in research and development. According to the data obtained from [2], in the period 2008-2017, EU member states fell behind other developed economies, such as the USA, Japan and South Korea in the area of research and development.

Foreign-owned companies invest twice as much as Czech-owned companies in research and development and other investments. Czech industry lacks innovation, the development of new technologies, and a focus on research, which limits the development of production with value added [3].

It is therefore necessary to find out what opportunities are lost or how much value-added economies may lose if they do not use innovations in industry and do not use all the subsidies intended for development. It is also necessary to determine the optimal value of investments for the EU28 economies in order to be competitive in the global value chain.

If investment in research and development is so important for creating corporate value added, why is the level of investment not determined by the real needs of companies and the investments possibilities of individual countries? How has investment in research and development evolved since the establishment of the EU and how has the COVID-19 pandemic affected the given field?

The objective of the contribution is to find out to what extent investment in research and development participated in the creation of value added for the EU28 countries in the period 2013-2017. We assume that a strong connection between

* Corresponding author: 26567@mail.vstecb.cz

https://doi.org/10.1051/shsconf/20219001008

the given items will be confirmed. As a result, special attention will be focused on those industries that create the highest value added – manufacturing and construction. The following research question was formulated:

What is the ideal ratio between value added and investment?

By answering the question, it will be possible to find out which member states invest inefficiently and therefore have room for better and more effective use of investment, as well as for achieving overall value added.

One of the most significant effects of globalization has been the increase in global value chains (GVC) [4]. Their modernization is explicitly connected with the long-term growth of companies and product innovations [5]. Globalization has accelerated the increase in innovation in many countries outside the European Union (EU). To therefore remain competitive, companies or individual EU economies support the development of innovations by investing in research and development (R&D).

The most popular definition of investment in scientific literature was created by [6], who defines investment as a current sacrifice for future prosperity, and defines the present as a relatively well-known variable, whereas the future is a mystery. That is why in this case, investment is seen as a certain sacrifice for uncertain profit. [7] defines innovation as new ideas which are in reality applied in various ways to generate new or additional value.

[8] proposed a complex investment framework – the Innovation Value Chain (IVC). In their work, the innovation value chain is seen as a gradual three-phase process: gathering knowledge, transformation of knowledge, and its application [9].

[10] point out that the purpose of innovation is much broader than the simple creation of value for a customer, competitive advantage for companies, and environment for a better quality of life. In accordance with the authors´ view, the objective of innovation should be the creation of a better, intelligent future.

The issue of the influence of investment in R&D was addressed by [11], who examined the effectiveness of research and development using the Malmquist index. The index indicated that 15 out of 28 EU countries evaluated showed an increase in the effectiveness of research and development between 2010 and 2015, with Italy and Germany showing the highest increase.

[12] analysed the relationship between investment in research and development and economic growth in 25 EU countries. Multiple regression analysis was used to confirm the strong impact of investment at the regional level for distances of more than 200 km. In their research, they highlight the importance of vicinity for the transfer of economically productive knowledge.

[9] focused on a comparative panel-data analysis of the innovation value chain in Irish and Swiss production companies. Panel-data analysis was used to examine the influence of innovations on productivity. They concluded that in both countries, the measures adopted to support the development of skills will have a positive impact on the innovation and productivity in both countries. However, there were differences in the case of the relationship between innovation and productivity, as in Ireland, process innovation is more important, while in Switzerland, product innovation is.

[13] investigated how the level of participation in global value chains (GVC) affects corporate innovation activities in Japanese companies. On the basis of the results, he emphasized the importance of GVC in supporting the innovative abilities of a company, which enables companies and countries to upgrade within the value chain.

[14] studied the relationship between the participation in (GVC) and innovative performance at country level. Their findings proved that participation in GVC positively correlates with the level of innovation within a country.

[15] investigated how the combination of approaches related to GVC and innovation systems may enable us to understand the trajectories of education and innovation in developing countries. They found that in some cases, there is an improvement in local innovative capacity with potential positive effects on overall competitiveness, while in other cases, there is only a small improvement or even loss of innovative capacity.

[16] examined the role of intellectual capital in the participation and re-valuation of value in GVC on a sample of European countries. They found that intangible assets other than R&D play a more important role than research and development, while the main driver of the participation is training. The evidence presented in the article also shows that some countries may be trapped in a vicious circle of low investment in activities that create high value added and low competitiveness in international markets.

[17] examined the relationship between innovation and the participation of companies in GVC on a sample of 90 countries. Their empirical estimation was made using a probit model which enabled them to estimate the participation probability of innovative enterprises in value chains. Empirical analysis revealed a positive and significant influence of innovation on companies´ participation in GVC.

[18] used the matching method to test whether the public funding makes Spanish industrial companies invest in research and development. The research results did not show any effect of public funding on investing in R&D. Public

funding is more efficient in the case of small companies and in companies operating in industries with a low level of technology. In their observations, González and Pazó also point to the delayed result in estimating the effect of R&D funding.

EU member countries, which are among the world´s economically most developed economies, should make significant profit from investment in research and development [19]. Since the EU is aware of the importance of these investments in the long term, a strategic plan was accepted in Barcelona in 2002, according to which the investment in innovations within R&D should achieve 3% of GDP per year until the year 2010, whereby two thirds of the investment should come from the private sector [20]. Unfortunately, the achievement of the 3% investment target is not feasible for all member countries because "investment in research and development in the EU declines from the north-west to the south-east" [21, p. 3]. According to Eurostat [22], at the time of the adoption of this strategic plan, the required level of investment was achieved only in Finland and Sweden. Other EU member states showed much lower levels of investment in R&D, e.g. in the Czech Republic, it was 1.1% of GDP. The goal to get close to the required value by 2010 was very ambitious and other EU countries did not manage to achieve it after overcoming the crisis of 2008. According to the latest data, for 2018, other member countries have since achieved the goal, e.g. Denmark, Germany, Austria, Slovenia. Investment in R&D is a driver of innovation. As innovation is synonymous with change, EU economies need significant changes in their budget policies to gradually strengthen their own value added by means of innovation.

To compare global investment processes among the members of the United Nations (UN), investment in R&D within the overall context of innovative development is monitored by means of Global Innovation Index (GII). In 2017, a total of 126 countries were evaluated within GII. The highest GII value was achieved by Switzerland, followed by the Netherlands, Sweden, Great Britain, Singapore, the USA, and Finland. The Czech Republic occupied the 24th position, while in 2018, it dropped to 27th position [23].

Every two years since 2002, the Czech Statistical Office has been conducting statistical surveys on the innovative activities of companies (see Figure 1). Innovative companies in the Czech Republic are mostly foreign controlled companies, which are mostly large companies operating in industry. At the same time, the share of foreign controlled companies is high, especially in terms of medium-sized and large companies (50 – 75%).

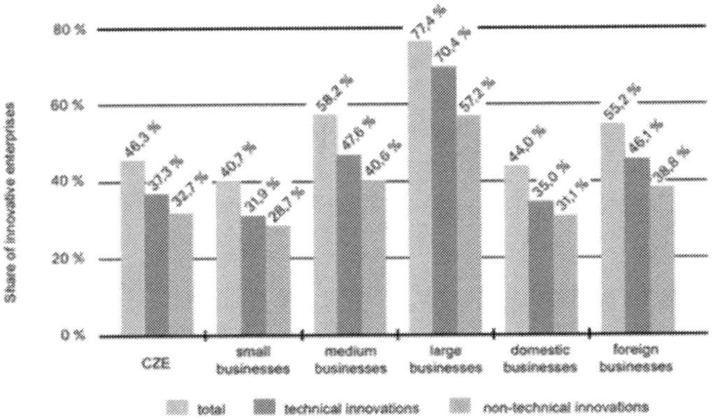

Figure 1. Share of innovative Czech companies by type of innovation (2014–2016)

Source: [23].

This indicates their importance for the innovative performance of the Czech Republic and the necessity for cooperation between the public sector and this segment of R&D [23].

For research and development to be a driver of economic growth in the EU, investment in this field should be properly distributed in order to transform the European economy as a whole [24]. People often point to the delay between economic activities and their consequences, referred to as time lag, which starts with the investment in research and development and ends with the actual economic return on investment [25]. From the macroeconomic perspective, EU funds are considered as an attractive instrument for funding investment opportunities, especially in times of crisis, when private investment declines [26]. The global crisis caused by the COVID-19 pandemic is growing at an exponential rate, and the associated growing uncertainty is leading to the loss of investment and escalating fluctuations in international trade [27]. In our opinion, the COVID-19 pandemic has undermined the overall economic ambitions of the world economies, and as research and development needed financial investment even before the pandemic outbreak, it will be even more difficult to achieve the goals set after the pandemic ends. The growth of investment can be assumed in sectors such as health, logistics and digital technologies, which will subsequently be reflected in their increased value added. However, sectors such as manufacturing, construction and logistics have become less attractive during the pandemic. This is evidenced by reduced production in large manufacturing companies, the cancellation of many flights due to the reduction of travel opportunities, etc. As [24] state, the achievement of national goals is not only a question of financial possibilities, but also of mutually beneficial research alliances between the public and private sectors. In the history of European investment in R&D, it has become commonplace that the EU budget includes funds for innovation; however, the number of patents is not enough to obtain support from the state. This problem is also associated with the lack of experts and researchers who might put forward and adapt innovations to the actual needs of states and according to EU priorities. [19] argue that realistic national goals may stimulate patent activities provided that there is a suitable environment and cooperation between public and private research activities. According to the authors, investment in research and development is of the highest priority in countries that lie in the north-west of the block (Scandinavia), whilst the priority is declining in those countries that lie in the south-west of the block (Mediterranean and Balkans). Growth potential is unevenly distributed among the EU member states; as a global player, the EU must ensure that the potential growth in R&D supports the economic recovery of the EU as a whole [24].

The theoretical framework shows that the issue of investment in research and development has a positive impact on the development of the value added chain. Moreover, studies have repeatedly confirmed the relationship between investment in R&D and economic growth, both at the level of states and regions, as well as companies.

The aforementioned studies included more than two factors that would affect the relationship between innovation and global value chains, which naturally enabled a more detailed description of this relationship. At the same time, however, the studies were not focused on determining the optimal value of R&D investment for individual industrial sectors. In our research, regression analysis is used to focus on two data time series, the use of which will make it possible to determine the optimal level of investment required to achieve the desired indicator of value added.

2 Methodology

The aim of this contribution is to analyse the investment of individual countries and the generated value added. The answer to the research question is the ideal ratio between investment and value added. To avoid and eliminate extreme values it is appropriate to use a regression model. As stated above, information on country-specific investment and the generated value added will be used for the analysis. For the purposes of this contribution, data based on published statistics of the European Commission for the period 2013-2017 were used:

1) Data on investment in research and development in the EU28 countries in the manufacturing, construction and sectors were used from published Eurostat data [1].

2) Data on the value added of individual countries were drawn from the Eurostat-OECD Entrepreneurship Indicator Programme [1].

The generated dataset included a total of 269,892 companies from across the European Union, of which 74,759 companies from the construction sector and 195,133 companies from the manufacturing sector.

For all the calculations, TIBCO's Statistica software (version 13.0) was used. For the analysis, artificial neural networks are used to examine the dependence of investment and value added in the construction and manufacturing sectors of individual EU countries in the period 2013-2017. At the same time, the relationship will be examined through the application of neural networks. Neural networks provide efficient image recognition, prediction, optimization and control of processes, and adaptation of linear and nonlinear functions [28].

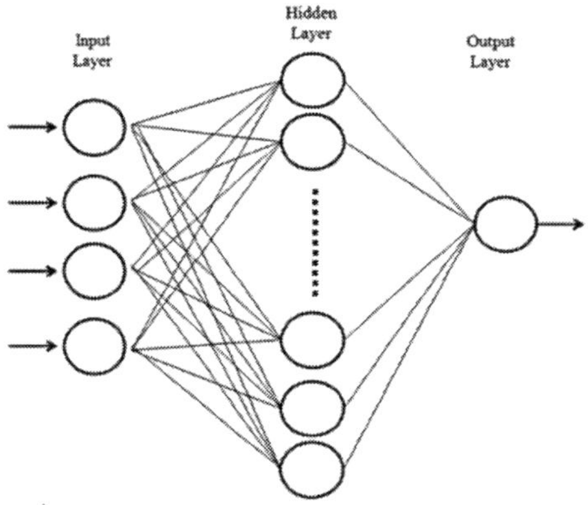

Figure 2. Neural networks

Source: [29].

Specifically, these will be multilayer perceptron networks (hereinafter MLP) and neural networks of radial basis function (hereinafter RBF). The data will be entered from the aforementioned sources, i.e. Eurostat. The Data mining tool neural networks will be used in the Statistica software, specifically, a section called Regression. First, the data is selected. The investments of a specific state are the independent variables. We understand the value added to be an dependent variable. Subsequently, the data is divided into three sets: training, testing and validation. The ratio is 70:15:15. Table 1 provides information on the basic statistics of the input data.

Table 1. Basic statistics of input data

Statistics	VA Total construction	Investment construction	VA Total manufacturing	Investment manufacturing
Minimum (Train)	616.8	10.55	707.0	4.54
Maximum (Train)	111850.8	53359.10	592027.2	58493.50
Mean (Train)	19048.2	4701.88	70990.9	4327.45
Standard deviation (Train)	28397.7	10861.67	110342.4	10008.83
Minimum (Test)	497.7	7.26	1883.1	23.55
Maximum (Test)	124704.8	58493.50	519792.5	49482.40
Mean (Test)	33307.8	6977.28	58409.4	4308.50
Standard deviation (Test)	41822.9	14455.32	122510.5	11770.55
Minimum (Validation)	850.6	16.50	4541.6	23.05
Maximum (Validation)	109303.8	10453.63	569863.5	53359.10
Mean (Validation)	19466.6	2271.15	97360.2	5983.40
Standard deviation (Validation)	27147.8	11656.49	125406.3	11247.96
Minimum (Overall)	497.7	7.26	707.0	4.54
Maximum (Overall)	124704.8	58493.50	592027.2	58493.50
Mean (Overall)	21199.3	4679.12	72960.6	4561.31
Standard deviation (Overall)	30397.8	10719.18	116861.0	10607.58

Source: Authors.

The input data concerns data on the amount of investment in research and development and value added in the construction and manufacturing sectors for the period 2013-2017. Subsequently, on the basis of the training dataset, suitable neural networks are identified that best describe the relationship between the independent and dependent variables. On the basis of the testing and validation datasets, it is then verified whether the identified neural structures are adequate and faithfully describe the situation.

10,000 neural structures are generated, of which the top 5 in terms of performance and error are retained. For MLP, a minimum of 2 and a maximum of 20 neurons in the hidden layer applies. To activate the inner and outer layers of neurons, the following functions are used:

- Identity;
- Logistics;
- Hyperbolic tangent;
- Exponential;
- Sine.

In the case of RBF, a minimum of 11 and a maximum of 30 neurons in the hidden layer applies.

3 Results and Discussion

Based on the procedure above, the following statistical calculations were made for the construction and manufacturing sectors

Construction sector. Table 2 provides an overview of the five most successful neural networks that were retained.

Table 2. The five most successful neural networks for the construction sector

Index	Network title	Training perf.	Test perf.	Validation perf.	Training error	Test error	Validation error	Training algorithm	Error function	Hidden activation	Output activation
1	RBF 1-16-1	0.741394	0.698510	0.886277	28507279	50464189	18596061	RBFT	SOS	Gaussian	Identity
2	RBF 1-18-1	0.853169	0.436614	0.895073	15854994	121745126	1167300	RBFT	SOS	Gaussian	Identity
3	RBF 1-18-1	0.805408	0.739256	0.902445	20644240	45729420	6074010	RBFT	SOS	Gaussian	Identity
4	RBF 1-21-1	0.804324	0.599527	0.893076	20956051	67124939	4037686	RBFT	SOS	Gaussian	Identity
5	RBF 1-20-1	0.721194	0.628954	0.898485	27961883	63696129	1210406	RBFT	SOS	Gaussian	Identity

Source: Authors.

The identified neural networks were selected by the programme from 1000 neural structures on the basis of performance. For these networks, the basic prediction statistics were subsequently calculated, as presented in Table 3.

Table 3. Basic prediction statistics for investment (construction sector)

Statistics	Predictions statistics. Target: Investment				
	1.RBF 1-16-1	2.RBF 1-18-1	3.RBF 1-18-1	4.RBF 1-21-1	5.RBF 1-20-1
Minimum prediction (Train)	-6462.8	-441.5	-159.9	-3475.1	-5514.1
Maximum prediction (Train)	34384.4	53756.8	32210.5	36954.3	45465.3
Minimum prediction (Test)	-6442.0	-247.9	-34.1	-355.5	-2311.1
Maximum prediction (Test)	29085.4	52616.3	28394.6	35761.2	14437.2
Minimum prediction (Validation)	-4767.2	-231.3	-56.3	-736.7	-1602.9
Maximum prediction (Validation)	23667.4	14158.2	22428.0	20886.3	14437.2
Minimum residual (Train)	-24575.8	-16872.4	-23547.6	-28330.4	-18554.6
Maximum residual (Train)	23624.4	33657.5	27201.1	23280.9	37952.0
Minimum residual (Test)	-13494.8	-36407.2	-12804.0	-20170.6	-11136.4
Maximum residual (Test)	34855.7	44342.2	36076.9	37634.2	44056.3
Minimum residual (Validation)	-13213.8	-3704.6	-11974.3	-10432.7	-3983.6
Maximum residual (Validation)	4821.8	1700.2	795.5	2911.1	2086.7
Minimum standard residual (Train)	-4.6	-4.2	-5.2	-6.2	-3.5
Maximum standard residual (Train)	4.4	8.5	6.0	5.1	7.2
Minimum standard residual (Test)	-1.9	-3.3	-1.9	-2.5	-1.4
Maximum standard residual (Test)	4.9	4.0	5.3	4.6	5.5
Minimum standard residual (Validation)	-3.1	-3.4	-4.9	-5.2	-3.6
Maximum standard residual (Validation)	1.1	1.6	0.3	1.4	1.9

Source: Authors.

The basic statistics for R&D investment predictions from Table 3 and their relation to VA are graphically presented in Figure 3. With the help of the Statistica software, the five most successful neural networks (RBFs) were identified that best described the logic of the regression function. The identified neural networks point in the graph where each of the EU28 countries is located with regards to the optimal amount of investment in R&D for the period 2013–2017.

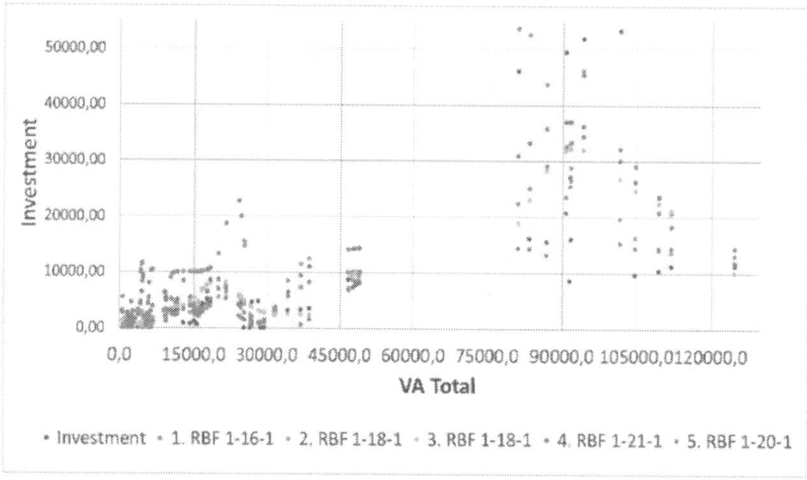

Figure 3. Relation between VA and R&D investment

Source: Authors.

A comparison of the prediction statistics and the real data from the five neural networks showed that neural network No. 1 (RBF 1-16-1) was the most successful, which is graphically presented in Figure 4. The searched values on the (x, y) axis are closest to the point (0, max), i.e. where the least waste occurs in the ratio between investment and value added.

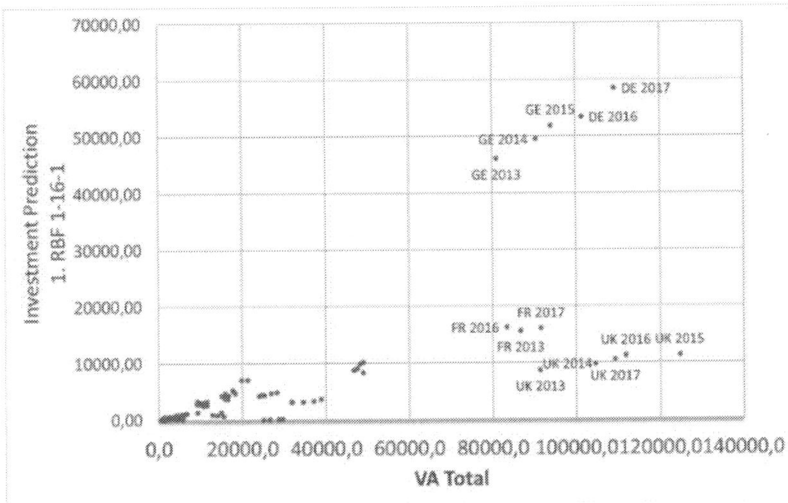

Figure 4. The most successful neural network No. 1 (RBF 1-16-1)

Source: Authors.

In this case, when analysing neural network No. 1 (RBF 1-16-1), it is clear from Figure 4 that for the construction sector, the most optimal ratio of R&D investment and value added was achieved by the United Kingdom during the monitored period (2013 - 2017). The most successful year for the United Kingdom as a member of the EU28 was 2015. This year, the value of investment in R&D reached EUR 11.3 billion.

France was another country that, according to neural network No. 1 approached the ideal point (0, max) on the graph. In 2013, investment in R&D in France amounted to EUR 15.6 billion, and in 2017, EUR 16.1 billion, an increase of EUR 500 million. Despite achieving approximately similar levels of value added, Germany invested more than the United Kingdom during the period under review, but did not reach similar optimum values.

Manufacturing sector. The five most successful neural networks that best describe the logic of the regression function are presented in Table 4.

Table 4. The five most successful neural networks for the manufacturing sector

Index	Network title	Training perf.	Test perf.	Validation perf.	Training error	Test error	Validation error	Training algorithm	Error function	Hidden activation	Output activation
1	MLP 1-5-1	0.985507	0.995239	0.997513	1428085	725669.2	1246492	BFGS 66	SOS	Tanh	Logistic
2	MLP 1-3-1	0.984483	0.994447	0.996925	1525709	769331.4	1505471	BFGS 748	SOS	Tanh	Logistic
3	MLP 1-3-1	0.984451	0.994435	0.996969	1528674	770448.9	1493403	BFGS 9999	SOS	Logistic	Logistic
4	MLP 1-6-1	0.985452	0.995294	0.997145	1434234	637001.4	870003	BFGS 140	SOS	Exponential	Exponential
5	MLP 1-8-1	0.987951	0.996519	0.997107	1226937	572504.8	561127	BFGS 51	SOS	Logistic	Exponential

Source: Authors.

Multi-layer perceptron (MLP) networks consist of at least three layers of nodes: input layer, hidden layer, and output layer. The time series are divided into three sets – training, testing and validation. In the hidden and output layers, the following activation functions were considered: Logistic, Exponential and Tanh. A total of 10,000 neural networks were generated, out of which 5 with the best characteristics were retained and used for the calculation of the basic prediction statistics (see Table 5 below).

Table 5. Basic prediction statistics for investment (manufacturing sector)

Statistics	Predictions statistics. Target: Investment				
	1.MLP 1-5-1	2.MLP 1-3-1	3.MLP 1-3-1	4.MLP 1-6-1	5.MLP 1-8-1
Minimum prediction (Train)	4.54	4.54	4.54	4.54	15.17
Maximum prediction (Train)	58493.29	58493.50	58493.50	59111.59	55018.81
Minimum prediction (Test)	4.54	4.54	4.54	4.54	20.60
Maximum prediction (Test)	47916.50	49546.50	49590.97	49677.17	51255.03
Minimum prediction (Validation)	4.54	4.54	4.54	4.55	42.35
Maximum prediction (Validation)	58492.41	58493.50	58493.50	56296.32	54627.74
Minimum residual (Train)	-4004.35	-4090.89	-4055.54	-3784.10	-2736.97
Maximum residual (Train)	5498.27	6048.29	6064.35	6163.58	5081.69
Minimum residual (Test)	-3366.29	-3614.50	-3626.81	-3478.81	-2616.57
Maximum residual (Test)	1921.24	2106.25	2090.51	1948.49	1986.08
Minimum residual (Validation)	-5133.31	-5134.40	-5134.40	-2937.22	-2855.58
Maximum residual (Validation)	1600.28	1694.08	1678.75	1212.40	825.34
Minimum standard residual (Train)	-3.35	-3.31	-3.28	-3.16	-2.47
Maximum standard residual (Train)	4.60	4.90	4.90	5.15	4.59
Minimum standard residual (Test)	-3.95	-4.12	-4.13	-4.36	-3.46
Maximum standard residual (Test)	2.26	2.40	2.38	2.44	2.62
Minimum standard residual (Validation)	-4.60	-4.18	-4.20	-3.15	-3.81
Maximum standard residual (Validation)	1.43	1.38	1.37	1.30	1.10

Source: Authors.

The relationship between the basic prediction statistics for investment in R&D (see Table 5) and value added are graphically presented in Figure 50. With the help of the Statistica software programme, the five most successful neural networks were identified (all of them MLP networks). In the graph below, each of the neural networks shows a point where the EU28 countries are situated with regards to the predicted optimal amount of investment in R&D.

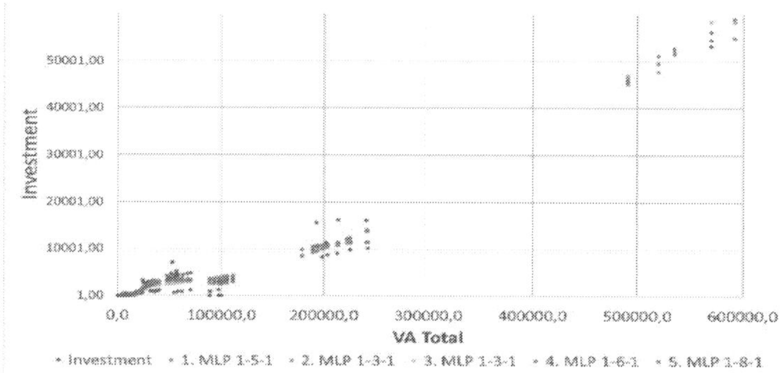

Figure 5. Relationship between VA and investment in R&D

Source: Authors.

A comparison of the prediction statistics and the real data obtained from the five neural networks showed that neural network No. 5 (MLP 1-8-1) was the most successful (see Figure 6).

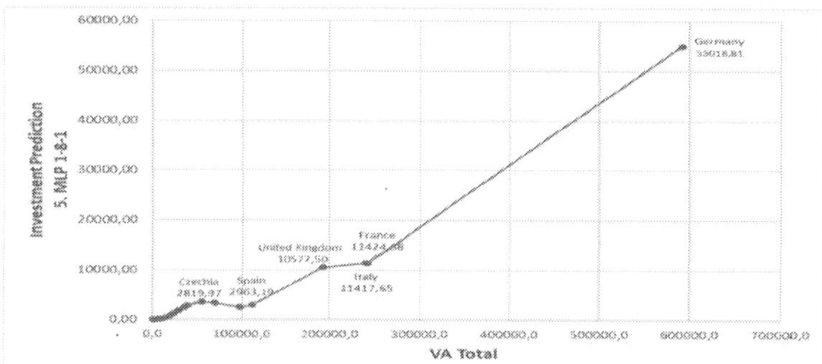

Figure 6. The most successful neural network No. (MLP 1-8-1)

Source: Authors.

As the VA for the monitored period 2013 - 2017 had similar values, for the clarity of the graphical representation of neural network No. 5 (MLP 1-8-1), the year 2017 was selected to make an evaluation of the EU28 countries.

As can be seen in Figure 6, in the manufacturing sector, the optimal ratio between value added and investment in R&D was achieved by France (~EUR 11.42 billion), Italy (~EUR 11.41 billion) and Great Britain (~EUR 10.6 billion).

Like in the construction sector, Germany is a country with high investment in R&D in the manufacturing sector. The funds invested are subsequently reflected in the amount of value added. Germany is undoubtedly a leader in the creation of VA in various industries. However, in terms of the 1:1 ratio (investments: value added), it does not achieve the same optimal results in the manufacturing industry as, for example, France, Italy and Great Britain, where the least waste occurs. In the graph, the Czech Republic achieves a value of EUR 2.8 billion, which, in terms of the 1:1 ratio regarding the optimality of the funds invested, is a positive result.

In the research part, the authors examined the ideal ratio between value added and investment in R&D in the EU28 countries in the manufacturing and construction sectors. Regression analysis of neural networks was used to analyse the most successful neural networks in these sectors. Scatter plots were used to find the point (0, max) at which the least amount of investment in R&D was wasted.

In the construction sector, the optimal values were achieved by Great Britain in the year 2015 (EUR 11.3 billion). For many years, Great Britain was one of the EU leaders in creating value added. It is clear that with Brexit at the beginning of 2020, the European Union lost its leader in the creation of value added in the European market. This was reflected in the global market where there is strong competition between Asian and European countries in terms of creating value added. Germany, as another EU leader in creating value added in the construction sector, shows a high level of creating VA during the monitored period, which can be explained by the high standard of living in the country, which is constantly developing itself, as a result of which the construction industry is prospering. Despite the high VA, which is the objective of all national economies, Germany invests high amounts of money in the given sector, but no longer reaches the optimal ratio between the examined variables, unlike, for example, Great Britain and France. The high investment in the given sector in this developed country can be explained by high wage costs and other factors, which are the highest in the EU.

In the manufacturing sector, the optimal ratio between VA and investment in R&D was achieved by France, Italy and Great Britain. In this sector, Germany did not achieve the optimum between the examined variables, despite the fact that it achieved the highest values. The ideal ratio between investment in R&D and the resulting VA are therefore the values of this ratio in which investments are not wasted.

4 Conclusion

The objective of this contribution was to find to what extent investment in research and development participates in the creation of value added for the EU28 countries. This objective was achieved by means of regression analysis. Neural networks were used to determine the optimal ratio of investment in R&D to value added. The ratio was shown in a scatter plot, which also showed that the sought optimum of invested funds in R&D is achieved if the amount of value added is higher than the investment in R&D.

The results of the research showed that optimal values in the construction sector were achieved by Great Britain during the monitored period (2013-2017); the most successful year for Great Britain as a member of EU28 was 2015. In this year, the value of investment in R&D reached ~EUR 11.3 billion. France, with investment of ~EUR 16.1 billion in 2017, came in second place. Despite achieving approximately the same amount of value added and investing more than Great Britain during the monitored period, Germany did not achieve similar optimal values.

In the manufacturing sector, the optimal ratio between VA and investment in R&D was achieved by France (~EUR 11.42 billion), Italy (~EUR 11.41 billion) and Great Britain (~EUR 10.6 billion). As in the case of the construction sector, Germany is a country that invests highly in R&D also in the manufacturing sector.

Limitations of the research. In this case, missing data on some countries in Eurostat was a limitation. The authors believe that the data published by the EU countries are correct and calculated using the same methodology.

Practical application. Although the EU countries use the same currency, the standards of living and the historical evolution of the sectors they specialize in makes it difficult to create and compare a dataset for the EU28 countries. However, such comparisons are very useful for the national economies to be able to judge their competitiveness in the European and global market.

Possible direction of further research. Further research could compare the influence of investment on value added using regression analysis, but with a time lag. This would, according to the authors, better describe the relationship between investment and value added.

When comparing the statistical data for the period 2013-2017 on the basis of the predictions made by the identified neural networks using the Statistica software programme as artificial intelligence, it was observed how the given programme was trying to predict the future development of investment in R&D. Its predictive abilities were able to provide a rough estimate that was quite close to the actual value; however, 2 years later. Using the assumption of the possible development, it was able to predict the development of investment in R&D. When dealing with the issue of investment, it is necessary to take into account the time lag with regards to returns, which is more than one year in most cases. The application of neural networks can therefore be considered a suitable method for predicting possible developments. Of course, no artificial intelligence is able to predict the development with complete accuracy if it does not include uncontrollable factors, such as economic crises, ecological disasters, or pandemics in its calculations. However, its predictive abilities can be useful for the purposes of strategic planning of companies in a given sector.

References

1. European Commission, *European Innovation Scoreboard 2020*. Luxembourg: Publications Office of the European Union (2020)
2. European Statistical Office (Eurostat), *Smarter, greener, more inclusive? Indicators to support the Europe 2020 strategy*. Luxembourg: Publications Office of the European Union (2020)
3. Technology Centre of the CAS, *Potenciál růstu soukromých investic ve výzkumu, vývoji a inovacích mající vliv na strukturu ekonomiky, zaměstnanost a veřejné rozpočty* [*Growth potential of private investments in research, development and innovation affecting the structure of the economy, employment and public budget*]. [online]. Available at: https://www.spcr.cz/aktivity/z-hospodarske-politiky/12541-cesko-potrebuje-zlepsit-podminky-pro-vyzkum-vyvoj-a-inovace (2018)
4. T. Ylömäki, Global Value Chain Upgrading. *The Research Institute of the Finnish Economy (ETLA), ETLA Working Papers*. **36** (2016)
5. C. Pietrobelli, R. Rabellotti, Global value chains meet innovation systems: Are there learning opportunities for developing countries? *World Development*. **39**, 1261–1269 (2011)
6. J. Hirshleifer, Investments Decision under Uncertainty: Choice - Theoretic Approaches. *The Quarterly Journal of Economics*. **79**, 509–36 (1965)
7. S. Lee, D. Olson, *Convergenomics: Convergenomics: Strategic Innovation in the Convergence Era*. Surrey: Gower Publishing (2010)
8. M. T. Hansen, J. Birkinshaw, The Innovation Value Chain. **85**, *Harvard Business Review*. 121–130 (2007)
9. S. Roper, S. Arvabitis, From knowledge to value added: A comparative, panel-data analysis of the innovation value chain in Irish and Swiss manufacturing firms. *Research Policy*. **41**(6), 1093-1106 (2012)
10. S.M. Lee, S. Trimi, Innovation for creating a smart future. *Journal of Innovation & Knowledge*. **3**(1), 1-8 (2018)
11. M. Halaskova, B. Gavurova, S. Korony, Change of EU28 countries research and development indicators between 2010 and 2015. *Economics and Sociology*. **13**(1), 230–248 (2020)
12. A. Rodríguez-Pose, R. Crescenzi, Research and development, spillovers, innovation systems, and the genesis of regional growth in Europe. *Regional studies*. **42**(1), 51–67 (2008)
13. S. Kergroach, National innovation policies for technology upgrading through GVCs: A cross-country comparison. *Technological Forecasting and Social Change*. **145**, 258–272 (2019)
14. L. Tajoli, G. Felice, Global Value Chains Participation and Knowledge Spillovers in Developed and Developing Countries: An Empirical Investigation. *The European Journal of Development Research*. **30**, 505–532 (2018)
15. R. Lema, R. Rabellotti, P. Gehl Sampath, Innovation Trajectories in Developing Countries: Co-evolution of Global Value Chains and Innovation Systems. *The European Journal of Development Research*. **30**, 345–363 (2018)
16. C. Jona-Lasinio, S. Manzocchi, V. Meliciani, Knowledge based capital and value creation in global supply chains. *Technological Forecasting and Social Change*. **148** (2019)
17. K. Reddy, R. Chundakkadan, S. Sasidharan, Firm innovation and global value chain participation. *Small Business Economics*. (2020)
18. X. González, C. Pazó, Do public subsidies stimulate private R&D spending? *Research Policy*. **37**, 371–389 (2008)
19. Z. Potuzakova, J. Ohm, R&D Investments, EPO Patent Applications and the Economic Heterogeneity within the EU. *Review of Economic Perspectives*. **18**(2), 177–191 (2018)
20. Archive of European integration, *Barcelona European Council 15-16 March 2002. Presidency conclusions*. [online]. Available at: http://aei.pitt.edu/43345/ (2002)

21. N. Albu, Research and Development Spending in the EU: 2020. Growth Strategy in Perspective. *SWP Working Paper.* **1**(8) (2011)
22. Eurostat. *Gross domestic expenditure on R&D* [online]. Available from: https://ec.europa.eu/eurostat/tgm/table.do?tab=table&init=1&language=en&pcode=t2020_20&plugin=1 (2020)
23. Office of the Government of the Czech Republic, *Innovation Strategy of the Czech Republic 2019—2030* [online]. Available at: https://www.vyzkum.cz/FrontAktualita.aspx?aktualita=867990 (2019)
24. A. Ruser, H. K. Anheier, The EU's Future Role on the Global Stage. *Global Policy.* **5**(1) (2014)
25. J. W. Lee, Lagged Effects of R&D Investments on Corporate Market Value: Evidence from Manufacturing Firms Listed in Chinese Stock Markets. *The Journal of Asian Finance, Economics and Business.* **7**(8), 69–76 (2020)
26. C. T. Albulescu, D. Goyeau, EU Funds Absorption Rate and the Economic Growth. *Timisoara Journal of Economics and Business.* **6**(20), 153–170 (2013)
27. P. K. Ozili, T. Arun, Spillover of COVID-19: Impact on the Global Economy. *SSRN Electronic Journal.* (2020)
28. I. S. Didych, O. Pastukh, Y. Pyndus, O. Yasniy, The evaluation of durability of structural elements using neural networks. *Acta Metallurgica Slovaca.* **24**(1), 82–87 (2018)
29. M. Vochozka, J. Horak, T. Krulicky, Innovations in Management Forecast: Time Development of Stock Prices with Neural Networks. *Marketing and Management of Innovations.* **2020**(2), 324–339 (2020)

Process Model of the Economic Efficiency of the Financial Administration of an EU Member State

Sebastian Kot[1], *Ján* Dobrovič[2*]

[1]The Management Faculty, Czestochowa University of Technology, Armii Krajowej 19B, 42-201 Czestochowa, Poland
[2]Institute of Technology and Business, Faculty of Corporate Strategy, Department of Management, Okružní 517/10, 370 01 České Budějovice, Czech Republic

Abstract: The current reform of Slovakia´s tax system is a part of the country´s comprehensive tax and levies reform. The reform should contribute to improving and making public finances more efficient. However, its implementation in practice is problematic due to the attitudes and problems of both the professional public and business entities. The aim of this contribution is to help students and business entities to understand the tax system reforms. The contribution is based on information from the main actors of the tax reforms. The data were obtained during 3 months (2019 – 2020) using the questionnaire method. These were processed using basic statistical methods of the descriptive type, followed by an analysis of the principal factors and factor analysis. The factor analysis was crucial for our contribution because it identified 5 significant indicators with regards to the assessment of the reforms and the functioning of the tax system. Subsequently, a correspondence analysis was conducted of the factor "Electronisation/computerisation - greater user comfort" to find out the differences in the perception hereof among the respondents. The contribution therefore provides a new perspective on the evaluation and functioning of the reforms of the Financial Administration's management system in the Slovak Republic.
Within the framework of the research into this matter, and with a view to increasing the efficiency of the system globally, we also looked at the existing functional organizational structures and tax administration systems in Hungary, Poland, Czech Republic and Slovenia. On the basis of trend analysis, we can assume that the upcoming reform of the Tax and Customs Administrations will significantly contribute to the increasing efficiency of the system and to the positive perception of taxes, which are currently viewed as a socially unpopular obligation.
Keywords: tax system reform; financial administration; taxes; factor analysis; principal components

1 Introduction

The management of the Tax Administration in Slovakia is governed by principles introduced after 1989 [9, 13]. Despite efforts in recent years to decrease the number of employees, the system of administration of taxes in Slovakia can be considered to be inefficient in terms of the competitiveness of the state [10]. At present, the general tendency is to complain about the rigid, unfriendly, bureaucratic apparatus, which brings about an unwillingness of citizens to comply with tax obligations and the stricter regulations of the state. The consequences of this are the aforementioned decrease in competitiveness of the state and a lack of credibility in the eyes of citizens [11, 19].

The aim of this contribution is to help students and business entities to understand the tax system reforms. The contribution is based on information from the main actors of the tax reforms.

1.1 Trends in the management of the Tax Administration in the Slovak Republic

The starting point here is the programme declaration of the government of 4 November 2002, which states, under the section "Economic Policy", the following objectives in the area of the administration of taxes: simplify tax legislation; amend parts of tax laws which allow for ambiguous interpretation; simplify the system of penalties; decrease direct taxes; shift the tax burden from direct taxes to indirect taxes; review and re-evaluate the application

* Corresponding author: 24574@mail.vstecb.cz

of property tax rates; unify income tax rates; analyse possibilities of introducing a single tax (rate); strengthen own tax incomes of municipalities; determine own tax incomes of higher territorial units; ensure strict, direct, just and efficient tax collection and decrease tax rates; reduce tax evasion; create a new system of horizontal financial settlement [1, 14, 22].

The Slovak Republic, not only thanks to the tax reforms of 2004, which saw the introduction of a single tax rate, has joined the progressive states of the European Union and has significantly strengthened its attractiveness and competitiveness [25]. From the point of view of the levels of managing taxes within the Slovak Republic, the current situation can officially be defined as a two-level management system. However, by transposing some of the competencies of the Tax Directorate of the Slovak Republic on to the Branch Offices of the Tax Directorate of the Slovak Republic (hereinafter "BOTD"), it is actually a three-level management system, based on the need to efficiently manage 102 local tax offices, which cannot be assured from one central point [1]. The organization of the Tax Administration in this way is not ideal from several reasons [2, 15]:

the performance of the main business processes is scattered throughout the whole territory, when each local tax office (whether small, middle-sized or large) carries out all the processes related to the administration, audit and enforcement of taxes, disabling thus the optimization of the performance of these processes as well as costs for their performance from the point of view of the tax administration as a whole [3, 29, 20];

the system of the distribution of local tax offices is not flexible enough because it does not enable adjustments to the allocation of the main organizational units to the needs of the taxpayers [4, 16, 27];

BOTD have become an administrative level of management within the current system of management, and for quite some time it has been clear that there is a need to concentrate the performance of certain processes (such as book-keeping, salaries) because these are unnecessarily split between TDSR and BOTD, which therefore increases the intensity of the administration and communications (e.g. demand excessive administration and communications) [5, 17];

in performing the work tasks of the employees of BOTD, some problems occur which are typical for organizations that apply other levels and types of management apart from linear management (such as project, technical, methodological, etc.) [2].

Based on the aforementioned, the outline of the planned reforms takes into account the principle of justice, neutrality, simplicity and unambiguity, efficiency and the elimination of double taxation. Analyses by the Institute of Financial Policy conducted between 2001 – 2004 show the reasons and obvious need for reforms [17]:

- complicated tax legislation – intelligibility;
- too many exceptions, exemptions and allowances leading to social ineffectiveness because the production and consumption is not governed by demand and supply but by tax advantages;
- variability of setting tax base enabling optimization on the side of the taxpayer is increasing administrative costs and decreasing the efficiency of control.

From the point of view of the management and organization of the Tax Administration, further reasons can be:

- complexity of organizational structure – ambiguity and duplicity of the functions and competences at the central and regional levels;
- costly administrative apparatus of the Tax Administration;
- non-transparent project management, decreased possibility of controlling processes;
- unwillingness of taxpayers to pay taxes.

The intention of the Slovak government, as stated in the previously referenced programme declaration is to implement reforms of the Tax Administration in such a way that will make it more effective, with the aim of providing methodological assistance to compliant taxpayers and detecting those taxpayers who are avoiding taxes [21]. The objective is to create the conditions necessary for the efficient coordination of public authorities, to guarantee access to the internet for all citizens and to secure the interconnection of public authority information systems. Another priority of the Ministry of Finance of the SR is to reform the Customs Administration, with the aim of unifying the processes for the collection of taxes, fees, duties and contributions. The reform should proceed in two phases. The first stage is the merger of the tax and customs administrations. The second stage will unify the collection of taxes, fees, duties and insurance contributions [18, 26].

2 Methodology

The results of the questionnaire survey and the presented analyses will enrich the knowledge and help people understand the reforms of the tax system in the Slovak Republic from the point of view of their main participants, namely entrepreneurs and financial administration offices.

When interpreting the results of the analyses, it should also be borne in mind that all the conclusions are based on the subjective self-evaluation of the respondents.

This contribution presents the views and attitudes of a third, very important group of the population - students. The main research objective is to determine the main indicators for assessing the reform of the management of the Financial Administration of SR. By deconstructing the main objective of the research, it is possible to determine partial objectives:

1. To determine and subsequently analyse the differences in points of views of the three main targets of the research - entrepreneurs, financial administration authorities and students.
2. To analyse the relationships and differences in opinions with regards to the identified key factors of the tax system reforms according to entrepreneurs, financial administration authorities and students.

In order to obtain the most relevant data to identify the main indicators of tax reform, a simple questionnaire survey built on the Likert scale was used. The categories were as follows:

1. Willingness of tax office employees
2. Competence of tax office employees
3. Tax collection – process/method and simplicity
4. Collection of tariffs/customs – process/method and simplicity
5. Collection of levies - process/method and simplicity
6. Administrative difficulty of the tax and customs agenda
7. Time savings
8. Availability
9. Electronisation/computerisation – greater user comfort
10. Transaction costs
11. Postage costs
12. Labour costs

Key criteria were selected based on the expected benefits of the reforms (if implemented). This represents both qualitative and quantitative research based on the facts associated with the reforms (as mentioned in the introductory part of this contribution). The aim of the questionnaire survey was to identify the needs and expectations of the people most affected by the reforms. In addition, our research spontaneously reveals the need for the reforms, their effectiveness and inefficiency. The analysis of individual segments is based on the experience of the respondents and the perception of the benefits of the tax and customs reforms from a citizen's point of view. The survey, as well as the selection of the representative sample, were carried out as follows:

Tax Reform and its Expectations – selected results of the survey [3]

- time schedule of the conducted survey: 01/11/2019 - 31/1/2020
- geographical structure: Prešov and surroundings, Košice and surroundings, Banská Bystrica – town, Bratislava – city
- age structure of respondents: 18 – 60 years
- employed as: clerks, businessmen, students, some did not disclose their occupation
- representative sample: 1,500 (see Table 1)
- number of collected and completed questionnaires: 2,963

- The representative sample (1,500 respondents) was selected by applying a random number generator to the 2,963 completed questionnaires to generate a sample set of roughly the same number of males and females, as well as geographical areas - the four largest towns/cities in Slovakia (Prešov and surroundings, Košice and surroundings, Banská Bystrica – town, Bratislava – city).

The actual processing of the data was carried out as follows:

1. Analysis of the main components;
2. Factor analysis;
3. Scatter analysis;
4. Correspondence analysis.

However, the focus of the analysis is on the results of the factor analysis. Factor analysis is a multidimensional statistical method that focuses on generating new variables and attempts to reduce the amount of data (data reduction) with the minimum loss of information. It is based on the assumption that the dependence between the observed variables is due to the effect of a smaller number of non-measurable variables in the background, which are referred to as common factors. These common factors are defined as a linear combination of original variables. The main objectives of factor analysis are to find out and use the structure of common factors that are considered to be the hidden causes of mutually correlated variables to explain the observed dependencies in the best way possible.

Table 1. Quantitative analysis by age of the respondents

Category	Frequency: Age			
	Frequency	Cumulative (Frequency)	Relative (Frequency)	Cumulative (Relative Freq.)
26 - 35 years old	369	369	24.60000	24.6000
36 - 45 years old	359	728	23.93333	48.5333
46 - 60 years old	388	1116	25.86667	74.4000
18 - 25 years old	384	1500	25.60000	100.0000
ChD	0	1500	0.00000	100.0000

Source: Authors.

3 Results and Discussion

3.1 Analysis of the research file

The respondents were divided into four categories (intervals) according to the classification of the Statistical Office of the Slovak Republic, based on their age: 18 - 25 year olds, 26 - 35 year olds, 36 - 45 year olds and 46 - 60 year olds. The first group (18-25 year olds) features 394 respondents, representing 26.27%. The second group (26-35 year olds) features 348 respondents, representing 23.20%. The third group (36-45 year olds) features 291 respondents, representing 19.40%, and the last group (46-60 year olds) features 467 respondents, representing 31.13%. A detailed breakdown of the distribution of respondents with regards to three basic identification criteria (employment, gender and geographical distribution) for the first and fourth age groups, i.e. 18-25 year olds and 46-60 year olds, is presented in Figure 1 [3].

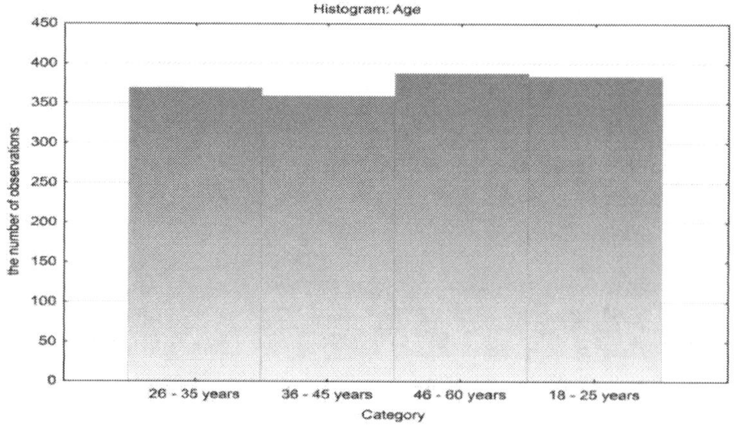

Figure 1. Distribution of respondents by age

Source: Authors.

Based on KMO test results, we can state that the achieved value 0.895 points to the suitability of using factor analysis for processing the research data. Bartlett's test of sphericity produced a value of 91.564, with 66 degrees of freedom. The corresponding p-value is 0.001, so the hypothesis that the realization of a selection correlation matrix with 12 considered variables is a unit matrix is rejected on the asymptotic level of significance 0.05. The assumptions for the application of factor analysis have therefore been fulfilled and its usage for the data analysis justified [1,18].

3.2 Results of the analysis

The Crombach alpha coefficient was used to assess the reliability of the research tool used. The calculated value was 0.856371. Based on this result, it is possible to say that it is not necessary to remove any variable to increase the alpha value because the change would be negligible. Since the value of Cronbach's alpha exceeds 0.7, it can also be said that the research tool is reliable and that it is possible to continue to work with the data obtained.

The aim of the principal component method is to simplify the description of a group of mutually dependent characters, namely the decomposition of the source matrix. Each of the principal components represents a linear combination of the original characters. The main components are ranked according to their importance - decreasing variance (see Table 2). It follows that most of the information about the variability of the original data is centred in the first principal component and the least information in the last principal component.

Table 2. Table of custom numbers from the source matrix

Value number	Eigenvalues of correlation matrix, and related statistics			
	Eigenvalue	% Total variance	Cumulative Eigenvalue	Cumulative %
1	1.793891	14.94909	1.79389	14.9491
2	1.148705	9.57254	2.94260	24.5216
3	1.086579	9.05482	4.02917	33.5765
4	1.046359	8.71966	5.07553	42.2961
5	1.001378	8.34482	6.07691	50.6409
6	0.960782	8.00651	7.03769	58.6474
7	0.940269	7.83558	7.97796	66.4830
8	0.911057	7.59215	8.88902	74.0752
9	0.876618	7.30515	9.76564	81.3803
10	0.822587	6.85489	10.58822	88.2352
11	0.818052	6.81710	11.40628	95.0523
12	0.593724	4.94770	12.00000	100.0000

Source: Authors.

Table 2 shows that the first principal component features 14.949% of the original data variance, the second principal component features 9.572%, the third 9.055%, the fourth 8.719% and the fifth 8.344%. These five principal components, whose own number is greater than 1, amass 50.649% of the variance of the original data of the research file. These results will be used to describe and identify factors within the factor analysis. The scatter plot (see Figure 2) suggests that the first principal component, Component 1, divides respondents' responses into two clusters separated by a vertical axis. These clusters are more apparent in the charts featuring components 3, 4 and 5. The scatter plot shows that the combination of components 2 to 5 does not form these clusters as the research objects are centralized approximately in the centre of the coordinate system. On the other hand, it is possible to identify outlying respondents in terms of their views on the reform of the tax system (e.g. respondent 716 shows different behaviours with regards to components 1 and 2; respondent 870 with regards to components 2 and 3).

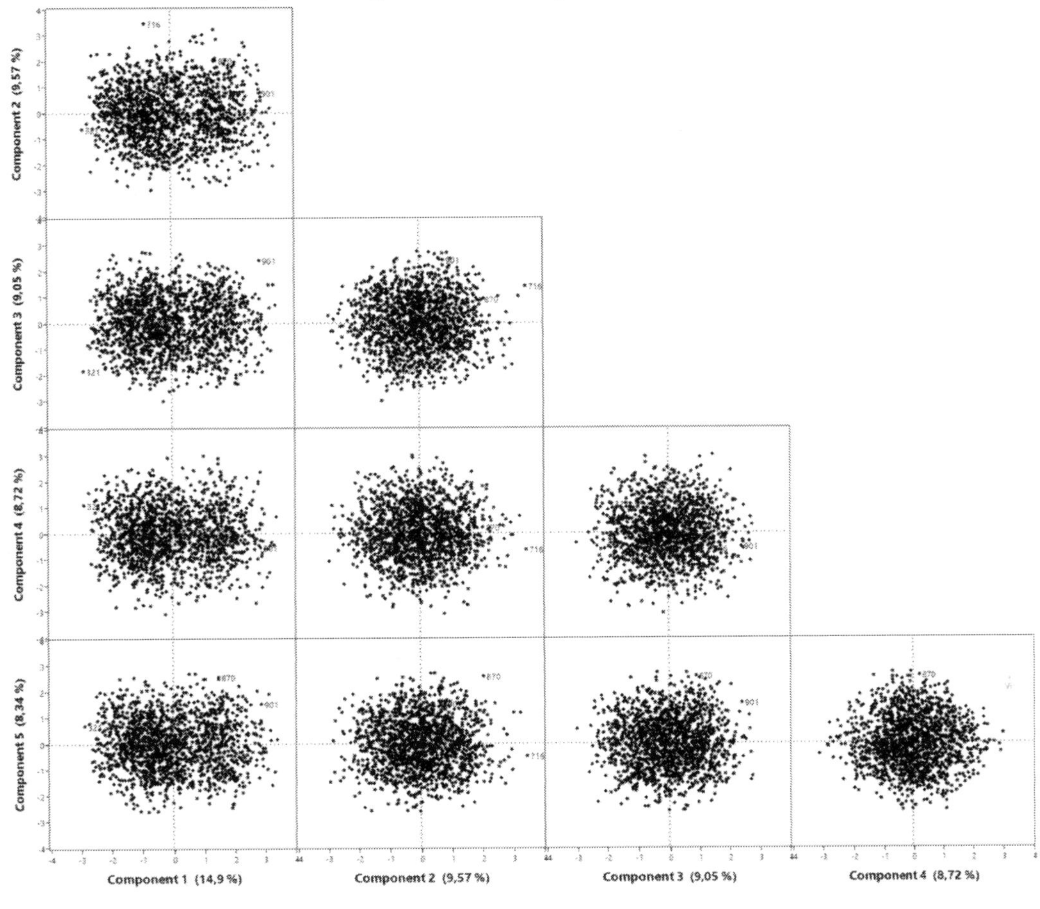

Figure 2. Scatter plot of components´ scores

Source: Authors.

From the scatter plots and the figure containing the component weights of the group of respondents identifying themselves as "entrepreneurs" and the first and second principal components (see Figure 3), the following conclusions can be drawn. A positive correlation for the group of "entrepreneurs" was identified for Collection of levies - process/method and simplicity and Availability, while no correlation was found with Postage costs. On the other hand, a negative correlation exists between Transaction costs and Postage costs. With regards to the second principal component, there is a positive correlation between the Collection of tariffs/customs - process/method and simplicity, Time savings and Labour costs. From the point of view of importance, Postage costs, Labour costs, Transaction costs and Electronisation/computerisation - greater user comfort are the most important factors for the group of entrepreneurs. It can be further inferred from the figure of the component weights that the least important factor for the entrepreneurs is the Administrative difficulty of the tax and customs agenda. The very last two conclusions are closely related - the importance of electronisation. It can be assumed that if the computerization of the tax system reaches the required level, in terms of compliance and convenience, then

Willingness of tax office employees will become a pointless indicator. From the scatter plot it is clear that the attitudes of entrepreneurs towards the items of the research tool are almost homogeneous. However, this can only be said when accepting the assumptions of the analysis without taking into account differences between men and women, age categories and geographical location. These assumptions form the basis for the conclusions drawn for the group of clerks and students (see Figures 4 and 5).

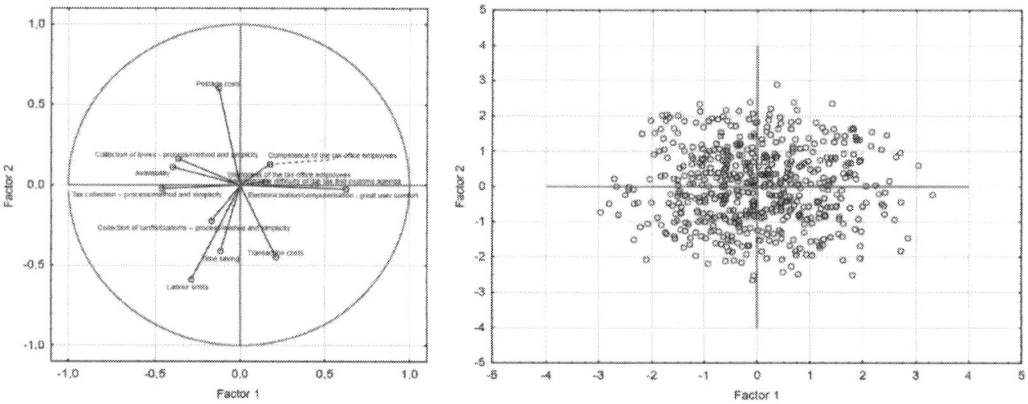

Figure 3. Component weights chart and scatter plot for the group "entrepreneurs"

Source: Authors.

As far as the group "clerks" and the first and second principal components are concerned (see Figure 4), the variables Time Savings and Competence of tax office employees correlate positively in the first quadrant of the chart. Another positive correlation is found between Labour costs and Electronisation/computerisation - greater user comfort and between Administrative difficulty of the tax and customs agenda and the Collection of levies - process/method and simplicity. At the same time, it is necessary to point out the indirect correlation between Electronisation/computerisation and the Administrative difficulty of the tax and customs agenda. Even though the item Electronisation/computerisation - greater user comfort is not significant for the group "clerks", the indirect correlation indicates that clerks are also of the opinion that if the electronic tax system is simplified, it would minimize administrative difficulties. In terms of the least important, clerks identify Tax collection - process/method and simplicity, Electronisation/computerisation - greater user comfort and Postage costs. On the contrary, the most important are Willingness of the tax office employees, Competence of tax office employees and Administrative difficulty of the tax and customs agenda.

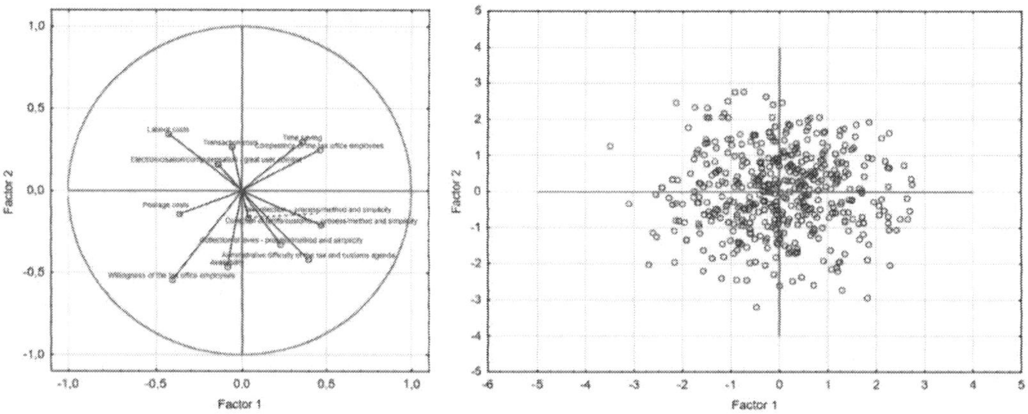

Figure 4. Chart of Component Weights and the Scatter Plot for the "Clerks "

Source: Authors.

As far as the group "students" and the first and second principal components are concerned (see Figure 5), a positive correlation exists between the Transaction costs and Electronisation/computerisation - greater user comfort, and between Administrative difficulty of the tax and customs agenda and Willingness of tax office

https://doi.org/10.1051/shsconf/20219001009

employees. In terms of the most important, students identified Time savings, Tax collection - process/method and simplicity and Postage costs. Students consider Administrative difficulty of the tax and customs agenda and Availability to be the least important. From these three graphs it is possible to deduce the differences in the perception of the importance of the research parameters among the three groups surveyed - entrepreneurs, clerks and students.

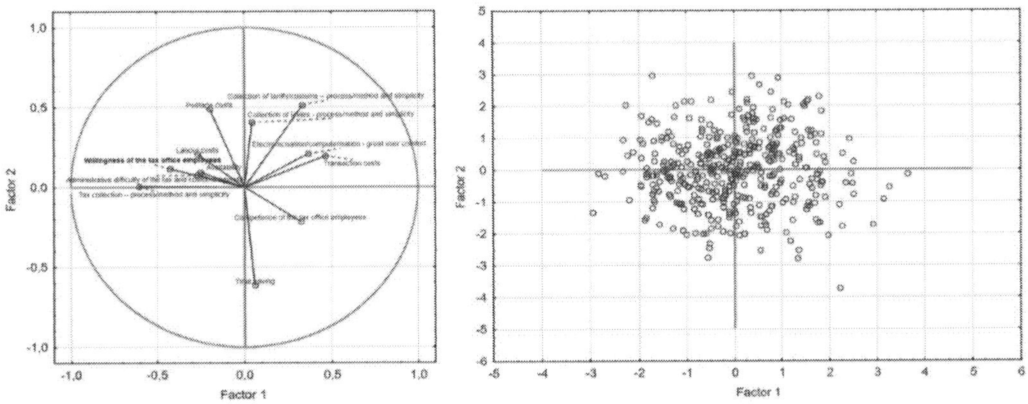

Figure 5. Component Weights Chart and the Scatter Plot for the "Students"

Source: Authors.

The basic indicators for determining the suitability of the use of factor analysis are KMO statistics and Bartlett's test of sphericity. Kaiser-Meyer-Olkin's statistics (KMO) are an index that compares the size of the experimental correlation coefficients to the size of the partial correlation coefficients. If the sum of the squares of the partial correlation coefficients is smaller compared to the sum of the square correlation coefficients, the KMO is close to 1. Low KMO statistical values indicate that factor analysis of the original characters is not suitable because the correlation between the pairs of characters cannot be explained by other characters. According to the value of KMO statistics (0.645) and according to Kaiser's definition, it can be said that the correlation rate is good and that the choice of factor analysis is justified. Bartlett's test of sphericity is a statistical test of the correlation between the original characters. It tests the zero hypothesis (H0: There is no correlation between the characters), so the correlation matrix is the unit matrix. The significance of the Bartlett test of sphericity $p = 0.000$ is less than the chosen level of significance $\alpha = 5\%$ and therefore the zero hypothesis is rejected. It is therefore possible to state that factor analysis is suitable for analysing the data concerning the reform of the tax system.

Table 3. Assumptions for the use of factor analysis

Kaiser-Meyer-Olkin measure of sampling adequacy.		.645
Bartlett's test of sphericity	Approx. Chi-Square	584.448
	df	66
	Sig.	.000

Source: Authors.

One of the basic tasks of factor analysis is the reduction of the original number of characters. The basic premise is to get the right combination of original characters to explain the scattering, and then to search for combinations that are increasingly smaller and smaller. This contribution focuses on the criterion of own numbers. The own number criterion - factors that match a number greater than 1 are, according to Keiser, considered statistically significant. On the contrary, factors whose own number is less than 1 are statistically insignificant. Based on Table 1 and Kaiser´s criterion of statistical significance, five factors can be identified whose own number is greater than 1. The cumulative of these five factors explains 50.6409% of the total variance. The first common factor represents 14.94909% of the scatter, the second 9.57254%, the third 9.05482%, the fourth 8.71966% and the fifth 8.34482% of the total scatter.

The first step in interpreting the results of the factor analysis is the factor matrix analysis (see Table 4), which gives us the initial number of factors. The factor matrix contains factor loads for each character and factor, and represents the best linear combination of original characters while including as many characters as possible. The first factor is always the most important because it represents the best linear relationship found in the original characters. The second factor is the second best representation of the linear relationship. However, there is a limitation to it - it must be orthogonal to the first factor. The factor load explains the role of each original character while defining a common factor. It is actually the correlation coefficient between each original character and the factor.

Table 4. Factor load table

Variable	Factor loadings extraction : Principal components (Marked loadings are >.550000)				
	Factor 1	Factor 2	Factor 3	Factor 4	Factor 5
Willingness of tax office employees	-0.490469	0.312143	0.035937	0.035268	-0.066297
Competence of tax office employees	**-0.756493**	-0.001618	-0.015697	-0.075119	0.114479
Tax collection – process/method and simplicity	0.028661	**0.620819**	-0.295487	0.132194	0.239475
Collection of tariffs/customs – process/method and simplicity	0.022983	-0.101647	0.288631	-0.521288	**0.699792**
Collection of levies - process/method and simplicity	0.047041	0.376630	-0.037996	**-0.569445**	-0.034917
Administrative difficulty of the tax and customs agenda	**0.670673**	-0.073099	-0.025161	0.040498	-0.137827
Time savings	-0.251409	-0.116226	**-0.655635**	-0.037791	0.105068
Availability	0.153056	0.457617	-0.198887	-0.350003	-0.343257
Electronisation/ computerisation - greater user comfort	-0.448133	-0.208580	-0.009483	-0.000949	-0.110170
Transaction costs	-0.092307	-0.308164	0.084271	-0.518963	-0.465836
Postage costs	-0.050492	0.376508	**0.642661**	0.166675	-0.092219
Labour costs	0.478705	-0.071952	-0.150750	-0.057563	0.223821
Expl.Var	1.793891	1.148705	1.086579	1.046359	1.001378
Prp.Totl	0.149491	0.095725	0.090548	0.087197	0.083448

Source: Authors.

From Table 4, it is clear that the first factor correlates significantly with Administrative difficulty of the tax and customs agenda. The factor load values reach 75.6493% for Competence of tax office employees and 67.0673% for Administrative difficulty of the tax and customs agenda. The analysis of Factor 1 shows that 57.223% of the variability of the Competence of tax office employees and 44.981% of the variability of the Administrative difficulty of the tax and customs agenda is explained by the first common factor. The second common factor correlates with the item Tax collection - process/method and simplicity of the research tool with the factor load value of 62.0819%, whereby 38.542% of the variability of this item is explained by the second common factor. The third common factor correlates significantly with Time savings and Postage costs. The factor load values are -65.5635% for Time savings and 64.2661% for Postage costs. Based on the analysis of Factor 3, we can state that the 42.986% variability of the Time savings item and the 41.301% variability of the Postage costs is explained by the third common factor. The fourth common factor correlates with Collection of levies - process/method and simplicity of the research tool with a factor load value of -56.9445%, whereby 32.427% of the variability of this item is explained by the second common factor. The fifth common factor correlates with Collection of tariffs/customs – process/method and simplicity with the load factor of 69.9792%, whereby 48.971% of the variability of this item is explained by the second common factor. In so ascertaining, the practical importance of the factors has been verified.

Based on the above, we can therefore postulate the factors that should be the main research objectives:

- Factor 1 – Competence and Administrative difficulty
- Factor 2 – Tax collection
- Factor 3 – Time savings and Postage costs
- Factor 4 – Collection of levies
- Factor 5 – Collection of tariffs

Factor analysis focuses mainly on factor model parameters. Common factor estimates, called factor scores, may be required. The values of the common factors for the selected objects or observations are not only a useful tool for data diagnostics, but also, where necessary, as inputs for further analyses. The factor score is an estimation

https://doi.org/10.1051/shsconf/20219001009

of parameters based on common sense because it estimates unobserved values. Estimates of the factor score for a given object can be represented as the coordinates of this object in the R-dimensional space.

Table 5. Factor score coefficients

Variable	Factor score coefficients rotation : Varimax normalized extraction : Principal components				
	Factor 1	**Factor 2**	**Factor 3**	**Factor 4**	**Factor 5**
Willingness of tax office employees	0.313260	-0.137702	-0.107087	0.133449	0.094002
Competence of tax office employees	0.413710	-0.001934	0.092446	-0.027742	-0.125672
Tax collection – process/method and simplicity	0.038782	-0.615576	0.074223	0.230054	0.023989
Collection of tariffs/customs – process/method and simplicity	-0.009761	-0.006138	-0.024719	0.019145	-0.902220
Collection of levies - process/method and simplicity	0.032871	0.037248	-0.004882	0.596863	-0.219229
Administrative difficulty of the tax and customs agenda	-0.377910	0.066310	-0.028692	0.019058	0.128076
Time savings	0.078301	-0.115548	0.619134	0.027501	0.052065
Availability	-0.023931	0.011181	0.017036	0.627254	0.185583
Electronisation/computerisation - greater user comfort	0.220404	0.193086	0.083401	-0.102637	0.064863
Transaction costs	0.035079	0.672051	0.034622	0.295748	0.038417
Postage costs	0.117976	-0.092137	-0.684152	0.011194	0.036032
Labour costs	-0.285506	-0.101049	0.158766	-0.017867	-0.173540

Source: Authors.

In line with the defined partial objectives of the research, the next part of this contribution analyses the opinions of the respondents represented by the factor scores in relation to the five extracted factors – factors of the tax system reforms using Fisher´s ANOVA (Analysis of Variance). In the analyses, consideration is only given to significant independent variables at the value of the relevant factor at the chosen level of significance $\alpha = 0.05$.

The analysis of scattering is a set of induction statistics procedures used to test hypotheses of mean values in a different, often complex configuration of experiments when testing more than two groups. The zero hypothesis in this case indicates that the test groups do not statistically significantly differ in the mean value.

Table 6. ANOVA for Factor 1 (Competence and Administrative difficulty)

Effect	Univariate tests of significance for Factor 1 : Sigma-restricted parameterization : Effective hypothesis decomposition				
	SS	Degrees of freedom	MS	F	p
Intercept	**18.6060**	**1**	**18.6060**	**46.5828**	**0.000000**
Age	1.1098	3	0.3699	0.9262	0.427321
Gender	0.3336	1	0.3336	0.8351	0.360948
Employment	**846.2751**	**3**	**282.0917**	**706.2562**	**0.000000**
Geographical distribution	0.7619	3	0.2540	0.6358	0.591955
Error	594.7339	1489	0.3994		

Source: Authors.

Competence and Administrative difficulty has a significant impact on Employment at a chosen significance level of 5%. Also, Intercept has a significant impact, integrating the influence of variables that have not been analysed in the research, namely age, gender and geographical distribution. If the factor score is considered to represent either agreement or disagreement with the given factor, the average value of the factor score in the group "entrepreneurs" is -0.58823 ± 0.05341, for the group "clerks" it is 1.07095 ± 0.055511, and for the group "students" -0.591651 ± 0.061451. Based on the average score value and Figure 6, it is possible to conclude that students and entrepreneurs perceive Factor 1, defined as Competence and Administrative difficulty, negatively,

 https://doi.org/10.1051/shsconf/20219001009

while clerks and respondents who did not report their job perceive it positively. At the same time, based on the Scheffe test, it can be stated that at the selected level of significance of 5%, the mean values of the factor scores for students and entrepreneurs can be considered identical (p = 0.999851) and statistically different from the group "clerks" (p=0.00001).

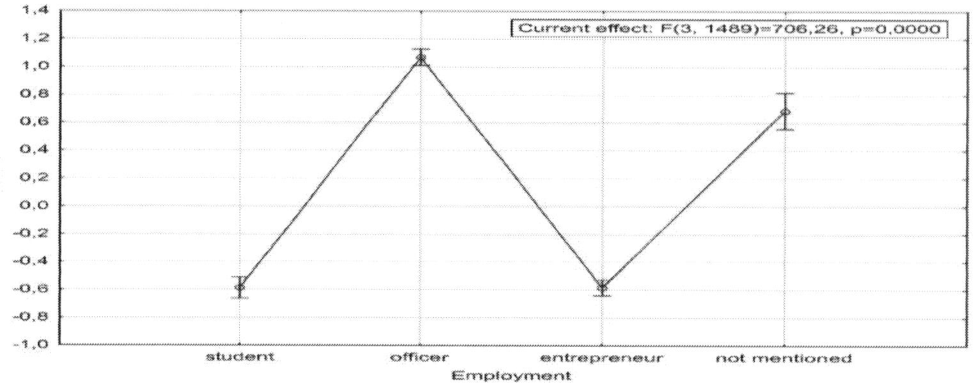

Figure 6. Factor score chart for Factor 1 and Employment

(Factor 1 - Competence and Administrative difficulty)

Source: Authors.

When considering Factor 2 (Tax collection), the interaction between Geographical distribution and Employment (p = 0.03518) is seen as a significant influence that affects the change in the factor score. The average value of the factor scores for students in Prešov is -0.01858, Košice 0.03811, Banská Bystrica 0.02730 and Bratislava-0.18959. Students from Prešov and Bratislava have a negative attitude towards Tax collection (Factor 2), with a higher rate for students from Bratislava. Students from Košice and Banská Bystrica are slightly positive in this respect, with the average value in both cases almost identical. In the group "clerks", the average value of the factor scores in Prešov is 0.02599, Košice -0.06077, Banská Bystrica 0.17242 and Bratislava -0.15917. The clerks in Banská Bystrica have a positive attitude towards Tax collection, while clerks from Bratislava perceive it negatively. The average value of the factor in the case of entrepreneurs from Prešov is 0.15264, Košice 0.04020, Banská Bystrica -0.13218 and Bratislava 0.07535. From the above average values, Tax collection is negatively perceived by entrepreneurs from Banská Bystrica, while entrepreneurs from Prešov see it positively. For more detailed information see Figure 7.

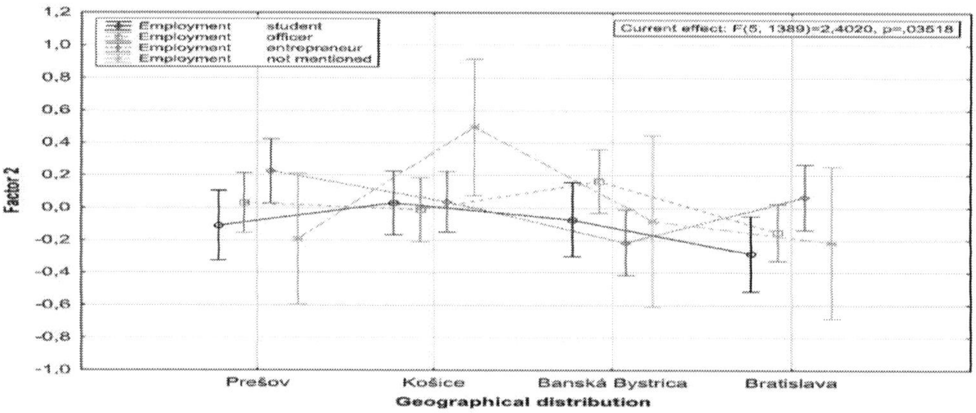

Figure 7. Factor score chart for Factor 2 and interaction between Geographical distribution and Employment

(Factor 2 - Tax collection)

Source: Authors.

Factor 3, defined as Time savings and Postage costs, is significantly influenced by Employment at the chosen significance level 5%. The average value of the factor score for the group "entrepreneurs" is -0.182388, for "clerks" 0.147971, and for "students" -0.035865. From the average factor score value and Figure 8, it can be concluded that Factor 3 is perceived by entrepreneurs negatively, whereas clerks, respondents who did not state their job and students perceive it positively.

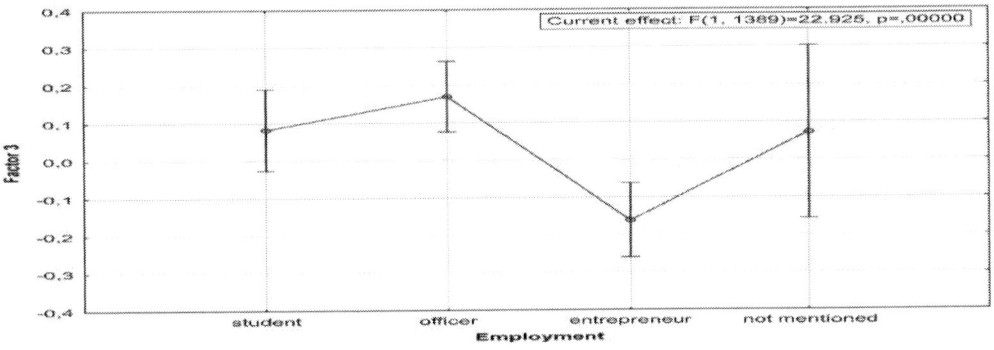

Figure 8. Factor score chart for Factor 3 and Employment

(Factor 3 - Time savings and Postage costs)

Source: Authors.

When analysing Factor 4 (Collection of levies), Geographical distribution and Employment (p = 0.03412) have a significant effect on the change of factor score. The average values of the factor scores for students in Prešov is 0.19017, Košice -0.113131, Banská Bystrica 0.04434 and Bratislava -0.08728. It follows that students from Košice and Bratislava have a negative attitude towards the Collection of levies (Factor 4), with a higher rate for students in Košice. Students from Banská Bystrica and Prešov are slightly positive towards Factor 4, while the average value of the factor scores is significantly higher for Prešov students. In the group "clerks", the average value of the factor scores in Prešov is 0.02391, Košice -0.16331, Banská Bystrica -0.26136 and Bratislava 0.08506. Clerks from Prešov and Bratislava have a slightly positive attitude towards Factor 4, while in case of Košice and Banská Bystrica the score revealed the slightly negative attitude of the clerks living there. The average factor score for the group "entrepreneurs" from Prešov is -0.04147, Košice 0.05416, Banská Bystrica 0.03768 and Bratislava 0.06978. From the above average values, it follows that Collection of levies is negatively perceived only by entrepreneurs from Prešov; entrepreneurs from other cities perceive this matter positively. The conclusions are shown in Figure 9.

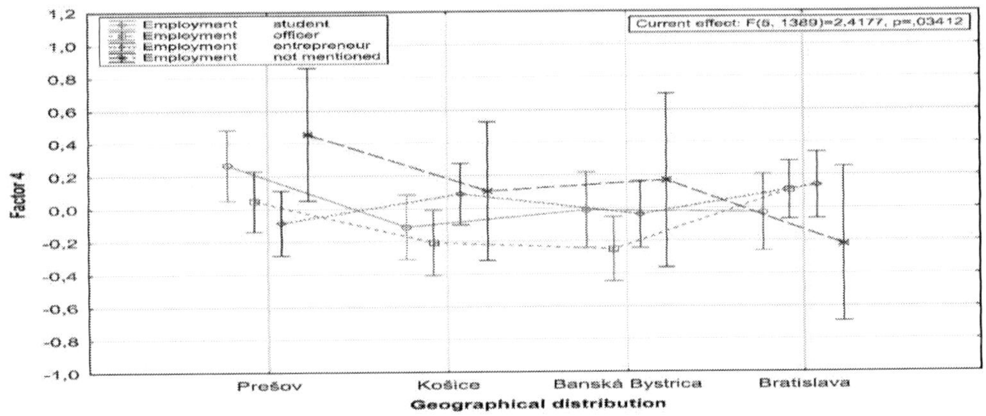

Figure 9. Factor score chart for Factor 4 and interaction between Geographical distribution and Employment

(Factor 4 - Collection of levies)

Source: Authors.

Factor 5, defined as Collection of tariffs, is significantly influenced by Age at the chosen significance level 5%. The average score for 18-25 year olds is 0.073782, 26-35 year olds -0.015995, 36-45 year olds 0.030348 and 46-60 year olds -0.133042. From the average score value and Figure 10, it can be concluded that Factor 5 is perceived by the respondents more negatively as they get older.

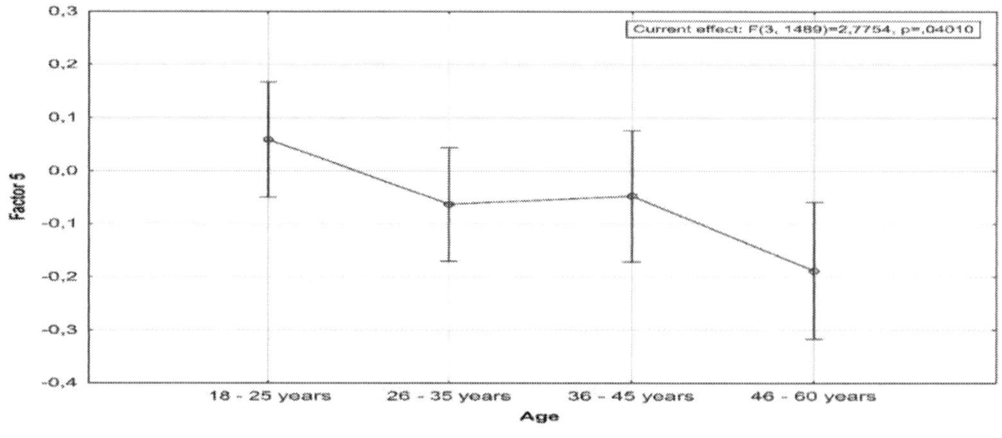

Figure 10. Factor score chart for Factor 5 and Age

(Factor 5 – Collection of tariffs)

Source: Authors.

An excessive and disproportionate administrative burden can have a real economic impact. Generally speaking, it is considered to be an irritating and disturbing element in business and is often cited as a priority objective of simplification efforts. The aim of the Financial Administration reforms is to use electronic services to the widest extent possible as it will decrease the amount of red tape placed upon tax subjects and streamline the functioning of the Financial Administration. The aim of electronisation is to introduce such information and communication technologies to secure paperless communication and automated processes first within the Financial Administration and later in relation to the tax subjects. The reform is based on the vision of the eGovernment from 2013, the aim of which is to achieve citizens' satisfaction with the public administration by providing services in an attractive and simple way, while increasing its efficiency, competence and reducing the costs of public administration. The implementation of the national project Electronic Services of Financial Administration – Taxes was officially completed on 21 October 2016. The project met two main goals. The first was to build a new internal financial reporting system for the Financial Administration, section taxes. This streamlined and partially restructured internal management processes. Full computerization and unification of these processes were necessary steps towards the second objective of the project - more effective communication with the public. In view of the above, this contribution also addresses the perception of the electronisation of the services of the Financial Administration by students, clerks and entrepreneurs. The analysis of responses of respondents in absolute numbers on the Likert scale is presented in Table 7.

Table 7. Absolute number for respondents' answers to the question on electronisation

	Observed table row variables: Employment(4) column variables : Electronisation/computerisation - greater user comfort(5)					
	1	**2**	**3**	**4**	**5**	**Total**
student	128	132	88	31	31	410
officer	78	101	73	101	109	462
entrepreneur	169	166	102	47	52	536
not mentioned	13	22	22	13	22	92
Total	388	421	285	192	214	1500

1 – absolutely dissatisfied, 2 – dissatisfied, 3 – I do not know, 4 – satisfied, 5 – absolutely satisfied

Source: Authors.

Based on the analysis, the application of the χ^2 test with a value of 146.753, and 12 degrees of freedom, it can be concluded that there is a significant relationship between the respondents' opinion on Electronisation/computerisation - greater user comfort and Employment (p = 0.0001) at the chosen significance level 5 %.

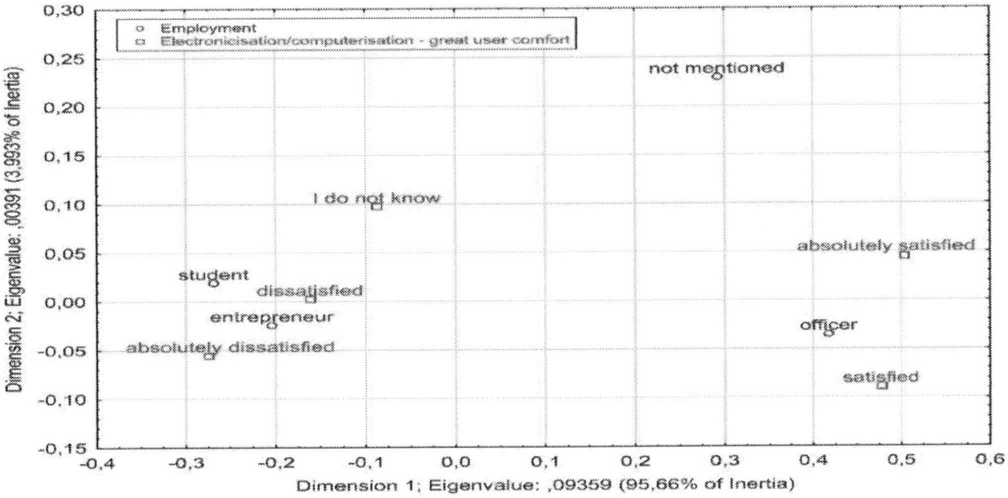

Figure 11. Correspondence map of Employment and Electronisation

Source: Authors.

From the correspondence map shown in Figure 11, which represents the relationship between Employment and Electronisation/computerisation - greater user comfort, it is clear that students and entrepreneurs are strongly inclined to disagree or strongly disagree, whereas clerks tend to agree or strongly agree with the opinion. Users of electronic financial management services, i.e. entrepreneurs and students, still perceive this system in a negative light. This negative attitude can be attributed to the frequent malfunctions and complexity of the entire electronic system, although it is necessary to say that, in view of the previous conclusions, these two groups perceive the process of electronisation very positively.

4 Conclusion

Slovakia, through the above mentioned process of tax reforms, is approaching what could be described as an effective tax system, one which will lead to an increase in the effectiveness and competitiveness of the state amongst other EU member states. The impact of the suggested changes can be divided into two main categories. The first category comprises the benefits of the reforms of the tax and customs administrations in terms of cost and time savings, growth of value added, efficiency of work, etc. The second category is represented by the expenditures used for the individual objectives of the reforms of the tax and customs administrations. Both of these categories can further be divided into the impact on the taxpayer, that is the client/user, and the impacts on the public administration. From the financial point of view, the greatest importance should be attached to those impacts of a permanent nature or that have a repeated effect [6, 28].

The research carried out and the subsequent processing of the data obtained in the form of factor analysis gave results similar to those of authors carrying out research in the field of social sciences. For example, research into activities for the unemployed using AZN methodology and factor analysis, identified 8 factors. Dobrovič et al. (2012) [1] used the method in the field of social sciences - the perception of security risks by citizens of a city - and outlined 11 factors. Dobrovič et al. [24] analysed 35 indicators, which gave rise to 6 factors. We note that the factor analysis carried out identified 5 new factors that were based on 12 indicators. These factors sufficiently reflect the views held about the Financial Administration reforms from the point of view of the main actors of these reforms, namely entrepreneurs and the Financial Administration. The analysis also included the third, less significant group represented by students. The analysis revealed different perceptions of key factors from the viewpoint of the three groups studied. However, these differences are particularly pronounced in the evaluation of electronisation, where similar opinions, i.e. dissatisfaction, are observed among students and entrepreneurs, while satisfaction was observed only in the group "clerks".

When Slovakia joined the EU in May 2004, it had to align and continue to align its tax policy with the tax policy of the EU Member States in order to harmonize the Slovak tax system with EU legislation. In some areas, the EU has exclusive competence, which in practice means that all important decisions or important laws are adopted by the European institutions, namely the Commission or the Council or the European Parliament. By joining the EU, Slovakia was obliged to eliminate certain incompatibilities with the fundamental EU law and thus to harmonize the tax system, value added tax, excise duties and tax on income (in particular for companies linked by common assets) with that of the EU. Had Slovakia failed to do so, the tax subjects concerned could sue it. It would also be sued by the European Commission for failing to implement the legal acts resulting from the founding treaties. The majority of Slovak legislation is based on so-called acquis communautaire of the European Union, which represents a set of common legal acts and obligations that have accumulated and been amended for over more than 50 years [since 1957 - The Treaty of Rome and the establishment of the European Economic Community (EEC) of six advanced Western European countries]. The EU Member States have some freedom in deciding to incorporate the acquis communautaire into their domestic law. For this reason, their impacts should be taken into account when implementing European legislation. The acquis communautaire is a set of rights, obligations, treaties, and is constantly expanding and developing. The latest developments in some EU Member States clearly show that the EU is going through very unfavorable economic developments, has no strong leaders, and that the issue of governance is raising a few eyebrows .

In connection with the process of European integration, tax harmonization primarily involves aligning Member States' legal and administrative regulations in order to prevent distortions in the common market. The aim of the effort is to remove existing tax boundaries within the EU. Tax harmonization can also be characterized as a process of convergence of the tax systems of the Member States on the basis of common rules. This contribution pointed out the difficulties of doing business based on certain factors of the Financial Administration in Slovakia. Therefore, it is necessary to consider that these confirmatory factors should be perceived as very serious for the collection of taxes and duties in Slovakia. The realization must come that unless the tax collection system is simplified, there will always be attempts to evade tax. The biggest shortcoming of the Financial Administration's system is its not very fortunate attempt at electronisation. However, unless we align our VAT with the EU or abolish it altogether in favour of a Turnover Tax, there will continue to be tax evasion, which will become increasingly sophisticated over time. That is why it is necessary to focus more on the following factors: Competence and Administrative difficulty; Tax collection; Time savings and Postage costs; Collection of levies; and Collection of tariffs, as well as diplomacy when negotiating with the EU countries on the terms for the reciprocal exchange of information. Alternatively, Slovakia should strive to revive 2009's negotiations on the abolition of Value Added Tax and the introduction of a Turnover Tax, but, above all, enforce the EU legislation applicable to tax subjects in the country where the tax base was created, which means eliminating the possibility of taxing tax subjects in the state they prefer.

References

1. J. Dobrovič, L. Sojka, M. Cehlárová, N. Svetozarovová, *Hodnotenie výkonnosti manažmentu efektívnej daňovej správy SR v kontexte regiónov pred reformou daňovej správy SR s návrhom jej zefektívnenia po reforme. [Performance Evaluation of Effective Tax Administration Management of the Slovak Republic in the Regional Context before the Tax Administration Reform of the SR with a Proposal for Its Higher Effectivity after the Reform*]. Prešov: University of Prešov (2015)
2. K. Kubátová, *Daňová teorie a politika.* [*Tax Theory and Policies*]. Praha: ASPI Publishing (2003)
3. J. E. Maede, The Theory of international Economic Policy. II. Trade and Welfare and Mathematical Supplement. In: *The Canadian Journal of Economic and Political Science.* **23**(4), 560–561 (1995)
4. R. A. Musgrave, P. B. Musgrave, *Public Finance in Theory and Practice.* 5th ed. New York: McGraw-Hill (1989)
5. P. Pudil, V. Vybíhal, L. Vítek, J. Pavel, J. et al., *Zdanění a efektivnost.* [*Taxation and Efficiency*]. Praha: Eurolex Bohemia s.r.o. (2004)
6. R. Rajnoha, J. Dobrovič, Simultaneos Management of Economics Business Processes by Added Value Knowledge. *E & M Ekonomie a Management.* **14**(1), 53–69 (2011)
7. R. Rajnoha, A. Sujová, A., J. Dobrovič, Management and Economics of Business Processes Added Value. *Procedia Social and Behavioral Sciences.* **62**, 1292–1296 (2012)
8. R. Rajnoha, D. Slivková, J. Dobrovič, Globatization and Transer Pricing in Multinational Corporations in Slovakia and OECD Countries – Analytical Study and Decision-Making Model on the Choice of Optimal Transfer – Pricing Method. *Ekonomický Časopis.* **62**(6), 609–630 (2014)
9. Ministry of Finance of Slovak Republic, *Tax and Customs Administrations Reform. Framework Analysis.* (2007)
10. Tax Directorate of the SR, *Tax Administration Reform. Framework Analysis.* (2002,2004)

11. Ministry of Finance of Slovak Republic, *Act No.595/2003 Coll. on Income Taxes* (2003)
12. R. B. Adams, D. Ferreira, Women in the boardroom and their impact on governance and performance. *Journal of Financial Economics*. **94**, 291–309 (2009)
13. T. D. Allen, Family-Supportive Work Environments: The Role of Organizational Perceptions. *Journal of Vocational Behavior*. **58**(3), 414–435 (2001)
14. R. M. Blackburn, J. Browne, B. Brooks, J. Jarman, Explaining gender segregation. *British Journal of Sociology*. **53**(4), 513–536 (2002)
15. I. E. Broder, Review of NSF Economics Proposals: Gender and Institutional Patterns. *The American Economic Review*. **83**, 964–970 (1993)
16. V. Cabinova, E. Onuferova, P. Gallo, P. Gallo, J. Gallo, A comparative analysis of modern performance methods in economic practice. *Montenegrin Journal of Economics*. **14**(4), 85–96 (2018)
17. A. Afonina, Strategic Management Tools and Techniques and Organizational Performance: Findings from the Czech Republic. *Journal of Competitiveness*. **7**(3), 19–36 (2015)
18. B. Gavurová, J. Belás, K. Kočišová, T. Klieštik, Comparison of Selected Methods for Performance Evaluation of Czech and Slovak Commercial Banks. *Journal of Business Economics and Management*. **18**(5), 852–856 (2017)
19. E. H. Mamdani, S. Assilian, An Experiment in Linguistic Synthesis with a Fuzzy Logic Controller. *International Journal of Man-Machine Studies*. **7**(1), 1–13 (1975)
20. P. Micheli, M. Luca, Theory and Practice of Performance Management. *Management Accounting Research*. **25**(2), 147–156 (2014)
21. P. Novák, B. Popesko, Cost Variability and Cost Behaviour in Manufacturing Enterprises. *Economics and Sociology*. **7**(4), 89–103 (2014)
22. J. Belas, L. Smrcka, B. Gavurova, J. Dvorsky, The impact of social and economic factors in the credit risk management of sme. *Technological and Economic Development of Economy*. **24**(3), 1215–1230 (2018)
23. I. Derun, H. Mysaka, Stakeholder perception of financial performance in corporate reputation formation. *Journal of International Studies*. **11**(3), 112–123 (2018)
24. J. Dobrovič, M. Lambovska, P. Gallo, V. Timkova, Non-financial indicators and their importance in small and medium-sized enterprises. *Journal of Competitiveness*. **10**(2), 41–55 (2018)
25. J. Dobrovič, M. Urbański, P. Gallo, E. Benková, V. Čabinová, Balanced scorecard concept as a tool of strategic management and its usage in the construction industry. *Polish Journal of Management Studies*. **18**(2), 59–72 (2018)
26. G. Farkas, The Effects of Strategic Orientations and Perceived Environment on Firm Performance. *Journal of Competitiveness*. **8**(1), 55–65 (2016)
27. B. Haltofova, Using Crowdsourcing To Support Civic Engagement In Strategic Urban Development Planning: A Case Study Of Ostrava, Czech Republic. *Journal of Competitiveness*. **10**(2), 85–103 (2018)
28. J. Kramoliš, A. Kotásková, The gap of importance of design in business between 2014 and 2016. *E & M: Ekonomie a Management*. **21**(2), 133–146 (2018)
29. Š. Slávik, B. Zagoršek, Relationship between Business Strategy and Business Model Studied in a Sample of Service Companies. *Journal of Competitiveness*. **8**(4), 72–84 (2016)
30. S. Szabo, B. Mihalčová, P. Gallo, M. Ivaničková, Evaluating efficiency in specialized hospital facilities – developing the model by way of the discriminant analysis. *E & M: Ekonomie a Management*. **21**(3), 88–106 (2018)

Human Capital Management and Industry 4.0

Alžbeta Kucharčíková[1,*], *Martin* Mičiak[1], *Aneta* Bartošová[1], *Monika* Budzeľová[1], *Simona* Bugajová[1], *Alžbeta* Maslíková[1], *Simona* Pisoňová[1]

[1]Department of Macro and Microeconomics, Faculty of Management Science and Informatics, University of Žilina, Univerzitná 8215/1, 010 26, Žilina, Slovak Republic

Abstract. Human capital (HC) is considered to be an important factor influencing the success of. The concept of human capital management (HCM) deals with the effectiveness of human capital, as the most precious intangible asset. However, with the advancement in technology and the introduction of Industry 4.0, the position of human capital in companies can alter with regard to the expected changes. On the other hand, new elements being brought in by Industry 4.0 will allow for the HCM process to be performed better, in favour of the improvement in the performance and competitiveness of companies. The aim of this article is to point out the interconnection between the concept of HCM and some elements of Industry 4.0, as well as to explain the findings from a questionnaire survey focused on how IT companies in Slovakia use information systems for the implementation of HCM, whether these systems enable them to analyse data related to HC, and whether these companies are prepared for the introduction of changes in this field in the future.
Keywords: human capital management; Industry 4.0; information systems; digitization; big data

1 Introduction

Employees are currently considered to be the most valuable asset of an organization [1]. Human capital (HC) is one of production inputs of companies. The increase in the human capital value is an important way to increase a company´s market value. Highly qualified HC ensures the strong competitiveness of companies. By investing in HC, a company increases the knowledge, skills and abilities of its employees, who can make the production of the company more efficient and generate profit.

Human capital is a constituent of intellectual capital, which represents the key, long-term knowledge of a company in the competitive environment. These pieces of knowledge are used in the strategic management of a company while being created as a combination of pieces of information and knowledge from the company's external surroundings [2]. Similarly to other production inputs, HC also needs to be managed, and with the intention of strengthening the competitiveness, new ways in which its effectiveness can be enhanced need to be sought after. This is what human capital management (HCM) is focused on.

1.1 Human Capital Management

HCM represents an approach to personnel management. The correct estimation of talent [3], motivation and the placement of the right people in the right positions can optimize the utilization of human resources. According to Lawler [4], the purpose of HCM lies in the retention of capable employees in a company, whose activity is reflected in the overall operations of the company. For companies, it is of vital importance to implement a policy for motivating employees at the right time, which helps achieve competitive advantage. Ghosh [5] understands the substance of HCM to be an overarching concept, including a set of practices, organizational needs, technologies and abilities with the aim of obtaining organizational capability for the acquisition, management and optimization of the workforce. HCM can also be defined as the management of employees in a company that ensures that each employee contributes to overall productivity through their skills and knowledge [6].

The Gartner company defines HCM as a set of procedures needed for the acquisition, management and optimization of the workforce within human resources management. The application of methods and procedures within the HC field can be performed in any organization, but it is especially suitable for those organizations where people play the most important role [7]. The Personio company [8] defines the main principle of successful HCM as the placement of the right person in a company in the right position at the right time. The consideration of talent and the personality of the individual, accompanied by their training and professional development, can help create a perfect candidate for a job vacancy.

[*] Corresponding author: alzbeta.kucharcikova@fri.uniza.sk

HCM is a modern concept of people management that is considered to be a supplement to human resources management. HCM represents a process consisting of seven phases, which is inevitably connected to a company's strategy. The basic elements of HCM are represented by key performance indicators [9-11], metrics for the expression of HC value, metrics for the effectiveness of HC utilization, and metrics for the effectiveness of investments in HC. Except for these elements, HCM is also a complex of corrective measures and reports on HC [12]. The HCM process is affected by many external as well as internal factors. The external factors include the situation in the labour market, economic policy measures [13], a company's suppliers, customers and competitors, as well as the technological advances and innovation environment in a country. The internal factors are represented, for example by the remuneration system, motivation program [14-16], corporate culture [17], employee turnover, and so on.

1.2 Industry 4.0 and the management of people

Another external factor influencing HCM is represented by the ongoing processes related to Industry 4.0 which is helping to develop and improve the third industrial revolution through intelligent and automated systems based on big data and machine learning [18]. Industry 4.0 and its implementation are closely related to technologies such as the Internet of Things, information and communication systems, business architecture, and so on [19]. It is estimated that Industry 4.0 and the gradual digitization of manufacturing will greatly affect human resources and their position in companies. Technological advances will change the structure of the workforce in a short period of time and will partially cause unemployment [20]. From a long-term perspective, the quality of the workforce will increase, alongside the level and quality of education [21]. Therefore, it can be presumed that, over time, the people employed in companies will be highly qualified individuals [22].

The fusion of human resources and human capital under Industry 4.0 has been the subject of several research studies. The main findings imply that Industry 4.0 would lead to a considerable decrease in standardized work tasks with low qualification needs, and to an increase in the tasks that require a high level of qualification. This is connected to the increase in demand for workers that can take care of planning, checking and the associated IT related tasks. According to Bonekamp and Sure [23], the majority of researchers expect an increase in the complexity of many job profiles as well as an increase in the continuous learning, professional training and development of employees so that they are able to adapt to new requirements.

It is also assumed that the transformation of industry, in connection with the business environment, will have an impact on the processes of human resources and human capital management. While technologically integrated business processes can cause the loss of employment for some low-qualified employees, the need for a highly qualified workforce will rise. However, there may be an issue with finding the highly qualified workforce required and with the subsequent complications of retaining these highly qualified employees [22].

Industry 4.0 requires human capital to be supported by continuous learning and the utilization of education systems that are more focused on knowledge beyond what is currently needed. It is not only about education in companies. According to Agolla [24] the effective integration of Industry 4.0 requires learning and development of HC from early childhood. The determinative role is not only played by higher education, but also by the right preschool, elementary, secondary or vocational education.

The implementation of HCM in an organization can change the overall approach to its management. Setting appropriate frameworks and technological systems can help with the acquisition of suitable talents and their right integration and retention [25]. For companies to be competitive today, it is not sufficient to just implement progressive methods in the management of people. These methods also need to be connected to the elements of the Industry 4.0 process, modern information systems (IS), digitization, the Internet of Things, advanced data analytics and so on. Only when the technologies, technological advances and innovations get gradually transformed from external factors into internal ones, can companies effectively use them in favour of their overall effectiveness and future competitiveness.

2 Methodology

Industry 4.0 gives companies a lot of different tools with which to increase the effectiveness of their processes. This also concerns the field of HCM. However, it is still questionable whether companies use at least some of the elements of Industry 4.0 and whether they are willing to consider changes in the future. The aim of this article is to point out the interconnections between the concept of HCM and elements of Industry 4.0 and to explain some of the findings from a questionnaire survey focused on whether IT companies in Slovakia use ISs for the implementation of HCM concept, whether these ISs enable them to analyse data related to HC, and whether the companies are prepared for the introduction of further changes in this field.

The article presents the application of several scientific methods. These include the analysis, comparison and systemization of pieces of knowledge within the field of human capital management, Industry 4.0 and the accompanying phenomena. In the next step, the results of an original extensive questionnaire survey are used. This survey was conducted in 2018 and the responses from 113 IT companies operating in Slovakia were collected. The attention was directed towards IT companies due to their presence in a knowledge-intensive industry of the economy. It can also be assumed that these companies will represent an example for others in relation to the introduction of new concepts for the management of people in connection with elements of Industry 4.0. For this article, the answers to three selected questions from the survey were processed and thoroughly analysed.

3 Results

3.1 Use of ISs in companies for human resources and HCM

This part of the questionnaire was focused on the current state of the possible informational support for the implementation of the concept of HCM. The respondents (HR managers of IT companies) were asked to disclose what IS they used in the company for the field of human resources or directly for HCM if they had already partially implemented the concept. The IS being used helps to identify what types of data are being collected, and with the use of what other technologies other types of data could be collected, processed and used while measuring and assessing the effectiveness of HC.

The question in the survey was open so that respondents were not limited by a predefined scale. Subsequently, the responses could be put into common categories, based on their frequencies. For the processing and evaluation within this question, only 70 usable responses were collected. The statistical description of the set of companies from which the responses were collected is presented in Table 1.

Table 1. Structure of the sample of companies

Company size	Number companies	%	Majority owner	Number of companies	%
micro-company	25	35.71	domestic	50	71
small company	15	21.43			
medium-sized company	15	21.43			
large company	15	21.43	foreign	20	29
Total	**70**	**100**	**Total**	**70**	**100**

Source: Authors.

The aggregate results include only those ISs that were stated at least by two respondents. Other ISs, stated only by a single respondent, were put into the category "other IS". The results processed in this way are depicted in Figure 1.

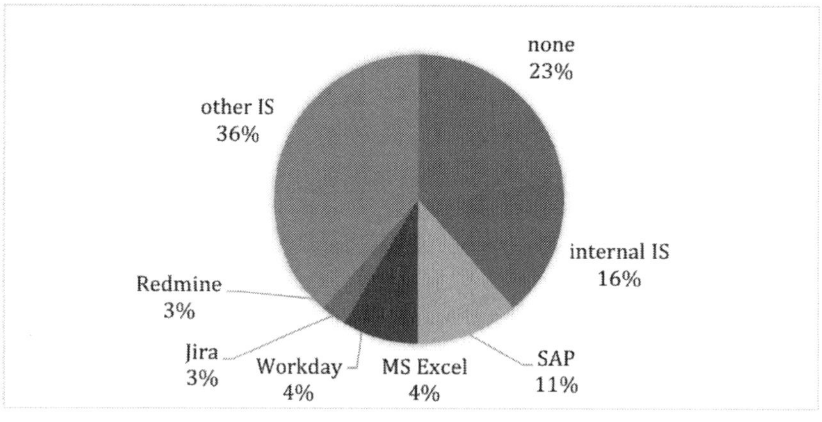

Figure 1. ISs used in the companies for HR and HC

Source: Authors.

The greatest portion of responses was represented by other IS (36%), followed by no specific IS (23%). The latter may be caused by the greater proportion of micro-companies within the sample. The third most popular response was the use of internal ISs developed and adjusted according to the company's needs (16%).

The IS SAP within the given field is used mainly by large companies (more than 33% of such companies) and by companies with foreign majority owners (30% of such companies). On the other hand, among the companies where the respondents stated that they did not use any IS for the field of HR or HC, there were mainly micro-companies (52% of such companies) and companies with domestic majority owners (32% of such companies).

 https://doi.org/10.1051/shsconf/20219001010

The results therefore revealed the use of a spectrum of ISs, ranging from spreadsheets for the creation and organization of the related records to sophisticated systems whose adjustment and extension is usually complicated and costly. The ISs identified by the respondents can be the object of further analysis for the evaluation of possibilities within the implementation of the concept of HCM and elements of Industry 4.0. The key issue here is to make sure that the implemented IS is suitable for the given. Company and that its application facilitates and supports the practical usage of the concept.

3.2 Support for the monitoring and analysis of data related to HC in the IS used

Another question in the survey was focused on the identification and description of how the IS used enables the HR managers to monitor and process the data related to HC.

This question was answered by 95 respondents. Therefore, the sample size for this question was bigger, while the structure of the sample in relation to company size and majority ownership remained comparable. The greater number of respondents means that the responses were given also by the managers who did not state a concrete name of the IS used in the previous question. Thus, the ISs used by them can be the same to those collected, but there can also be other specific software products. For this reason, the additional analysis of the information support for HCM can also be focused on other software products. This can be done using the information on websites of IS providers or the experience of users of these ISs captured in case studies.

The closed question focused on the perceived support for the monitoring and analysis of data related to HC. Respondents could choose from: the IS used allowed them to fully monitor and evaluate HC data based on all the metrics and indicators they use; the system allowed them only to monitor the qualifications of the employees and the completed learning and development activities; the system did not provide any monitoring of HC at all. The results captured in the form of relative frequencies are presented in Figure 2.

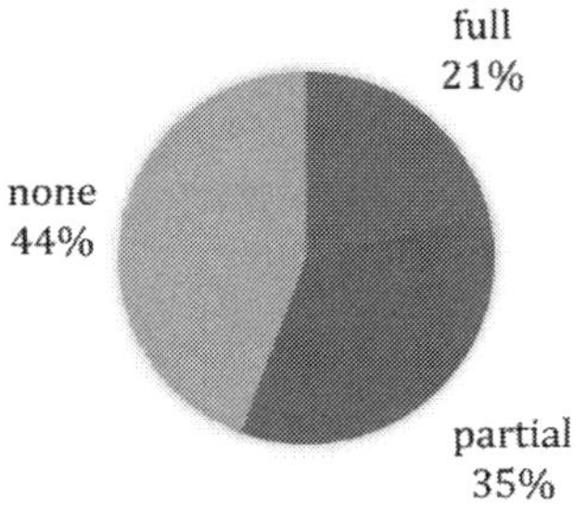

Figure 2. Informational support for the monitoring and analysis of HC data

Source: Authors.

The results show that the informational support for the concept of HCM is still relatively low, with only 20% of companies stating that their IS provided full support for all the metrics and indicators for HC. In contrast, up to 44% of companies stated that the IS they used did not allow them to monitor the metrics and indicators needed for the management of HC at all. This needs to be put into relation with the results from the previous question where many respondents, mainly from smaller companies, stated that they did not use any IS for the field of HR and HC. To confirm such a tendency with regards to the perceived informational support for HCM in the ISs, the results were further statistically analysed.

In relation to company size, statistically significant differences were detected, as supported by the following test results: $\chi^2(6) = 32.379$; at the critical value of C = 12,592; p-value <0.001; significance level of $\alpha = 0.05$. These statistical test results are reflected in terms of absolute and relative frequencies according to company size in Table 2.

Table 2. Differences in informational support for HC according to company size

HC support in the IS	Micro-company	% within the group	Small company	% within the group	Medium-sized company	% within the group	Large company	% within the group
Full	2	5.00	5	25.00	4	22.22	9	52.94
Partial	9	22.50	7	35.00	10	55.56	7	41.18
None	29	72.50	8	40.00	4	22.22	1	5.88

Source: Authors.

This means that with regards to the perceived informational support for HCM in the IS used, again, higher values of partial or full informational support were recorded for larger companies.

This is followed by the statistical testing of the differences based on majority ownership. The identification of statistically significant changes is supported by the following test results: $\chi^2(2) = 14.585$; at the critical value of C = 5,991; p-value <0.001; significance level $\alpha = 0.05$. The reflection of the results in terms of absolute and relative frequencies can be seen in Table. 3. Overall, higher perceived support for HCM is present in companies with foreign majority owners.

Table 3. Differences in informational support for HC according to majority ownership

HC support in the IS	**Domestic owner**	**% within the group**	**Foreign owner**	**% within the group**
Full	10	13.89 %	10	43.48 %
Partial	23	31.94 %	10	43.48 %
None	39	54.17 %	3	13.04 %

Source: Authors.

3.3 Expected introduction of changes leading to increased HC effectiveness

The purpose of the question in relation to the above was to reveal whether the companies planned to introduce changes, including the application of ISs and enhanced, modern data processing, to increase the HC effectiveness.

This question was answered by 108 respondents (again, with comparable structure in relation to company size and majority ownership). The aggregate results expressed in terms of absolute frequencies are presented in Figure 3.

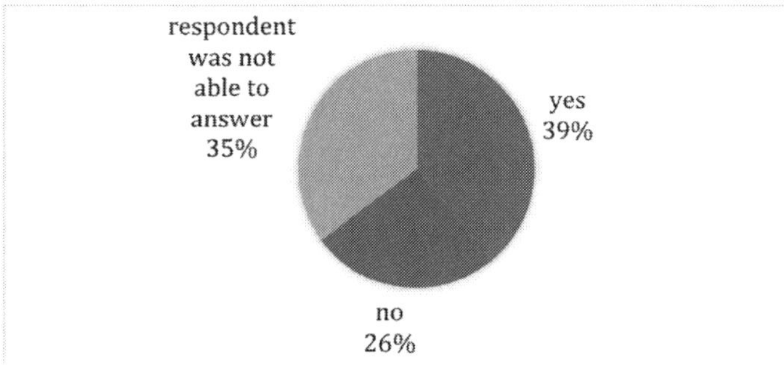

Figure 3. Expected changes leading to increased HC effectiveness

Source: Authors.

The results show a dominance of positive answers. This outcome supports further research in this field focused on the creation of practically applicable solutions the use of which will make it possible to precisely evaluate the effectiveness of HC utilization its changes over time and in relation to the measures taken as well as the business outputs and the introduction of elements of Industry 4.0.

When applying the statistical analysis, only statistically significant differences based on company size were detected. These are supported by the test results: $\chi^2(6) = 20.221$; at the critical value of C = 12.592; p-value = 0.003; significance level $\alpha = 0.05$. The structure of respondents' answers according to company size categories is presented in Table 4.

Table 4. Differences in expected changes regarding increased HC effectiveness according to company size

Expected changes	**Micro-company**	**% within the group**	**Small company**	**% within the group**	**Medium-sized company**	**% within the group**	**Large company**	**% within the group**
Yes	8	17.78	11	47.83	15	68.18	8	44.44
No	19	42.22	4	17.39	2	9.09	3	16.67
Respondent was not able to answer	18	40.00	8	34.78	5	22.73	7	38.89

Source: Authors.

 https://doi.org/10.1051/shsconf/20219001010

Micro-companies did not expect changes to lead to the higher effectiveness of HC, whereas in the other company size categories such changes and measures were expected.

A deeper understanding of these findings was achieved via an analysis of reasons stated by the respondents. Of the respondents who expected the changes described, 25 of them also added reasons for this. These reasons support further research on the topic. They included a diverse combination of external and internal impacts. Among the reasons stated was the desire for continuous improvement, pressure from competitors, as well as recognition of the importance of HC and its effect on overall business results.

Of the respondents who did not expect changes in HC effectiveness due to IS and data analytics, 16 respondents stated reasons. These reasons included insufficient capacity or satisfaction with the current level of effectiveness. However, this creates a contradiction with the philosophy of continuous improvement, whereby such a managerial decision can lead to stagnation, which can be very dangerous for a company's existence in relation to Industry 4.0. Another reason stated was that the company was planning to focus on different priorities in the next period. Such a reason implies that HC is not a high priority in the company, which sends negative signals to current and potential employees and other stakeholders. A reason that can be perceived more positively was that effective utilization of HC was included as one of the long-term goals of a company. This means that in this company no sudden changes were expected in this field since it was being dealt with on a systematic basis. Finally, another respondent stated that such changes had already been implemented and that this can be positive as long as it concerns the appropriate implementation of the entire concept of HCM and not just isolated efforts, and the evaluation of the effects there of on the company's operation.

Finally, 5 respondents stated a reason for them not being able to provide an answer to the question on future changes leading to the increased HC effectiveness. Once again, managers stated they were satisfied with the current situation. This can be a consequence of neglecting to focus on continuous improvement in a field that is critically important for IT companies. Another reason stated was that the respondent had not seen sufficient and clear outcomes that can be achieved via the implementation of HCM. This is therefore a reflection of the need to study this field even deeper, in connection with business practice, and to edify the managers regarding the successful introduction of HCM and elements of Industry 4.0. It is only then that the interest of the managers in HCM can be increased so that its implementation can help them achieve better results related to performance, efficiency, competitiveness, and employee satisfaction.

4 Discussion

The survey among the IT companies has shown that within the implementation of HCM, these companies use a lot of different information systems (36% of companies using some other software product), 16% of them use their own internal systems, 11% use SAP. This system is mostly used by foreign companies. However, up to almost a quarter of companies do not use any IS in this field, while 52% of them are micro-companies.

The survey revealed that 20% of the companies had a system supporting the analysis of data related to HCM, but up to 44% of companies could not fully monitor, evaluate and report the data needed for modern HCM. This was recorded despite the fact that even greater emphasis is being put on the use of elements of Industry 4.0 such as digitization and Big Data analytics, in connection with HCM.

On the one hand, it is understandable that the companies might not have the necessary funds for this, but on the other hand, the benefits from digitization considerably surpass the investment needed. This transformation should be as accurate as possible, respecting the mutual relationships between the processes, objects, phenomena and so on. The information describing the events in the past that are being organized in the databases during digitization does not need to be laboriously sought after later when they are used, and they can be utilized as experience for future decisions on similar issues [26]. Due to the digitization of processes as a whole, the paper form of keeping records has been substituted by the digital one. Digitization enables the better classification and selection of data, as well as the application of various structures for using the data in a better way. For now, digitization is mostly being connected to manufacturing companies that are digitally transforming their whole production processes. In relation to people, this applies, for example to the digitization of employee performance records.

Microsoft Excel has become the most frequently used tool for management and planning. Within the original survey presented in this article, 4% of IT companies are still using it instead of more complex and more expensive ISs. Even though this software was introduced more than 30 years ago, it is a successful tool for data processing in many small as well as larger companies. Its greatest advantage is its ease of use. A production worker with a paper version of a document is one of the most unstable elements in the process. Digitization eliminates the instability of the system, enables the real-time monitoring of the physical processes and can increase the reliability of the information. Via digital transformation, companies´ can relieve themselves of a huge pile of papers, thereby creating a more interactive environment [27].

After different departments of companies including marketing, finance, sales or security, have undergone digitization, the challenge of digitization in the fields of HC and human resources becomes accepted [28]. The transition from paper to information systems and analytics allows companies to implement automation and support strategic growth and create a competitive advantage [29]. The willingness of companies to introduce changes can still be quite low at the moment. In the original survey, up to 40% of the companies did not express a willingness to implement changes related to HCM, the utilization of ISs and data analytics [30]. At first sight, it might seem that these companies would need to spend additional funds, time and other resources. However, the aim of digitization in the management of people is the automation of processes that are done repeatedly, so the time saved can be spent on strategic causes [31]. Instead of performing routine tasks, an employee can join the process of planning [32]. Kenny & Company stated that with digitization, the data profiles of individual employees extended, which became one of the most important data sources within the organization. The information system then provides a wide range of data sets for the analysis, planning and decision-making in this field. The collected and aggregated data allows companies to identify trends related to the workforce together with the criteria for the job vacancies and the acquisition and retention of talents [29].

Industry 4.0 creates many challenges for companies with regards to the implementation of HCM as a modern concept of people management. Digitization, advanced analytics, Big Data and artificial intelligence have already begun changing the structure of work of HR specialists. Big Data are considered to be information assets, characterized by their large quantity, speed and variety, requiring specific technology and analytical methods for their transformation into value. Via Big Data analysis, a company can make better decisions [33], also in relation to the acquisition of the right talents for a certain job position.

Mauro et al. [34] deal with the use of Big Data in relation to people and their management. According to them, we live in a world where pieces of information are available everywhere. During the last few decades, exponential growth has been recorded in information flows, which has elicited the current integration. Pieces of information are now much more personal, in part because they are being directly consumed by the people using their personal devices. Technologies are becoming more and more accessible and more powerful when it comes to the processing of information. The efficiency of this processing is caused by fast calculations and accessibility of faster networks [35]. For the transformation of information into knowledge, it is necessary to use appropriate analytical methods and techniques. The latest analytical techniques used in the analysis of Big Data include cluster analysis [36], genetic algorithms, natural language processing, social network analysis, predictive models, data visualization, and others. Big Data analysis affects a wide range of fields including peoples' lives, culture and science. The effect can be positive as well as negative. According to Davenport, companies using Big Data can lower their costs, manage their processes more efficiently, and be more productive than their competitors. The negative impacts of Big Data analysis are related to the potential disruption of privacy and personal data [34].

Big Data analysis can be mostly utilized by companies with a larger number of employees because using these analytics enables them to find better solutions to problems connected with hiring, talent management, and so on. Via HR analytics, a company can cover up to nine fields in which it can increase its efficiency. These include the analysis of key employee skills, talent, workload, employee turnover, corporate culture, channel usage, leadership, as well as the individual and team performance of employees [37].

Big Data analysis is also related to the saving of the costs incurred by the employment of new people. If a company employs an unsuitable worker who leaves shortly after being hired, employee turnover rises, as do the overall personnel costs. If the companies use Big Data analysis before filling a job vacancy, they will be able to better predict who is suitable for the position and why, which will lower the turnover rate and the related costs. Besides that, Big Data analytics enables the monitoring of trends that can indicate how long the employee will stay in the company [38].

5 Conclusion

With the launch of the technologies related to Industry 4.0, companies become able to obtain and analyse more information than in the past. This amount rises every year due to modern software tools. For the field of human resources and HC, this presents a challenge in the form of investment in efficient ISs for HCM, which is also related to Big Data analysis. Via this, some large international companies already can, and medium-sized and small companies will in the future be able to, improve their business operations [39, 40]. For example, they can increase their employee retention rate and decrease the costs incurred as a result of employee turnover via better profiling of the requirements for a job position. The hiring of talents will be improved as well because the combination of Big Data analysis and an appropriate IS will improve the company's decision-making on hiring and the filtering of suitable candidates based on the keywords for a given position. Big Data analysis can also serve for the monitoring of work performance and the interaction between the best-performing employees and teams. The data can also serve to determine the optimal productivity of employees. Employee analytics will help companies analyse gaps in knowledge, directing the selection of the right training programs for the knowledge and skills that need to be developed. Logistic regression and predictive models will help with the estimation of the probability of the flu or other illnesses among employees, as well as help determine what variables will cause someone to leave. [41] found that the variables that affect employee turnover include higher education, number of years of working for the previous employer, changes in salary, history of promotions, and so on.

A survey conducted by Accenture in 2019 revealed that more than half of employees (64%) are afraid of losing their privacy. The use of Big Data analysis requires a company to be transparent and have the consent of the persons affected. Also, company ISs should be protected so that the loss of personal data or the violation of an employees' privacy does not occur [42].

Via Big Data analysis, a company can enhance its productivity and fill job vacancies appropriately. According to Lipkin [43], Big Data analysis related to HR that is transparently focused on how individual activities directly affect the company's performance can positively change the relationship between the employer and the employees [44]. For this reason, Hamilton and Sodeman [42] think that the application of Big Data analysis in the field of human resources can help strategically manage HC and bring higher individual performance as well as stability [45].

Since several studies came to the conclusion that Industry 4.0, besides technological advancement and business automation, will bring unemployment, it is necessary for company managers to focus their attention on the identification of new skills and abilities that will be needed in the future. The purpose of HCM lies mainly in the retention of capable employees in a company whose activity is reflected in its overall operations. Therefore, it is in a company's best interests to strive to protect current employees, adapt to more and more demanding conditions, and increase employees' qualifications.

Beyond the efforts to retain and train current employees, it will also be necessary to attract, retain and integrate new talents in companies. HR managers should therefore prepare themselves for a new definition of human resources, HC and hiring.

It will also be necessary for managers to focus on new ways of learning and development, so that new as well as current employees maintain their relevance and qualifications. One of the transitions within Industry 4.0 is artificial intelligence, leading to a change in the hiring of employees from the data management to the intelligence kind. It is predicted that within the next two years, a large number of HR teams will rely on machine learning and artificial intelligence in their HC activities [46].

Acknowledgement

This work was supported by project VEGA 1/0382/19 Building a sustainable relationship with stakeholders of enterprise through value creation using ICT.

References

1. R. Kumar, *Human Resource Management: Strategic Analysis Text and Cases*. New Delhi: IK International Publishing House (2011)
2. J. Porvazník, Celostná manažérska kompetentnosť – potreba, prístupy a metódy jej ohodnocovania [Holistic managerial competence - need, approaches and methods of its evaluation]. In *Zborník vedeckých prác z vedeckého seminára*. Bratislava: Ekonóm (2013)
3. L. Ližbetinová, M. Hitka, Selection of Most Suitable Candidates for the Talent Pool in a Furniture Manufacturing Company. *Drvna Industrija*. **67**(4), 333–340 (2016)
4. E. E. Lawler, Make human capital a source of competitive advantage. *Organizational Dynamics*. **38** (1), 1–7 (2009)
5. P. Ghosh, *What is Human Capital Management (HCM)? Definition, Process, Platforms (tools) with Examples* [on-line]. Available at: https://www.hrtechnologist.com/articles/performance-management-hcm/what-is-human-capital-management/ (2019)
6. MSG: MANAGEMENT STUDY GUIDE, *Human Capital Management - Meaning and Important Concepts* [on-line]. Available at: https://www.managementstudyguide.com/human-capital-management.htm (2019)
7. M. D'Angelo, *Beyond Human Resources: 4 Ways to Improve Human Capital Management*. [on-line]. Available at: https://www.businessnewsdaily.com/8145-human-capital-management.html?fbclid=IwAR19BmAI3xhyeO8xWRB1ONJkOXgGrg6gD0v9hugzKHPaknTQAua1oWNN3DI (2019)
8. Personio, *Human Capital Management* [on-line]. Available at: https://www.personio.com/hr-lexicon/human-capital-management-hcm/ (2019)
9. M. Ďurišová, E. Tokarčíková, F. O. Virlanuta, Z. Chodasová, The corporate performance measurement and its importance for the pricing in a transport enterprise. *Sustainability*. **11**(21), 1–17 (2019)
10. E. Malichová, M. Ďurišová, E. Tokarčíková, Models of application economic value added in automotive company. *Transport Problems*. **12**(3), 93–102 (2017)
11. M. Nadanyiová, J. Majerová, J. Oláh, 2019. Financing of adverts and its impact on the brand value. *Littera Scripta*. **12**(2), 1–12. (2019)
12. A. Kucharčíková, M. Mičiak, Human capital management in transport enterprises with the acceptance of sustainable development in the Slovak Republic. *Sustainability*. **10**(7), 1–18 (2018)

13. I. Kmecová, J. Stuchlý, L. Polanecký, M. Šuta, Analysing structure of employed and unemployed population of Czech Republic as part of human capital on labour market. *Littera Scripta*. **12**(1), 1–21 (2019)
14. M. Hitka, S. Lorincová, L. Ližbetinová, J. Schmidtová, Motivation preferences of Hungarian and Slovak employees are significantly different. *Periodica Polytechnica Social and Management Sciences*. **25**(2), 117–126 (2017)
15. L. Ližbetinová, M. Hitka, C. Li, Z. Caha, Motivation of employees of transport and logistics companies in the Czech Republic and in a Selected Region of the PRC. *MATEC Web Conferences*. **134**(8) (2017)
16. K. Stachová, Z. Stacho, J. Blštáková, M. Hlatká, L. M. Kapustina, Motivation of Employees for Creativity as a Form of Support to Manage Innovation Processes in Transportation-Logistics Companies. *Nase More*. **65**(4), 180–186 (2018)
17. Z. Stacho, K. Stachová, M. Hudáková, R. Stasiak-Betlejewska, Employee adaptation as key activity in human resource management upon implementing and maintaining desired organisational culture. *Serbian Journal of Management*. **12**(2), 305–315 (2017)
18. B. Marr, *What is Industry 4.0? Here's A Super Easy Explanation For Anyone*. [online]. Available at: https://www.forbes.com/sites/bernardmarr/2018/09/02/what-is-industry-4-0-heres-a-super-easy-explanation-for-anyone/#7fa47df49788 (2020)
19. Lu, Y., Industry 4.0: A survey on technologies, applications and open research issues. *Journal of Industrial Information Integration*. **6**, 1–10 (2017)
20. K. Kuba, F. Milichovský, Industry 4.0 and its impact on employees' age. *Littera Scripta*. **12**(1), 1–12 (2019)
21. Z. Kozubíková, The 4th industrial revolution: a challenge for the education system. In *International conference on informatization of economic and management processes*. 217–222 (2016)
22. O. Bayraktar, C. Atac, The Effects of Industry 4.0 on Human Resources Management. In: Y. Ertugrul, C. Hamza, *Globalization, Institutions and Socio-Economic Performance*. Berlin: Peter Lang GmbH. 337–360 (2018)
23. L. Bonekamp, M. Sure, 2015. Consequences of Industry 4.0 on Human Labour and Work Organisation. *Journal of Business and Media Psychology*. **6**(1), 33–40 (2015)
24. J. E. Agolla, Human Capital in the Smart Manufacturing and Industry 4.0 Revolution In: A. Petrillo, R. Cioffi, F. De Felice, *Digital Transformation in Smart Manufacturing* (2018)
25. HCM, *4 Features of a Human Capital Management Solution* [on-line]. Available at: https://www.paychex.com/articles/hcm/4-features-of-a-human-capital-management-solution (2015)
26. ATPJOURNAL. *Papierové riadenie je brzdou firiem, ich rozvoju pomôže digitalizácia* [*Paper management is a brake on companies, their development will be helped by digitization*]. [online]. Available at: https://www.atpjournal.sk/hlavna-stranka/slovenska/rubriky/podujatia/papierove-riadenie-jebrzdou-firiem-ichrozvoju-pomoze-digitalizacia.html?page_id=27071 (2018)
27. A. Bendová, *Meniaci sa svet výroby volá po digitalizácii [The changing world of production is calling for digitization]*. [online]. Available at: http://industry4.sk/magazin/industry-4-0/meniaci-sa-svet-vyroby-vola-po-digitalizacii/ (2018)
28. M. Šikýř, V. Sekerin, A. Gorokhova, Managing Human Resources Using the Best Practice. Best Fit Approach. *Journal of Applied Economic Sciences*. **13**(1), 113–122 (2018)
29. P. Daheb, *The Digitization of Human Resources*. [online]. Available at: https://michaelskenny.com/points-of-view/digitization-of-hr/ (2020)
30. L. Trnka, M. Gogola, Analysis of the tire contact area pressure on the road. *Perner's Contacts*. **14**(2), 49–59 (2019)
31. R. Eller, P. Alford, A. Kallmünzer, M. Peters, Antecedents, consequences, and challenges of small and medium–sized enterprise digitalization. *Journal of Business Research*. **112**, 119–127 (2020)
32. P. Lalwani, *What Is HR Digital Transformation? Definition, Strategies, and Challenges*. [online]. Available at: https://www.toolbox.com/hr/hr-innovation/articles/what-is-hr-digital-transformation-definition-strategies-and-challenges/ (2020)
33. L. Bartuška, J. Hanzl, L. Ližbetinová, Possibilities of Using the Data for Planning the Cycling Infrastructure. *Procedia Engineering*. **161**, 282–289 (2016)
34. A. De Mauro, M. Greco, M. Grimaldi, P. Ritala, Human resources for Big Data professions: A systematic classification of job roles and required skill sets. *Information Processing & Management*. **54**(5), 807–817 (2018)
35. N. Shah, Z. Irani, A. M. Sharif, Big data in an HR context: Exploring organizational change readiness, employee attitudes and behaviors. *Journal of Business Research*. **70**, 366–378 (2017)

36. M. Hitka, S. Lorincová, L. Ližbetinová, G. P. Bartáková, M. Merková, Cluster Analysis Used as the Strategic Advantage of Human Resource Management in Small and Medium-sized Enterprises in the Wood-Processing Industry. *BioResources.* **12**(4), 7884–7897 (2017)

37. Parsers, *How and why HR uses Big Data: big data technologies in human resources management* [online]. Available at: https://parsers.me/how-and-why-hr-uses-big-data-big-data-technologies-in-human-resources-management/ (2019)

38. insideBIGDATA, *6 Ways Companies Are Using Data Analytics to Reduce Expenses* [online]. Available at: https://insidebigdata.com/2019/02/24/6-ways-companies-are-using-data-analytics-to-reduce-expenses/ (2019)

39. M. Zarembo, *6 Ways Big Data Will Change HR.* [online]. Available at: https://www.aberdeen.com/hcm-essentials/6-ways-big-data-will-change-hr/ (2020)

40. V. Kumar, *The Role of Big Data in Human Resource Management.* [online]. Available at: https://info.aiim.org/aiim-blog/role-of-big-data-in-human-resource-management (2018)

41. S. Dil, *Big Data Can Help Cut Employee Turnover Costs.* [online]. Available at: https://www.predictiveanalyticsworld.com/machinelearningtimes/big-data-can-help-cut-employee-turnover-costs/6565/ (2020)

42. R. H. Hamilton, W. A. Sodeman, The questions we ask: Opportunities and challenges for using big data analytics to strategically manage human capital resources. *Business Horizons.* **63**(1), 85–95 (2020)

43. J. Lipkin, *Sieving through the data to find the person: HR's Imperative for Balancing Big Data with People Centricity.* [online]. 1–7. Available at: https://digitalcommons.ilr.cornell.edu/cgi/viewcontent.cgi?article=1079&context=chrr (2020)

44. S. Lorincová, M. Hitka, P. Štarchoň, K. Stachová, Strategic Instrument for Sustainability of Human Resource Management in Small and Medium-Sized Enterprises Using Management Data. *Sustainability.* **10**(10), 3687 (2018)

45. J. Blstakova, Z. Joniakova, Z. Skorkova, I. Nemethova, R. Bednar, Causes and Implications of the Applications of the Individualisation Principle in Human Resources Management. *AD ALTA: Journal of Interdisciplinary Research.* **9**(2), 323–327 (2019)

46. S. Libu, *Recruitment in the Era of Industry 4.0* [online]. Available at: https://www.highereducationdigest.com/recruitment-in-the-era-of-industry-4-0/ (2019)

Comprehensive Concept Model for Management Auditing in SMEs

Romana Kušnirová[1,*]

[1]Institute of Technology and Business, Faculty of Corporate Strategy, Okruzni 517/10, 37001 Ceske Budejovice, Czech Republic

Abstract. The aim of this contribution is to present managers and owners of SMEs with a comprehensive concept for the preparation and implementation of a systemic approach to management auditing. The concept is presented in the form of a model called the "Systemic Approach to Management Audits", as developed by the author. The principle behind the model is to conduct individual audit activities according to a predetermined order, as represented by a sub-model called "Advised Activities during Management Audits". The model, or the application of a purposeful, standardized and, above all, systemic approach, was compiled on the basis of a two-phase questionnaire survey conducted in 2015/2016 and 2018/2019 among managers of SMEs, as well as guided interviews with professional auditors. The partial aim is to also highlight the relatively poor knowledge and implementation of management auditing in SMEs, and to encourage the use of the presented methodology by them.
Keywords: SME; management audit; management audit phase; audited areas; management auditing activities; systemic approach

1 Introduction

A management audit, in comparison with other audits (e.g. personnel, financial forensics, social, internal, innovation, etc.), evaluates and analyses an enterprise as a whole, specifically focusing on a detailed assessment of its internal environment. As a result, the implementation and processing thereof brings many advantages, as confirmed by several authors [16, 20]. The most significant advantage is that it secures effective company management, followed by the establishment of the appropriate use and combination of resources, knowledge of the company´s strengths and weaknesses, determination of the suitability of the organizational structure and corporate strategy, and the assessment of the style and principles of company management as such.

1.1 Defining management audit

A number of company managers, as well as some professional auditors, associate a management audit with a personnel audit, i.e. an audit of managers and their activities, characteristics and/or skills. However, this is a misconception because a management audit deals with the issue of analysing and evaluating effective company management. This is substantiated, for example, in the detailed definition of a management audit given by [9]. They state that a management audit is a comprehensive, systematic, independent and objective verification of the management environment, goals, strategies and activities of a company in order to identify problematic areas or actualities, to recommend and propose a plan of action that would improve the performance of the company´s management.

1.2 Implementation of management auditing

Perhaps the best manner in which to consider the use and implementation of management auditing, and a systemic approach to it, is characterized by [11]. They define the approach as a logically arranged procedure of individual steps, phases, stages or activities that must be performed in a predetermined order because the individual steps (phases, activities) follow one another, as is also confirmed by [6]. Authors [4, 12] agree, recommending managers, as well as professional auditors, apply an implementation procedure that consists of 4 phases, i.e. planning (plan), organization (field work), data management (reporting), results (follow-up), as confirmed by [19].

The areas of focus of a management audit are suggested by [10], who perceives a management audit as an independent review of company management carried out by experts in the field of management and auditing, specializing only in this

* Corresponding author: pichova@mail.vstecb.cz

https://doi.org/10.1051/shsconf/20219001011

particular type of review. The review covers all areas related to managing company operations, including production, control, marketing, management, sales, finance, personnel, etc.

1.3 Goal and purpose of a management audit

According to [3], the main goal and purpose of a management audit is to identify and evaluate the actual (current) state of a particular system/style/principle of company management used so far, which is reflected in finding and characterizing problematic area(s) in a company that prevents it from being managed effectively. According to [18], it may be argued that the goal of a management audit (based on resulting outputs obtained from its implementation) is to help managers and company owners to effectively manage companies and meet their set objectives (operational, tactical and strategic) and vision. Likewise, [3, 18] claim that on the basis of the evaluated information from a management audit, its goal and purpose is to assist managers and company owners achieve effective results while performing managerial functions, i.e. managing a company as a whole.

2 Methodology

The aim is to present company managers with significant evidence of the advantages of implementing management audits. The methodological model presented here includes suggestions on how to proceed with the implementation of management auditing, as well as which areas need to be analysed, whereby the focus is on small and medium-sized enterprises in the Czech Republic. Based on [13], the method of experimental modelling was selected to achieve the objective.

In relation to experimental modelling, a specific model called "Systemic Approach to Management Audits" was compiled as part of the comprehensive concept for the implementation of management auditing. The model consists of several sub-models in sequence (see Results and Discussion), and was designed on the back of the results of a two-phase questionnaire survey (involving managers of SMEs) and guided interviews with 12 professional auditors. The first phase was carried out in 2015/2016 and resulted in the creation of a key sub-model ("Management Model of Decision-Making"). The second phase, was conducted in 2018/2019, and in combination with the guided interviews with professionals (with a minimum of 10 years´ experience), led to the development of a model concept for a systemic approach to management audits, including the areas that should be evaluated. The concept was generated only on the basis of the answers submitted by those respondents who prepare and implement such audits in their true sense, i.e. an assessment of the current state of the system of company management. The following table shows the results from both questionnaire surveys with regards to which participating SMEs (according to company size) perform or do not perform management audits or confuse the audit with other audits e.g. personnel, internal, etc.

Table 1. Questionnaire survey results - management audits by company size

Questionnaire survey 2015/2016				
Company size	Perform management audits	Do not perform management audits	Confuse management audits	Number of respondents
Micro-	1	151	9	161
Small	33	129	62	224
Medium-sized	33	121	71	225
Total number of respondents	610			
Questionnaire survey 2018/2019				
Company size	Perform management audit	Do not perform management audit	Confuse management audit	Number of respondents
Micro-	5	28	8	41
Small	17	45	16	78
Medium-sized	33	71	32	136
Total number of respondents	255			

Source: Author.

https://doi.org/10.1051/shsconf/20219001011

The data in the above table shows that the participating SMEs do not perform management audits very often. Based on this finding, the following hypotheses were formulated:

- Hypothesis 1 (H1): The preparation and implementation of management audits is dependent on company size.
- Hypothesis 2 (H2): When preparing and implementing management audits, less than half of companies use a systemic approach.
- Hypothesis 3 (H3): When preparing and implementing management audits, more than half of companies evaluate more areas, not only the area of management.

H1 was evaluated on the basis of correlation analysis, as recommended by [8]. The following formula was used to assess the correlation analysis.

$$r = \frac{\sum_{i=1}^{n}(x_i - \bar{x})(y_i - \bar{y})}{\sqrt{\sum_{i=1}^{n}(x_i - \bar{x})^2 \sum_{i=1}^{n}(y_i - \bar{y})^2}}$$

Figure 1. Pearson correlation coefficient

Source: Author

According to [7] and [8], H2 and H3 were evaluated on the basis of the application of a binomial test, for which the following formula was employed:

$$\frac{\dfrac{\text{Absolute expression } f1 = 1 - f0 \approx f1 > f;\ f1 < f;\ f1 = f}{\text{Relative expression } f1 = 100 - f0 \approx f1 > f;\ f1 < f;\ f1 = f}}{(f1 + f0 = 1;\ 1 - f1 = f0)}$$

Figure 2. Principle of binomial test calculation

Source: Author.

3 Results and Discussion

3.1 Principle of using systemic approach to management auditing

The questionnaire survey revealed that SMEs process management audits sporadically. Of the 865 survey participants, only 122 (14.1%) conduct management audits within the true sense of the meaning. Due to the relative unfamiliarity with management audits, previously created models that determine the main essence of a management audit are presented here. These indicate how to proceed with implementing a management audit and which areas to concentrate on.

In earlier publications, [5, 14 and 15], the author´s own proposals on how to handle management audits as efficiently as possible were introduced in the form of a model, namely the "Management Model of Decision-Making" (see Figure 3). The model consists of three sub-models that complement one another, and which follows on from previously known and frequently used models (McKinsey 7S model, IFE matrix, VRIO Framework). The latter have been expanded and modified by the author for the purposes of management audit implementation.

The main goal of the first sub-model ("7S Adaptation Model") and the second sub-model ("Application IFE 7S Matrix") is to give an in-depth evaluation of the current company management system, whereby managers should concentrate on choosing an appropriate management style (i.e. effective company management as such). The third sub-model ("Resource Model") evaluates company resources as a set of inputs, whose suitable allocation managers should focus on.

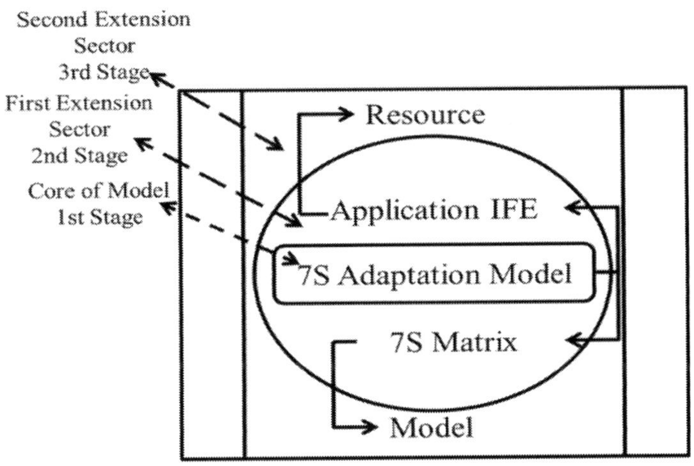

Figure 3. Stages of Management Model of Decision-Making

Source: Author.

In addition, in order to enable the implementation of management audits, the author generated a standardized procedure (see Figure 4) for the appropriate use of a systemic approach to them, as expressed by the model "Systemic Approach to Management Audits" (for more information see [15]). In order to fulfil the essence of management audits, i.e. to evaluate the current state of the system of company management, as well as the principle of a systemic approach to them, managers must be made aware of what needs to be analysed as part of a management audit. Given that purpose, the so-called "Managerial Questions" (see Table 2) were drawn up.

Table 2. Managerial Questions

Examined Questions	Factors patterned on the 7S Model
What does the company want to achieve and how does it want to achieve it?	Strategy
What (composition of) staff are required in the company?	Staff
What abilities and skills does every employee need to have?	Skills
What will be the hierarchical structure of the selected employees?	Structure
Which style of management should be used in the company?	Style
What systems should be used in the company?	Systems
Are all employees fully aware of company events and happenings?	Shared values

Source: Author.

Answering the above questions is the first step (and simultaneously the basis) of the entire "Systemic Approach to Management Audits" model (note 1), and is also part of the first stage of the "Management Model of Decision-Making". To make this even clearer, a specific list of questions (see Table 3) that makes up the structure of the "7S Adaptation Model" is given below. The evaluation of the questions can be performed separately by SMEs as part of their routine evaluation of the internal environment, whereby the recommendation to managers and owners of SMEs is to adhere to the original intention hereof and hence evaluate the questions both in terms of efficacy and efficiency.

Table 3. Evaluation questions

Factors	Evaluation Questions
Structure	Is the organizational structure of the company flexible? Does the number of management levels correspond to the specifics of the company? Are the superior-subordinate relationships in the company clearly defined? Is (Are) the set degree(s) of centralization / decentralization of powers to subordinates sufficient? Is the number of subordinates to individual managers optimal, i.e. is the span of management in the company appropriately used?
Systems	Are suitable communication methods and techniques used in the company? Does the company use an optimal combination of resources? Are appropriate and modern management methods and techniques used in the company? Is the company able to correctly respond to the changes in its corporate environment? Are appropriate information systems used in the company?
Management Style	Do managers use an appropriate style of management with their subordinates in the company? Do managers adapt the style of company management according to the current situation in the company? Are the responsibilities of managers and their subordinates in the

	company clearly defined? Do managers use the possibility of delegation of powers to their subordinates, or is there abuse of their power over subordinates? Is teamwork used?
Staff	Does the company have an optimal (sufficient) number of employees? Are the employee qualifications adequate for carrying out the job responsibilities? Do employees have any opportunities for career growth and progression? Are the job descriptions of individual employees, including their responsibilities, defined? Do managers use appropriate style(s) and method(s) of motivation and evaluation of their subordinates?
Skills	Are managers at all hierarchical levels able to independently plan, organize, manage and control? Are managers at all hierarchical levels able to compile plans so that these can complement and follow one another? Are managers and their subordinates able to use a self-management method? Do managers at all hierarchical levels use a management control system and feedback control? Are the skills of managers and their subordinates sufficient for the performance of their job?
Strategy	Has the company appropriately and clearly articulated its strategy? Is the company strategy sufficiently and appropriately developed (divided) into short-term procedures and activities (operations)? Has the company developed any concept of how to best meet the articulated strategy and implement it? Has the company adequately formulated and established its business goals and developed any concept of how to best achieve them? Are the goals, visions and strategies realistic and feasible?
Shared Values	Are managers and their subordinates familiar enough with the company policy, goals, vision, mission, tasks and strategy? Are managers and their subordinates sufficiently and informed in good time about the company affairs and actions? Do managers and their subordinates act in accordance with the established goals and values of the company? Are the competences of all company employees clearly defined and are employees mutually aware of them? Has the company formed its corporate culture that all employees are familiar with?

Source: Author.

Furthermore, the aforementioned systemic approach model was designed with respect to the principles of the PDCA cycle (as suggested by Deming), interviews with professional auditors (and their recommendations), observations made by [12], and was eventually arranged as four repeating elementary phases and related key management audit activities:

- P: Plan – planning of activities, i.e. planning management audit implementation, which also includes defining audit objective(s);
- D: Do – putting the plan into practice, i.e. management audit implementation, which is to evaluate the current company management system by responding to the "Managerial Questions" and processing the "Management Model of Decision-Making", see [15];
- C: Check – checking the result, i.e. comparing the current and desired state of the company management system and proposing measures (preferably 3 to 5) leading to the improvement of the system;
- A: Act – implementation of improvements, i.e. implementation of the proposed measures leading to a more effective company management system followed by re-evaluation of the current state of the system based on an assessment of the "Managerial Questions", as well as questions from the "Management Style" factor included in the "7S Adaptation Model".

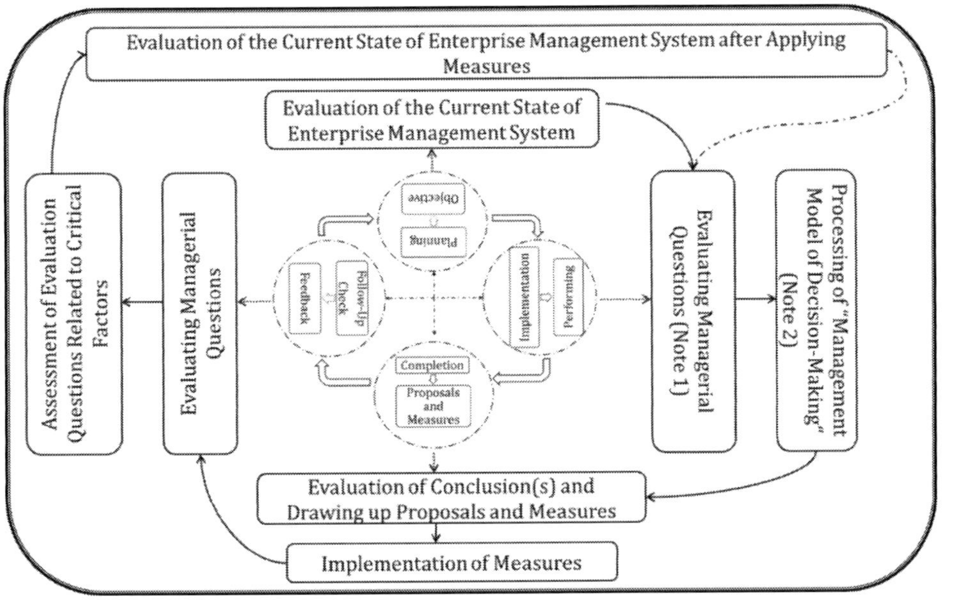

Figure 4. "Systemic Approach to Management Audits" model

Source: Author.

Furthermore, it may be added that the "Systemic Approach to Management Audits" model works on the assumption that a management audit is conducted in a specified order and time sequence, as presented in Figure 5 and previously in Figure 4, whereby the phases represent the basis for the model. As for the application of the model, the phases are characterized by specific activities that need to be carried out, as modified according to [17].

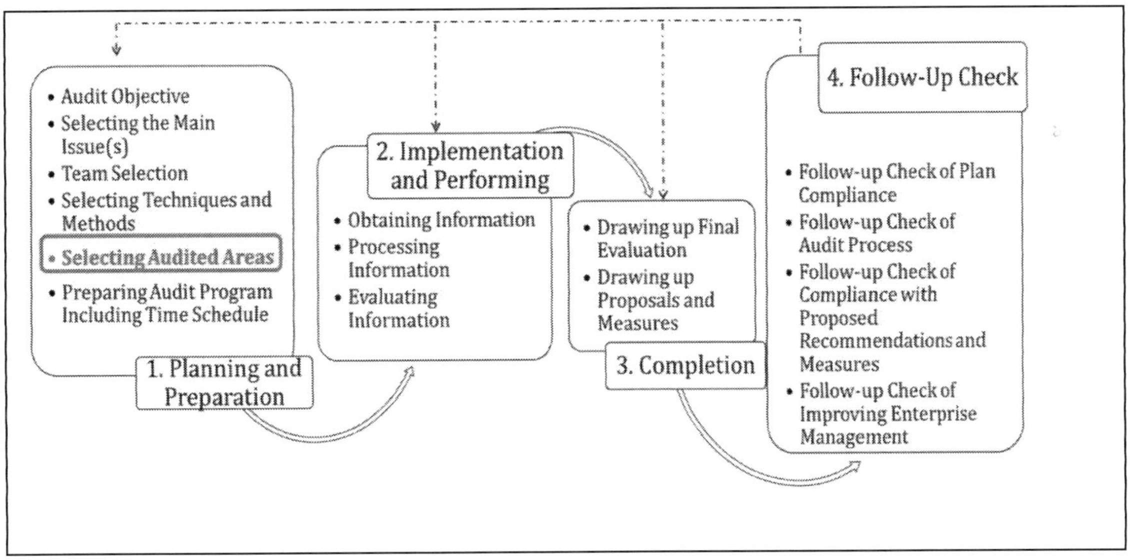

Figure 5. Advised activities during management audits

Source: Author.

The whole process must subsequently be extended with a "Six Area Auditing (6A Auditing)" model, as shown in Figure 6, which recognizes those areas that are suitable for auditing, as revealed by the survey. There are two prerequisites for the model. The first is that, in terms of management audits, companies tend to have multiple areas (not only the area of management) audited. The second is that the number of audited areas is affected by company size. The model therefore reflects both prerequisites and conforms with the requirements of managers of SMEs.

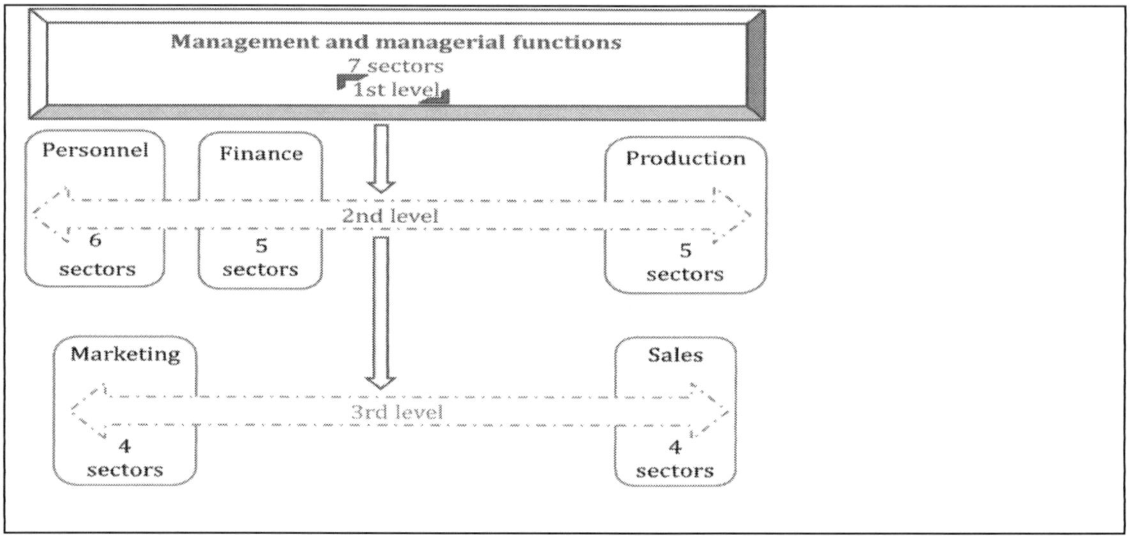

Figure 6. 6A auditing

Source: Author.

The above figure incorporates 3 levels, whereby the 1st level is essential and functions as a detailed assessment of the management area. On the basis of the "7S Adaptation Model", a total of 7 sectors have been allocated to "Management and Managerial Functions", which interconnects both models and puts them in sequence, in line with the stated aim of the author. The 2nd level (Personnel, Finance and Production) and the 3rd level (Marketing and Sales) should be viewed as added elements of a management audit, whereby certain consideration is only given to areas that habitually influence the management of a company. Besides the 1st level, small companies (11-50 employees) are advised to also audit areas of the 2nd level, whilst medium-sized companies (51-250 employees) are recommended to have a total of 6 areas audited, i.e. all 3 levels. In order to make the processing of the audit more feasible, numbers and names of sectors (see Table 4) have been allocated to each area in which their individual analysis is to be performed, since they mainly influence the given area as well as the management of a company as such.

Table 4. Areas and sectors of auditing

Areas	Number of Sectors	Sectors
Management and managerial functions	7	Structure; Systems; Management Style; Staff, Skills; Strategy; Shared values
Personnel	6	Work environment; Employee qualifications; Employee performance ratings; Motivation and stimulation of employees; Completed tasks; Workplace safety
Finance	5	Attitude to funding; Funding principle; Financial reporting; Financial stability and profit; Investing activities
Production	5	Production practices; Continuity of production process; Production technologies; Products; Product range
Marketing	4	4P (Product, Price, Place, Promotion); Direct marketing, 4C (Customer solution, Cost, Convenience, Communication); Customers
Sales	4	Specification; Services; Logistics and expediting; Sales

Source: Author.

3.2 Evaluation of hypotheses

When evaluating the hypotheses, only those respondents from both surveys that conduct management audits in the true sense of the meaning were taken into consideration. The total number of respondents amounted to n = 122 (micro-companies n = 6; small companies n = 50; medium-sized companies n = 66).

First, hypothesis 1 (H1) was evaluated on the basis of correlation analysis and p-value. The correlation analysis involved determining the rules for assigning the evaluation coefficients presented in Table 5. The test set included only those companies in which it was determined that there was a competency to implement and process management audits in the true sense of the meaning.

Table 5. Principle for assigning evaluation coefficients

Frequency	Evaluation coefficient
Multiple times a year	1.00
Once a year	0.8
Once in 2 years	0.6
Once in 3 years	0.4
Longer time span or according to own needs	0.2

Source: Author.

The correlation analysis evaluation, including the p-value (i.e. confirmation/refutation of H1), is presented in Table 6.

Table 6. Correlation analysis - H1 evaluation

Variables	Resulting values
Pearson correlation coefficient; r interval <-1;1>	0.1956179
p-value $\alpha = 5$ %; (p-value $\leq 0{,}05$)	0.1523

Source: Author.

The correlation analysis based on the Pearson coefficient calculation points to the fact that utilizing the competency to implement and process management audits does not depend on company size. The resulting coefficient value of r = 0.1956179 indicates a direct but weak dependence between the variables. The direct, weak dependence is clearer in Figure 7.

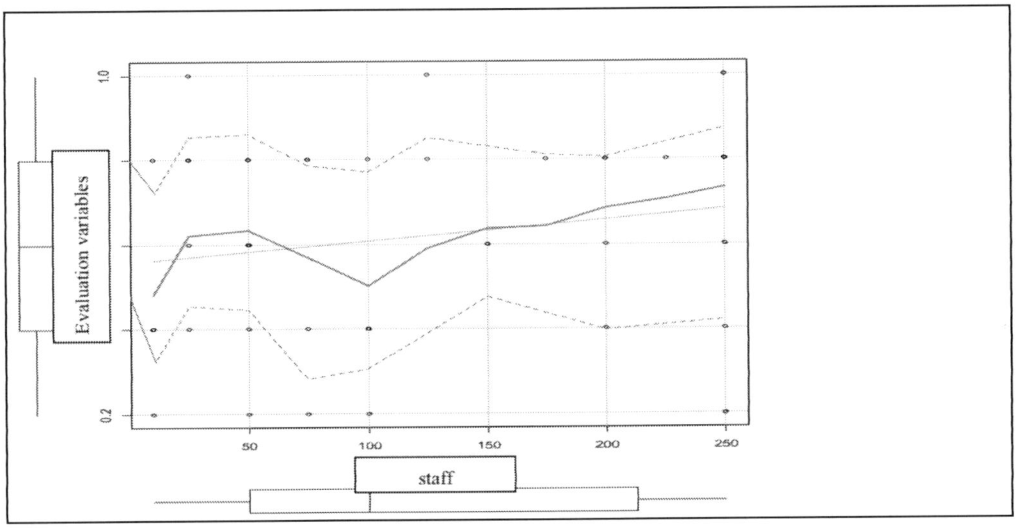

Figure 7. Graphical representation of correlation analysis

Source: Author.

The resulting p-value for hypothesis H1 is higher than the specified level of significance of 0.05, so it cannot be said that the correlation of the specified variables is significant. Therefore, hypothesis H1 (the preparation and implementation of management audits is dependent on company size) can be refuted.

The evaluation of hypothesis H2 was carried out in three stages. The first stage was to determine general criterion (1) and specific criteria (4). The general criterion involved conducting management audits in at least 4 phases, including the above audit content. The specific criteria were as follows: 1st Criterion – Phase 1: Audit Planning (Content: Defining audit objectives); 2nd Criterion – Phase 2: Audit Implementation (Content: Audit processing); 3rd Criterion – Phase 3. Audit Completion (Content: Drawing up proposals and measures to improve management); 4th Criterion – Phase 4: Follow-Up Check (Content: Feedback). The second stage involved assigning 0s and 1s (as the numerical values) in accordance with the categories specified in Table 7. The third stage involved a numerical assessment of the binomial test and the drawing of a conclusion. On the whole, the actual evaluation is the result of defining a desired frequency value (f), which was set at f = 0.5 as the absolute expression value, which corresponds to f = 50% as the alternative for the relative expression value. If the resulting value (f1) is <u>higher</u> than f (see Figure 2), i.e. f1 > f 0.5 (50%), the evaluation is <u>negative</u>, whereas if f1 is <u>less than or possibly equal</u> to f, i.e. f1 ≤ f 0.5 (50%), the evaluation is <u>positive.</u>

Table 7. Principle of assigning categorical numerical values

Category – audit implementation according to determined criteria		
Procedure according to determined criteria		Non-procedure according to determined criteria
Categorical numerical values – Code assignment		
Category	Non-procedure according to determined criteria	Procedure according to determined criteria
Code	0	1

Source: Author.

The evaluation of the binomial test is presented in the following table (Table 8).

Table 8. Evaluation of binomial test

Categorical numerical value		**N**	**Evaluation**	**Frequency**
Non-procedure according to determined criteria	0	88	1-f0 < 0.5	f0 = 0.721
Procedure according to determined criteria	1	34		f1 = 0.279
Total		122	1-0.721= 0.279	f∑ = 1.00

Source: Author.

On the basis of the calculated binomial test value, hypothesis H2 can be answered positively – YES (When preparing and implementing management audits, less than half of companies use a systemic approach), since the frequency value f1 = 0.279 is lower than f = 0.5.

Hypothesis H3 was also evaluated in three stages. The first stage involved establishing specific categories for assigning categorical numerical values. The second stage, once again, involved assigning 0s and 1s (as the numerical values) in accordance with the categories specified in Table 9. The third stage involved a numerical assessment of the binomial test and the drawing of a conclusion. On the whole, the actual evaluation is the result of defining a desired frequency value (f), which was set at f = 0.5 as the absolute expression value, which corresponds to f = 50% as the alternative for the relative expression value. If the resulting value (f1) is higher than f (see Figure 2), i.e. f1 > f 0.5 (50%), the evaluation is negative, whereas if f1 is less than or possibly equal to f, i.e. f1 ≤ f 0.5 (50%), the evaluation is positive.

Table 9. Principle for assigning categorical numerical values

Category – Analysis of areas		
Analysis only of the area of management		Analysis of multiple areas
Categorical numerical values – Code assignment		
Category	Analysis only of the area of management	Analysis of multiple areas
Code	0	1

Source: Author.

The evaluation of the binomial test is presented in the following table (Table 10).

Table 10. Evaluation of binomial test

Categorical numerical value		**N**	**Evaluation**	**Frequency**
Analysis of the area of management	0	25	1-f0 < 0.5	f0 = 0.205
Analysis of multiple areas	1	97		f1 = 0.795
Total		122	1-0.795= 0.205	f∑ = 1.00

Source: Author.

On the basis of the calculated binomial test value, hypothesis H3 can be answered positively – YES (When preparing and implementing management audits, more than half of companies evaluate more areas, not only the area of management), since the frequency value f1 = 0.795 is higher than f = 0.5.

4 Conclusion

A clear conclusion may be drawn from the two-phase survey conducted. Although the managers or owners of the SMEs concerned are aware of management audits, they perform them very rarely in their true sense, i.e. an assessment of the current state of the system of company management. This finding was evaluated on the basis of processing obtained survey data, where 198 (23%) of the total number of respondents (n = 865) confuse management audits with personnel audits, and 545 respondents (63%) do not perform management audits at all.

The results of the evaluations of the hypotheses shows that the preparation and implementation of management audits does not depend on company size (hypothesis H1 was therefore rejected), although it is clear that management audits are more often processed by medium-sized companies. Based on the confirmation of hypothesis H2, it can be claimed that the SMEs do not use a systemic approach to implementing management audits, which indicates that they do not have fixed rules on how to proceed with the implementation thereof. Out of the total number of respondents conducting management audits (n = 122), only 34 respondents (28%) use such an approach. Based on the confirmation of hypothesis H3, through which the author wanted to determine whether the SMEs evaluate more areas than just management and managerial functions, it was discovered that 97 out of 122 (80%) apply the procedure.

In light of the evaluations of hypotheses H1 and H2, it can be argued that the SMEs are missing out on an opportunity to streamline the principles and systems for company management. It is for this reason that the comprehensive "Systemic Approach to Management Audits" model, as outlined in this contribution, was proposed in order to encourage a systemic approach to the implementation of management audits. Based on the evaluation of hypothesis H3, the model was supplemented with specific areas, for which advice had been given to evaluate and analyse them as part of such an audit, thereby taking into account company size.

Finally, it is recommended that the managers and owners of SMEs pay closer attention to the implementation and processing of management audits, since the output thereof is an intention/prediction of how to manage a given company as effectively as possible in the future. On top of that, any company wanting to succeed against their competitors must first focus on their internal stability and constantly analyse their internal environment in order to achieve one of their main business goals, which is the continuous improvement of the system of company management. The author also agrees with opinions of [1, 2, 16].

References

1. E. S. S. Alzoubi, Audit committee, internal audit function and earnings management: evidence from Jordan. *Meditari Accountancy Research*. **27**(1), 72–90 (2019)
2. J. Belás, P. Bartoš, A. Ključnikov, L. Kozubíková, Významné determinanty kvality podnikatel´ského prostredia malých a stredných firem [*Significant determinants of the quality of the business environment of small and medium-sized companies*]. *Scientific Papers of the University of Pardubice, Series D*. **35,** 5–17 (2015)
3. Ch. D. Bose, *Principles of management and administration*. 2nd ed. New Delhi: PHI Learning Pvt. (2012)
4. L. D. Crumbley, *U.S. Master Auditing Guide*. 3rd ed. Chicago: CCH INCORPORATED (2004)
5. P. Gallo, R. Píchová, A. Senkova, D. Matusikova, J. Mitrikova, Techniques and analysis of management audits. *Cbu international conference proceedings 2017: innovations in science and education*. **5**, 132–137 (2017)
6. N. A. Hassan, S. H. Mohamad Zailani, H. A. Hasan, Integrated internal audit in management system: A comparative study of manufacturing firms in Malaysia. *TQM Journal*. **32**(1), 110–126 (2019)
7. J. Hendl, *Kvalitativní výzkum: základní teorie, metody a aplikace* [*Qualitative research: basic theories, methods and applications*]. 4th ed. Praha: Portál (2016)
8. R. Hindls, *Statistika pro ekonomy*. 8th ed. Praha: Professional Publishing (2007)
9. P. T. Kotler, K. L. Keller, *Marketing Management*. 15th ed. Boston: Pearson (2016)
10. J. Law, *A dictionary of business and management*. 5th ed. Oxford: Oxford University Press (2009)
11. P. Leung, P. Coram, B. Cooper, P. Richardson, *Modern Auditing & Assurance Services*. 6th ed. Hoboken: John Wiley (2015)
12. R. R. Moeller, *Brink's modern internal auditing: a common body of knowledge*. 8th ed. Hoboken: John Wiley (2016)
13. Z. Molnár, S. Mildeova, H. Řezanková, R. Brixi, *Pokročilé metody vědecké práce*. Praha: Profes Consulting (2012)
14. R. Píchová, D. Raušer, Management Audit as a Tool for Managing Small and Medium-Sized Enterprises. In *7th International Scientific Conference on Managerial Trends in the Development of Enterprises in Globalization Era (ICoM)* [online]. Available at: https://spu.fem.uniag.sk/fem/ICoM_2017/files/international_scientific_conference_icom_2017 (2017)
15. R. Píchová, D. Raušer, Principle of using systemic approach to management audit implementation by small and medium - sized enterprises. *Scientific Papers of the University of Pardubice, Series D*. **26**(3), 209–220 (2018)
16. P. Psar, V. Kupec, Innovative controlling and audit-opportunities for SMEs. *Problems and Perspectives in Management*. **17**(3), 184–195 (2019)

17. R. Stefko, R. Pichova, P. Gallo, D. Rauser, A Systemic Approach to Implementation of Management Audit. In: *6th International Scientific Conference on International Business and Management, Domestic Particularities and Emerging Markets in the Light of Research* [online]. Available at: https://www.unipo.sk/public/media/3948/ZBORNIK%202016 (2016)

18. T. L. Wheelen, D. J. Hunger, *Strategic management and business policy: concepts and cases.* 13th ed. Upper Saddle River: Prentice Hall (2012)

19. A. V. Zepeda, A. V. Ochoa, New public management based on rankings: from planning to evaluation. *OPERA-COLOMBIA.* **21**, 79–94 (2017)

20. Y. Zhaokai, K. C. Moffitt, Contract analytics in auditing. *Accounting Horizons.* **33**(3), 111–126 (2019)

An Analysis of Products Communicated on Instagram by the Most followed Czech Influencers

Kateřina Kvěchová[1,*]

[1]Institute of Technology and Business, Faculty of Corporate Strategy, Department of Tourism and Marketing, Nemanická 436/7, 370 10 České Budějovice, Czech Republic

Abstract. The aim of this contribution is to determine and analyse the types of products the five most followed Instagram influencers in the Czech Republic promoted during the period September 2019 - September 2020. The most followed influencers in 2020 were identified on the basis of the highest number of followers (Leoš Mareš, Petr Čech, Jirka Král, Karolina Kurková and Anna Šulcová). The next step was to calculate the frequency of product-related posts by each influencer according to the individual product types. Based on the data, it can be concluded that the most frequent product-related posts are associated with clothing and accessories, followed by magazines, cosmetics, e-shops, sports, travel, culture, children's goods or restaurants.
Keywords: influencer marketing; influencer; product-related posts; communication; Instagram

1 Introduction

One of the influencer agencies states that the amount of funding spent in this area in 2018 was $ 4-7 billion [1]. Other available sources predict further growth in investment in this area. This prediction is confirmed by the fact that in 2018 the number of Instagram users exceeded one billion for the first time. Influencer marketing is part of online marketing, which is one of the fastest growing forms of marketing. In other sources, it has been proven that advertising inspires only 30% of users, whereas the trust followers have in influencers is 95%. It is clear from past developments in the industry that the gap between these numbers will widen even further in the future [2].

It is stated that the Internet population of the Czech Republic (older than ten years) reached 7.8 million people in September 2019, of which 5.3 million users connected to the Internet via a mobile device [3]. According to the CZSO [4] the total volume of social network users in the Czech Republic in 2019 was 4.735 million. Various studies have shown that Czechs spend the most time on the Facebook (59% of respondents), followed by Seznam, YouTube, Google and Instagram [5].

Data from the Czech Republic show that in September 2019 Instragram was used by 2,275,000 users, of which the largest group (730,000 users) were people aged 18-24 years [6].

The above data show that social networks atttract many users from different segments, so there is great potential for a broad marketing communication mix. In 2019, in the US, over 60% of companies increased their influencer promotion budgets [7]. The worldwide trend is to move advertising from Facebook to Instragram. The advantage of Instragram is the higher interaction on product-related posts, as well as the wide community network of users. Current trends show that consumers are more likely to determine their purchasing behaviour based on the reviews and references of influencers that they follow [8].

Based on the highest number of followers of foreign influencers, the worldwide trend in communication on Instragram is mainly focused on clothing, cosmetics, catering, travel, culture and sports.

In the last few years, taking advantage of influencers on social networks has become an increasingly popular advertising strategy for brands looking to promote products, especially when targeting younger demographic groups. The results show that the informative value of influencer-generated content, influencer credibility, attractiveness, and similarity with followers positively affects their confidence in the product-related posts, which in turn affects brand awareness and purchasing intentions [9].

It has been shown that there should be ongoing cooperation between companies and influencers, in order to achieve better creative results [10]. Kumar, Aksoy and Donkers [11] proposed four components of the perception of the value of the customer's commitment to a company. These are the lifetime value of the customer (customer shopping behaviour),

* Corresponding author: kvechova@mail.vstecb.cz

the value of the customer's recommendation, the value influencing the customer (which includes the customer's behaviour with regards to influencing other customers), and the knowledge value for the customer. The findings reveal that Instagram product-related posts about highly trusted brands have a positive impact on its credibility, access to advertising and intention to purchase by followers [12].

The developing influencer community therefore has considerable power over brand perception, and society can greatly affect its reputation. These new features require communication professionals to seek continuous collaboration through various social web channels [13]. Jin, Mugaddam and Ryu [14] have even shown that consumers exposed to Instagram celebrity product-related posts perceive the source as more credible, show a more positive attitude towards the supported brand, feel their social rank improves, compared to consumers exposed only to celebrities.

Influencers who publish recommendations on original products and therefore have no commercial relationship with the brand are also successful if they explicitly state this [15]. The acquired credibility of influencers clearly shows significant and positive relationships to the purchasing intention [16].

In addition, they found that the reduction in interest was more pronounced in influencers than in branding. These facts suggest that the commercial orientation of a promotion has a strong negative impact on how it is perceived, reduces confidence in the influencer and also reduces the credibility of the product-related posts [17].

Chetioui et al. [18] demonstrate that perceived credibility, perceived behavioural control, perceived subjective norms, perceived expertise, and perceived compliance positively influences attitudes toward fashion influencers. Conversely, for organic products, for example, influencers want folllowers to believe in the ecology of a product only on the basis of a statement they make in their account [19]. On the contrary, in tourism, an important factor in an influencer's success is being loyal to one's self, the brand and their audience [20]. For fashion influencers, success factors are not only credibility but also the social and physical attractiveness of the influencers [18]. Like for fashion product-related posts, similar requirements apply to influencers in the cosmetics sector, whereby Instagram is becoming an increasingly important tool for placing products or services [21].

Curiel [22] examined the impact of the communication model on digital social media users in policy-oriented product-related posts. He performed a quantitative-qualitative comparative content analysis and confirmed the impact of strategic marketing tools on the response rate of social media users. The same research method was applied to the field of fashion. The results of the study showed that the application of strategic tools can restore a company's image and brand popularity [23].

The data examined clearly show that influencer marketing has a positive effect on consumer involvement in brand communication, influence on brand choice, shaping brand perception and improving customer willingness to publish brand promotional content [24].

2 Methodology

The aim of this contribution is to determine and analyse the types of products the five most followed Instagram influencers in the Czech Republic promoted during the period September 2019 - September 2020. Based on the aim, two research questions were determined:

1. *In the last year, what were the ten most frequently promoted products via product-related posts on Instagram by the five most followed influencers in the Czech Republic?*
2. *Were the product-related posts of the five most followed influencers in the Czech Republic targeted at the relevant consumer segment?*

The most followed influencers in 2020 were identified on the basis of a published list [25]. The analysed influencers were identified on the basis of the highest number of followers. The football player Petr Čech has the highest number of followers (2,062,470), followed by presenter Leoš Mareš (1,079,795), YouTuber Jirka Král (902,903), the model Karolína Kurková (810,965, who was born in the Czech Republic but lives in the USA, and Anna Šulsová (829,927). Furthermore, for each monitored influencer, the history of product-related posts from September 2019 to September 2020 was analysed. The next step was to calculate the frequency of product-related posts for each influencer according to individual product types. Based on the frequency of product-related posts for a given product type by the analysed influencers, the ten most frequently promoted products were determined.

After determining the most frequently communicated products by each influencer, heuristic methods were applied to determine whether the products communicated by the influencer corresponded to their target segment, i.e. their followers. The follower segment was determined according to research conducted by a marketing agency [26]. Heuristic methods provided an approximate solution based on the author's expert judgment and experience.

3 Results and discussion

The analysis of the product-related posts of the most followed influencers took place over a one year period (from September 2019 to September 2020), whereby each valid influencer post had to be product related.

At the time of analysis, Petr Čech had the most followers (2,062,470). Petr Čech posted a total of 23 product-related posts during the monitored period. Most of the posts were focused on sports equipment (13), in particular for football (Table 1).

Table 1. Petr Čech – product-related posts

Petr Čech (2,062,470 followers)	
Product type*	**Frequency of appearance**
Sport	13
Clothes, accessories	1

Source: Author.

(*List of represented brands: Chelsea, Euro 2020, 21 Euro, Premier League (13), Puma (1))

Table 2. Leoš Mareš – product-related posts

Leoš Mareš (1,079,795 followers)	
Product type*	**Frequency of appearance**
Retail	24
Restaurants, cafes	10
Foodstuffs	5
Clothes, accessories	2
Cosmetics	2
Sport	9
Magazines	20
Travel, culture	20
Children´s goods	5
TV, Radio	12

Source: Author.

(*List of represented brands: Penny (23), Thestorepraque (1), If Cafe (5), My Chef Kitchen (4), Yuniku bbq (1), Mumm (2), Bigboybutter (2), Tostiny (1), Linda Farow (1), LV (1), Whitepear001 (2), Slavia Praha (7), Praqueplayoff (2), Forbes (3), Esquire (9), iDnes (1), Reflex (1), Elle (1), Proč ne magazin (5), The Chedi Muscat (1), Flymeto.world (4), Letiště Václava Havla (1), Ticketstream, Ticketportal (14), Trihorse (5), Mall TV (1, Evropa 2 (11))

Based on the number of followers, Leoš Mareš (1,079,795) is the second most followed influencer. Table 2 shows that during the monitored period, he had a total of 142 product-related posts promoting various products. Leoš Mareš has a wide range of specializations on Instragram. Penny Market, the supermarket chain, was the most frequently communicated brand (5.7%). This was followed by magazines and journals, travel and culture.

The third most followed influencer is Jirka Král (902,903). IT technologies were the most promoted type of product in his Instragram posts. Beyond this, his posts also featured sports, restaurants and coffee shops (Table 3). Jirka Král is more active on YouTube, so the frequency of his product-related posts on Instragram is lower.

Table 3. Jirka Král – product-related posts

Jirka Král (902,903 followers)	
Product type*	**Frequency of appearance**
IT	10
Sport	2
Restaurants, cafes	3

Source: Author.

(*List of represented brands: Asus (5), GoPro (5), MMA (1), NHL (1), Z Kotva (2), Mc´Donalds (1))

The fourth most followed influencer is Karolína Kurková (810,965). During the monitored period, she had a total of 168 product-related posts. Among the most common products communicated to followers were cosmetics, clothing and accessories and magazines (Table 4). Karolína Kurková lives in the USA, but her product-related posts are also focused on the Czech consumer.

Table 4. Karolína Kurková – product-related posts

Karolína Kurková (810,965 followers)	
Product type*	**Frequency of appearance**
Magazines	40
Children´s goods	10
E-shops	15
Travel, culture	1
Clothes, accessories	42
Cosmetics	43
IT	1
Foodstuffs	2

Source: Author.

(*List of represented brands: Elle (3), Harpers bazaar (12), Forbes (1),Voque (9), Museum magazine (1), Parents (2), Stella (2), Number magazine (1), Ocean drive magazine (4), W magazine (2), Vanity fair (1), Instyle (1), Madam magazine (1) Cybex (10), Aboutyou (15), Editions (1), Billieblooms (15), Chanel (1), David Jones (3), Gucci (1), Versace (1), Tods (2), Victoria Secret (4), Prada (1), Valentino (2), Dolce a Gabbana (1), Dior (2), J. P. Gaultier (2), Donna Karan (1), Rocket (1), Oscar de la Renta (2), Hermes (1), Levis (1)Joannaczech (2),Gryphanddivyrose (36), Inner beauty (1), Theanycosmetics (1), Myskinbuddy (2), Daylisleeper (1) Leica camera (1), Beyodmeet (1), Dayilyharvest (1))

The fifth most followed influencer is Anna Šulcová (829,927). This influencer had 54 product-related posts. Table 5 shows that e-shops and clothing and accessories were most often mentioned on her profile.

Table 5. Anna Šulcová – product-related posts

Anna Šulcová (829 927)	
Product type*	**Frequency of appearance**
E- shops	24
Clothes, accessories	21
Cosmetics	1
Travel, culture	2
Cars	3

Source: Author.

(*List of represented brands: Zalando (23), Kup kosmetiku (1), Tezenis (11), DW a FP (3), JVD Forher (4), Honor (1), Prada a Max Mara (2), Batiste (1), Visitbudapest (1), Saintjohnresort (1), Hyundai (3))

Table 6. Summary of product-related posts by influencers

Order	**Product type**	**Frequency of appearance**	**Occurrence (%)**
1.	Clothes, accessories	66	16.3
2.	Magazines	60	14.8
3.	Cosmetics	46	11.3
4.	E-shops	39	9.6
5.	Sport	24	5.9
6.	Travel, culture	23	5.7
7.	Children´s goods	15	3.7
8.	Restaurants, cafes	13	3.2
9.	IT	11	2.7
10.	Foodstuffs	7	1.7
11.	Other	78	19.2
Special	Retail	24	5.9

Source: Author.

Table 6 shows the most frequently promoted products according to the frequency of product-related posts for the given product type. For each product type, it was desirable for the product type to be promoted by more than one influencer. For example, Leoš Mareš' product-related posts concerning Penny Market (retail trade) did not meet the aforementioned condition. The influencer was the only one to communicate this type of product, but the frequency was relatively significant (5.7%). Tabel 6 shows that the most frequent product-related posts were for clothing and accessories (16.3%), followed by magazines (14.8%), consumer cosmetics (11%), e-shops (9.6%) and sports, travel and culture (~6%). Although cars were promoted in product-related posts by at least two influencers, the frequency of this collaboration was low (4 out of 406 posts). The latter point also applies to other product types which were promoted by only one influencer and where the frequency of occurrence was at most in the order of units.

Research by a marketing agency [26] determined that the most active users of Instagram are the age group 18-24 years, followed by 25-34 years. Petr Čech's product-related posts are most often focused on sports, especially football. These product-related posts are interesting for both these segments of followers. Of all his product-related posts, Leoš Mareš most frequently mentioned retail chains, for example Penny Market. In this case, the targeting was not appropriate. For the age groups which are the most frequent followers of influencers, Penny Market is not an object of interest. His other product-related posts focused on products such as magazines, travel, restaurants and cafes, which are suitable for this type of communication. Jirka Král's product-related posts were most often focused on IT technologies, which is a matter of course for both age groups. For Karolina Kurková, product-related posts on Instragram were most often focused on cosmetics, clothing, accessories and magazines. All the products target the younger generation, so these product-related posts are appropriate. Anna Šulcová mostly communicated about e-shops, clothes and accessories, which are also appropriate to both target groups.

Tafesse and Wood [27] examined how elements of the content of product-related posts and influencers involvement strategies contribute to the engagement of followers on Instagram. In the research they conducted, the focus of influencers' product-related posts was also determined. Based on a comparison of the results of both analyses, we can state that the most followed Czech influencers focus on similar types of products, especially clothing, cosmetics, travel, children´s goods and food. Tafesse and Wood's research [27] also mentioned the following product-related posts on Instragram: weddings, luxury goods and design.

4 Conclusion

The aim of this contribution was to determine and analyse the types of products the five most followed Instagram influencers in the Czech Republic promoted during the period September 2019 - September 2020. The influencers were identified on the basis of the highest number of followers. The identified top five were football player Petr Čech with 2,062,470 followers, presenter Leoš Mareš with 1,079,795 followers, YouTuber Jirka Král with 902,903 followers, Karolína Kurková, a model born in the Czech Republic but living in the USA, with 810,965 followers, and Anna Šulcová with 829,927 followers. Leoš Mareš had the widest range of communication on Instragram. During the monitored period, he had a total of 142 product-related posts that promoted various products. Model Karolina Kurková had 54 product-related posts promoting mainly e-shops, clothing, accessories and cosmetics. Petr Čech posted a total of 23 product-related posts mainly focused on sports equipment, especially football. The most active users of Instagram are the age group 18-24 years, followed by 25-34 years. The product-related posts of Petr Čech are interesting for both groups of followers. Of all the product-related posts, Leoš Mareš mostly mentioned retail chains, in particular Penny Market. In this case, the targeting was not appropriate. His other product-related posts were appropriate to the target groups and focused on products such as magazines, travel, restaurants and cafes. Likewise, Jirka Král's product-related posts were most often focused on IT technologies, which is a matter of course for both age groups. Anna Šulcová most often communicated about e-shops, clothes and accessories, both of which are appropriate to the target groups.

References

1. Get Boost - Influencer marketing & Social media marketing agentura. *Influencer marketing* [online]. Available at: https://www.getboost.cz/influencer-marketing/ (2020)
2. PROMOPRODUKCE CZ. *5 věcí, které musíte vědět o influencer marketingu* [*5 things you need to know about influencer marketing*]. [online]. Available at: https://www.webfin.cz/5-veci-ktere-musite-vedet-o-influencer-marketingu (2020)
3. SPIR. *Internetová populace ČR dosáhla v březnu nového maxima: 7,77 milionu reálných uživatelů* [*The Internet population of the Czech Republic reached a new maximum in March: 7.77 million real users*]. [online] Available at: https://www.spir.cz/internetova-populace-cr-dosahla-v-breznu-noveho-maxima-7-77-milionu-realnych-uzivatelu (2020)
4. ČSÚ. *Sociální sítě* [Social networks]. [online]. Available at: https://www.czso.cz/csu/czso/7-vyhledavani-vybranych-informaci-na-internetu-491bfx37fh (2020)

5. KANTAR – Česká republika. *KANTAR – Česká republika* [KANTAR – Czech Republic]. [online]. Available at: https://cz.kantar.com/ (2020)

6. BUSINESSGRAM. *Počet lidí na českém a slovenském Instagramu k září 2019. Jsme jednička ve školení firemního Instagramu na českém trhu* [Number of people on the Czech and Slovak Instagram as of September 2019. We are number one in corporate Instagram training on the Czech market]. [online]. Available at: https://businessgram.eu/pocet-lidi-na-ceskem-a-slovenskem-instagramu-zari-2019/ (2019)

7. eMarketer. *Is Everyone on Instagram an Influencer?*[online]. Available at: https://www.emarketer.com/content/is-everyone-on-instagram-an-influencer (2020)

8. Marketing & Media. *Trendy a data na sociálních sítích podle Socialbakers* [Trends and data on social networks by Socialbakers]. [online]. Available at: https://mam.cz/marketing/2019-12/trendy-a-data-na-socialnich-sitich-podle-socialbakers/ (2020)

9. L. Chen, S. Yuan, Influencer marketing: how message value and credibility affect consumer trust of branded content on social media. *Journal of Interactive Advertising*. **19**(1), 58–73 (2019)

10. L. Casaló, C. Flavián, S. Ibáñez-Sánchez, Influencers on Instagram: Antecedents and consequences of opinion leadership. *Journal of Business Research*. **117**, 510–519 (2018)

11. V. Kumar et al. Undervalued or overvalued customers: capturing total customer engagement value. *Journal of service research*. **13**(3), 297–310 (2010)

12. S. Lee, E. Kim, Eunice. Influencer marketing on Instagram: How sponsorship disclosure, influencer credibility, and brand credibility impact the effectiveness of Instagram promotional product-related post. *Journal of Global Fashion Marketing*. **11**(3), 232–249 (2020)

13. J. Vodák, et al. Influencer Marketing as a Modern Phenomenon in Reputation Management. *Managing Global Transitions*. **17**(3), 211–220 (2019)

14. S. V. Jin, A. Muqaddam, E. Ryu, Instafamous and social media influencer marketing. *Marketing Intelligence & Planning*. **37**(5), 567–579 (2019)

15. M. De Weirman, V. Cauberghe, L. Hudders, Disclosing sponsored Instagram product-related posts: the role of material connection with the brand and message-sidedness when disclosing covert advertising. *International journal of advertising*. **39**(1), 94–130 (2020)

16. K. Sokolova, H. Kefi, Instagram and YouTube bloggers promote it, why should I buy? How credibility and parasocial interaction influence purchase intentions. *Journal of Retailing and Consumer Services*. **53** (2020)

17. F. J. Martínez-López, R. Anaya-Sánchez, M. F. Giordano, D. Lopez-Lopez, Behind influencer marketing: key marketing decisions and their effects on followers' responses. *Journal of Marketing Management*. **36**(7–8), 579–607 (2020)

18. Y. Chetioui, H. Benlafgih, H. Lebdaoui. How fashion influencers contribute to consumers' purchase intention. *Journal of Fashion Marketing and Management: An International Journal*. In print (2020)

19. M. Söderlund, J. Mattsson, Consumers reactions to unsubstantiated claims about ecological products. *Journal of Consumer Marketing*. **37**(5), 569–578 (2020)

20. M. L. Wellman, R. Stoldt, M. Tully, et al. Ethics of Authenticity: Social Media Influencers and the Production of Sponsored Content. *Journal of Media Ethics*. **35**(2), 68-82 (2020)

21. E. Villena Alarcón, M. Jesus Fernandez-Torres, Relaciones con los públicos a través de Instagram: los influencers de belleza como caso de estudio/Beauty on Instagram: Relations between influencers and the stakeholders. *Revista Internacional de Relaciones Públicas*. **10**(19), 111–132 (2020)

22. C. P. Curiel, Trend towards extreme right-wing populism on Twitter. An analysis of the influence on leaders, media and users. *Communication & Society*. **33**(2), 175–192 (2020)

23. C. P. Curiel, P. Sanz-Marcos, Branding, influencers and new target in the communication of fashion and luxuary the Gucci tendency on Instragram. *Prisma Social*. **24**, 1–24 (2019)

24. C. Stubb, AG NystrÖm, J. Colliander, Influencer marketing. *Journal of Communication Management*. **23**(2), 109–122 (2019)

https://doi.org/10.1051/shsconf/20219001012

25. Mzone. *TOP 10 českých influencerů: Kdo má kolik sledujících na Instagramu? Někteří překvapí!* [*TOP 10 Czech influencers: Who has how many followers on Instagram? Some will surprise you!*]. [online]. Available at: https://www.mzone.cz/instagram-top-10-ceskych-influenceru/ (2020)

26. Dantoa. *Facebook vs. Instagram - proč se většina konverzí odehrává na Facebooku?* [Facebook vs. Instagram - why do most conversions take place on Facebook?]. [online]. Available at: https://www.dantoa.cz/facebook-vs-instagram-proc-se-vetsina-konverzi-odehrava-na-facebooku/ (2020)

27. W. Tafesse, P. B.Wood, Followers' engagement with instagram influencers: The role of influencers' content and engagement strategy. *Journal of Retailing and Consumer Services*. **58** (2021)

Economic Growth, Investment, Population Growth and Unemployment in Ethiopia

Betgilu Oshora[1,*], *Tiblets* Nguse[1], *Maria* Fekete-Farkas[2], *Zoltan* Zeman[2]

[1]Doctoral School of Economic and Regional Science, Szent István University, 1 Páter K. Str., Gödöllő, 2100, Hungary
[2]Faculty of Economics and Social Sciences, Szent István University 1 Páter K. Str., Gödöllő, 2100, Hungary

Abstract. The study examines the correlation between economic growth, investment, population growth and unemployment in Ethiopia. For the purposes of this study, secondary time series data collected from the National Bank of Ethiopia, IMF and World Bank databases were used. The study extracts the perceived relationship between the variables through principal component (PCA) analysis. Both the Kaiser-Meyer-Olkin measure of sampling adequacy and Bartlett's test of sphericity were used to determine the appropriateness of the dataset for PCA. The results revealed that an increase in unemployment is positively correlated, but in the opposite direction to economic growth, investment, total population and the working-age population. On top of the rapid spread of COVID-19, the present protests as a result of the political instability and ethnic problems across the country will lead to an increase in unemployment and the loss of many lives. Policymakers therefore need to emphasize and assure an increase in economic growth and investment in order to create more jobs in line with the increasing demand for jobs, particularly by young people. Moreover, the government must address the ongoing ethnic problems and political instability before it hits the economy adversely.
Keywords: COVID-19; Ethiopia; economic growth; population growth, investment, unemployment

1 Introduction

Inclusive and sustainable economic growth is very crucial for the success of the implementation of the Sustainable Development Goals (SDGs). Although the global economy continues to grow, it is doing so at a slower rate and unemployment is growing. Hence there is a need for more progress in creating more work that is decent for all. One fifth of the world's youth were not in education, employment or training [1]. Decent work is crucial to achieving inclusive economic growth, since such work is more than just a source of income and social protection. Also, across all country income groups, young people aged 15–24 years' experience a higher unemployment rate than the rest of the working-age population [2]. For instance, of the 5% global unemployed in 2018, youth were three times more likely to be unemployed than adults [1]. Furthermore, the study by [3] shows that unemployment can result in depression, which is a major public health problem among unemployed young adults in Ethiopia. Moreover, it is widely recognized that youth employment and economic empowerment are essential components of strong societal foundations [4].

According to [5], the unemployed are people of working age who are without work, are available for work, and have taken specific steps to find work. At present, unemployment remains unacceptably high at the global level and the majority of employed people are in informal employment, with many of them living in extreme poverty. For instance, according to the 2015 Employment, Skills and Human Capital Global Challenge Insight Report, more than 200 million people are out of a job globally and the numbers are forecasted to rise to 215 million jobseekers by 2018. In order to match the growing labour force, it is estimated that nearly 42 million new jobs need to be created annually [6]. In 2018, out of the 5.7 billion working-age people, i.e. 15 years or older, 3.3 billion people were in employment and 172 million unemployed [7]. Moreover, 2 billion people make their living in the informal economy at the global level according to [8] and almost 77% of the world's young workers are in informal jobs, compared with around 60% of adult workers [9].

The results of many empirical studies show that workers in the informal economy face a higher risk of poverty than those in the formal economy. For instance, in emerging and developing countries, more than 300 million

* Corresponding Author: betgilu2002@gmail.com

workers live in extreme poverty (<$1.90/day). Furthermore, unemployment in developing countries is expected to increase by half a million per year in both 2018 and 2019, with the unemployment rate remaining at around 5.3% [10]. Out of 1.3 billion people living in Africa, 764 million people are aged 15 years or older, of which 4.3% of those of working-age are unemployed. This may seem like a very small fraction compared with the 60% who are employed, but many people have no choice but to take up informal jobs of poor quality in order to meet their basic needs and escape poverty. Being in employment does not always guarantee a decent living. For instance, in 2018, more than one quarter of workers in low- and middle-income countries were living in extreme or moderate poverty [7].

In contrast to the issue of high global unemployment, [6] shows that employers continue to complain about the difficulties they face in finding talent for vacant positions, pointing to a mismatch between available and needed skills. It is the shared opinion of the authors that it is important to recognize that this issue is a common and serious problem mainly in developing countries where the educational curriculums for many disciplines have a theoretical orientation rather than a practical one. This problem is also reflected at the university level, which exacerbates the problem. In addition, industries also do not allow students that live close by to gain practical experience.

According to [11], young people are three times as likely as adults (25 years and older) to be unemployed. At the global level there are approximately 1.3 billion young people between the ages of 15 and 24, of which about 68 million are unemployed. Moreover, about 126 million (30%) of employed youth remain in extreme or moderate poverty despite having a job. In addition, over three-quarters of young workers are engaged in informal employment. Similarly, the ILO (2015) report on five facts about the informal economy in Africa shows that the quality of jobs is questionable and nearly 8 out of 10 employed persons in Sub-Saharan Africa were in vulnerable forms of employment. This indicates that being employed does not guarantee young people a decent life. For this reason decent work is given due attention in the United Nations 2030 Agenda as a specific goal (SDG 8), with many targets [12]. The sub-Saharan region needs to raise growth to create the additional 20 million jobs per year needed to absorb new entrants to the labour markets. More growth, more jobs policies are needed to create more jobs in the coming years [13].

The issue of unemployment has become a severe challenge for all African countries, regardless of their different socio-economic development. People often do not have a choice with regards to their employment due to the lack of available opportunities and are therefore underemployed, i.e. their job is not their profession or is irrelevant to their field of study. Ethiopia is the second most populous nation in Africa after Nigeria, and the fastest growing economy in the region. The economy experienced strong, broad-based growth averaging 9.9% a year from 2007/08 to 2017/18, compared to a regional average of 5.4%. However, it is also one of the poorest, with a per capita income of $790 [14]. Despite reducing the extreme poverty rate from about 46% in 1995 to 23.5% in 2016, Ethiopia still has more than 25 million people living in poverty [15].

Despite the country's strong economic growth over the last 15 years, statistical data for 2015 on the unemployment rate in urban areas shows it to be 17.5%, which is an enormous challenge when it comes to inclusive economic development [16]. In Ethiopia, the lack of economic opportunities is the primary driver of youth frustration, with more than 25% youth unemployment. However, the number of underemployed youth is likely to be much higher [17]. More importantly, Fantaw & Bitga also state that young people claim that their voices are not being heard when it comes to community-level decision-making processes, which will drive unrest and instability in Ethiopia over the next decade and beyond. Furthermore, the current political uncertainty, plagues of locusts and COVID-19 pandemic will create a serious challenge for economic growth and will contribute to increasing numbers of people living below the poverty line and requiring humanitarian assistance. At present, different countries around the world are experiencing demographic changes, unemployment and the shutdown of economic activities to save human lives [18]. Due to the nature of globalization and economic interconnectedness, [19] states that it is difficult to mitigate the pandemic.

The Federal Democratic Republic of Ethiopia has developed a comprehensive job creation strategy to reduce unemployment in a market that sees about 2 million new workforce entrants each year and which aims to create 14 million jobs by 2025 [20]. However, the rapid spread of COVID-19 will harm the implementation of the strategy unless a new strategy can be put in place that minimizes the risks to the implementation of the wide ranging and comprehensive strategy. Moreover, the lockdown due to the pandemic is expected to hurt companies that hire the majority of employees, in particular small private companies. As a result, it will more likely hurt the 25% of the population already living in extreme poverty and whose income, even before the pandemic, was more dependent on daily labour and employment in informal jobs. The main aim of this study is to examine the status of unemployment and COVID-19 in Ethiopia and more specifically to:

1. Explore the correlation between unemployment, economic growth, investment, total population and the working-age population;

2. Examine the current status of the COVID-19 pandemic and its effect on employment.

1.1 Conceptual framework

The conceptual framework is depicted in Figure 1. This takes into account four independent variables for examining the correlation between the variables. An increase in the working-age population will increase the numbers joining the labour market. As a result, unless job creation increases in line with the increase in numbers seeking employment, unemployment will rise. In contrast, an increase in real GDP growth rate and investment are likely to contribute positively to the reduction of unemployment. Unfortunately, the current COVID-19 pandemic will enhance the existing rate of unemployment due to lockdown and stay-at-home measures.

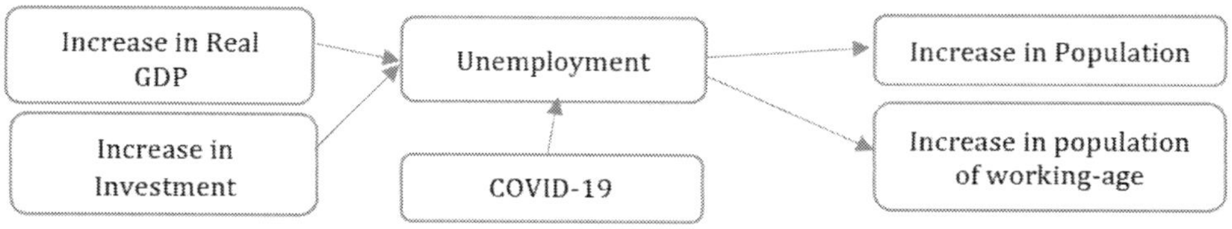

Figure 1. Conceptual Framework

Source: Authors.

2 Methodology

For the purposes of this study, we extracted 20-year time-series data from 2000 – 2019 from the National Bank of Ethiopia annual reports to examine the correlation between the variables under study, such as unemployment, total population growth, working-age population growth, investment and economic growth. To process the collected data, a multivariate statistical analysis, in the form of Principal Component Analysis (PCA), was applied using Minitab software to examine the relationships and significance in and between the variables. To verify that the dataset is suitable for PCA, Bartlett's test of sphericity and the Kaiser-Meyer-Olkin (KMO) were applied. Moreover, the eigenvalue rule was applied to determine the number of components to be considered for further investigation.

To support the results, we collected secondary data from the IMF, WHO and World Bank databases for descriptive analysis. This involved examining the trends for the variables over the past few decades and presenting them with the use of MSExcel spreadsheets. Furthermore, to develop a comprehensive understanding of the subject, empirical evidence was sought from different reports, websites, the Web of Science, Google Scholar, Research Gates, Academic.edu and Mendeley, under keywords like COVID-19, Ethiopia, growth and unemployment.

3 Results and Discussion

3.1 Descriptive (trend) analysis

In this section, the trends for the variables are presented and compared with other countries on the same continent. The main aim of reflecting the trend of the determinants and the status of the dependent variable is to triangulate the results obtained from the statistical methods using factor analysis.

As is evident in Figure 2 below, investment and real GDP growth for the country over the last two decades have been incremental, with exception to the sharp decline in real GDP growth rate in 2003. Despite the promising increase in economic growth and investment across the country, the unemployment rate remains 2 percent on average above over the same decades. This is mainly due to the increase in the population and of the working-age population joining the labour market, as illustrated in Figure 3 below. Over the last two decades, the total population and working-age population increased by more than 2 million and 1.3 million on average, respectively. As a result, with too few jobs available, the annual increase in the working-age population joining the labour market, plus the existing unemployed, is a major challenge for the government.

https://doi.org/10.1051/shsconf/20219001013

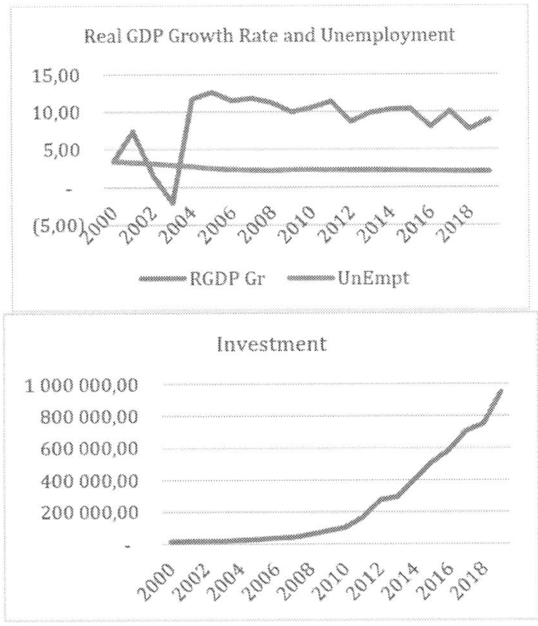

Figure 2. Economic growth, unemployment and investment trends

Source: Authors.

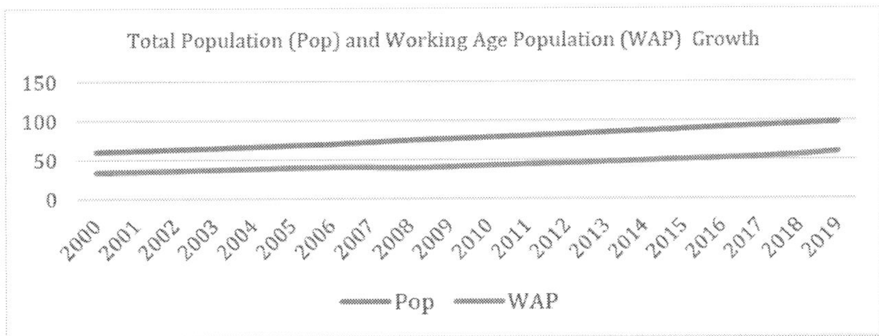

Figure 3. Total population and working-age population growth.

Source: Authors.

As is evident, the Ethiopian economy has been growing at a remarkable rate for the last few years. However, it is important to examine how that growth has translated into more jobs that are of a better quality and are inclusive. The statistical figures show that more than 70% of Ethiopians are under 30 years of age, with urban youth unemployment standing at 22% compared to an overall unemployment rate of 17%. Moreover, more than a quarter of all urban jobs are informal [21]. Given the large number of university and college graduates, it is a challenge for a developing country like Ethiopia to create and maintain productive employment. For instance, the authors of [22] reveal that educated unemployed youth are at the epicentre of the current protests, complaining about the fact that they have university degrees but can only find employment 'breaking cobblestones' for pedestrians sidewalks. There is therefore a need for a clearer understanding of the remaining challenges based on further empirical analysis to provide all stakeholders with urgent guidance on how to achieve SDG 8.

Moreover, as depicted in Figure 4 below, Ethiopia is also a country with the fastest population growth rate on average in sub-Saharan Africa, among low-income countries, and the world. This shows that unless employment opportunities increase equally, unemployment is expected to increase during the next few decades, thereby bringing into doubt the implementation and attainability of the SDGs. This will be exacerbated by the current COVID-19 pandemic, which will cause significant harm the global economy and consequently the realization of the SDGs. According to [23], COVID-19 will be a real challenge for the UN with regards to its efforts to eradicate poverty (SDG1) by 2030 and a major challenge for developing countries because the number of people living

 https://doi.org/10.1051/shsconf/20219001013

under the three poverty lines is estimated to increase to between 420–580 million. Therefore, even though no one knows how long the pandemic will last, the achievement of the 2030 Agenda, and in particular, no poverty and zero hunger (SDG 1) and decent work for all (SDG8), is under considerable threat. The impact of COVID-19 will more likely hit the poorest regions, especially SSA and South Asia, where the potential new poor, i.e. under the US$1.90/day and US$3.20/day poverty lines, are estimated to represent 80–85% of the total poor [23].

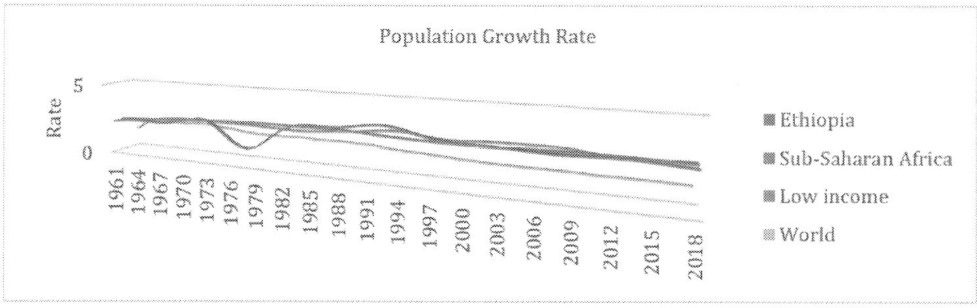

Figure 4. Ethiopian population growth compared with other populations

Source: Authors.

A comparison of the economic growth of the country with East Africa and sub-Saharan countries, reveals that Ethiopia is one of the fastest growing economies on the continent (see Figure 5 below). Like the global economy, lockdown measures due to COVID-19 are expected to have a negative impact on the country's economic growth in the coming years.

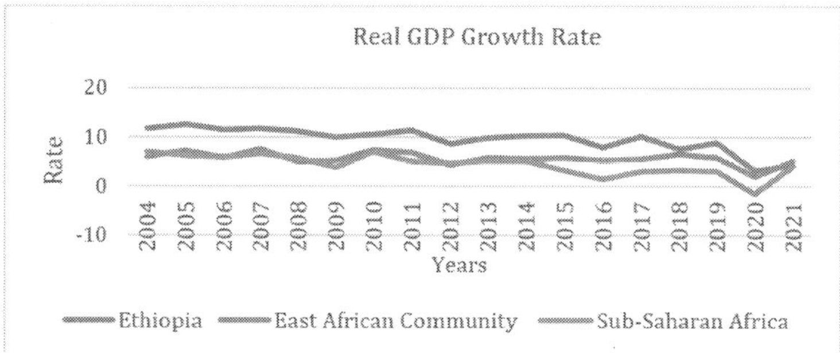

Figure 5. Economic growth compared with SSA and East Africa

Source: Authors.

3.2 Status of COVID-19 pandemic in Ethiopia

According to recent Worldometer data, Ethiopia is the second most populous nation in Africa after Nigeria and is ranked twelfth in the world, with a total population of 115,260,592 as of Wednesday, 5 August 2020. Moreover, the population is growing more rapidly than expected, which is evident when the numbers are compared to the UN's mid-year estimated figure for the total population of 114,963,588 [34].On 13 March 2020, Ethiopia joined the rest of the world by reporting the first case of COVID-19 and immediately called for several restrictions on human mobility in a bid to contain the pandemic.

The Ministry of Health (MOH) and Ethiopian Public Health Institute (EPHI) in collaboration with partners have intensified response efforts to prevent the spread and severity of the COVID-19 pandemic in Ethiopia [24]. Since the first case was reported in Ethiopia on 13 March 2020, the pandemic has spread rapidly throughout the country. As of writing this article, there were a total of 18,706 confirmed COVID-19 cases and 310 deaths [35]. As depicted in Figure 6 and 7 below, the number of new cases and deaths is increasing exponentially, in particular over the last month. In the last month, new cases and deaths accounted for 69% and 67% of the total cases, respectively. Moreover, despite the positive aspects of the country's early response, its young population, low population density in rural areas, and experience with handling large scale crises, there are also negative aspects such as a weak health system, poor nutritional status, lack of access to proper hygiene and sanitation and densely populated urban areas [25].

Authorities have closed borders, closed schools, ordered the shuttering of nightclubs and entertainment outlets, announced social distancing measures, and called in retired and in-training medical personnel. In addition, all people entering Ethiopia from another country are subject to a mandatory 14-day quarantine at designated hotels at the traveller's expense. The authorities postponed elections, which were scheduled for 29 August, for as long as 12 months. On 8 April, the Prime Minister declared a state of emergency under Article 93 of the constitution, which allows it to impose more stringent measures. The measures put in place to ease the spread of COVID-19 are pushing millions of people in Ethiopia into severe hunger and unemployment. Inactive economic activity can lead to unemployment, loss of wages, and consequently, loss of income. Yet now, the federal government and many humanitarian organizations are distributing food to ease the pain of the lockdown imposed to stop the spread of the coronavirus pandemic. However, this is unsustainable for a long period, leading to questions about the imposed measures.

Food prices have increased dramatically because of movement restrictions and border closures. However, according to [26], "the impact of COVID-19 on food insecurity is much weaker compared to that in upper-middle-income countries". Globally, food insecurity is becoming a very serious issue, challenging the economic development of the world. Of the 7.6 billion people that make up the world´s population, 17% and 9% are living under moderate and severe food insecurity. This situation is even more severe on the African continent, where, according to FAO data [27], of the total population of Africa, 31% and 21% are living in moderate and severe food insecurity. Highly crowded urban populations (particularly in capital city Addis Ababa), limited medicinal and pharmaceutical products, and weak economies that are unable to sustain health and lockdown costs, are just a few of the reasons that will make the country more vulnerable. For instance, the economic growth of Ethiopia is expected to fall to a 17-year low of 3.2% this year— down from 9% in 2019 [28].

Despite the hopes of Ethiopians [29] about democracy following the election of Abiy Ahmed as prime minister in 2018, hundreds of people have been killed in protests across the country over the last two years. The increasing intensity of the protests and demonstrations, particularly in the Oromia region and in Addis Ababa will enhance the spread of the pandemic. Moreover, failure to adhere to stay-at-home and physical distancing measures, the extreme poverty in which a large part of the population lives, and other preventive measures to minimize the risk of spread remain a challenge for the government. The shortage of hospital beds, the inadequacy of quarantine sites, the shortage of testing facilities, Personal Protective Equipment (PPE) for health professionals, the challenges of the track and trace system, the shortage of health professionals and so on are also hurdles that need to be jumped. The lack of health professionals and poorly equipped facilities relative to the total population will influence the country´s ability to combat the pandemic [30]. All the stakeholders therefore need to practice all of the imposed preventive measure to protect themselves and those around them in order to stay healthy.

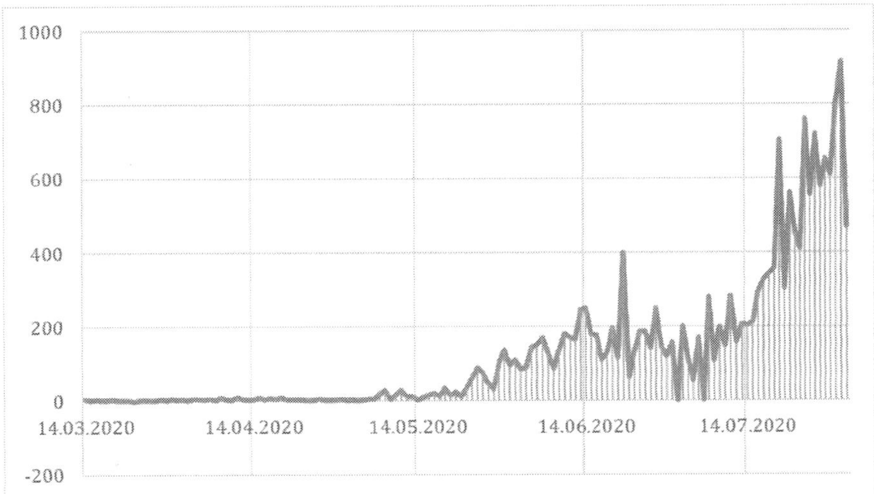

Figure 6. COVID-19 daily new cases in Ethiopia

Source: Authors.

 https://doi.org/10.1051/shsconf/20219001013

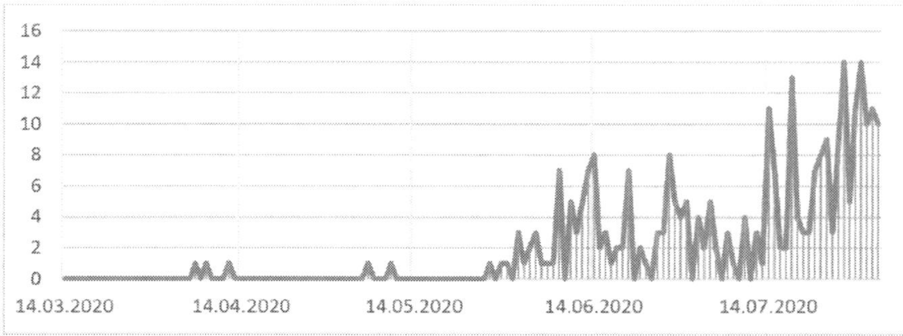

Figure 7. COVID-19 daily deaths in Ethiopia

Source: Authors.

As is the case in other countries, the increasing public health crisis, falling commodity prices and mitigation measures are having a significant effect on the economic activity of the country. Informal sectors are highly exposed to risk due to the impact of COVID-19, more so than other sectors despite hiring more daily wage earners. Hence the majority of wage workers will be highly exposed to the risk of food insecurity. For this reason, the mitigation measures are unlikely to be effective in the absence of sufficient government support to convince people to stay at home. Due to the lack of testing capacity to detect and quarantine people carrying COVID-19, we are highly concerned about the capacity of the country to deal with this outbreak. Taking into account that the majority of people are living under severe and moderate poverty, it will be difficult to implement travel and movement restrictions, lockdowns, border closures, bans on public gatherings, school closures, and so on. The government must therefore ensure that if stay-at-home measures are to be effective in combating the pandemic, they must not be at the expense of the human rights.

3.3 Testing of data adequacy

We conducted two tests to measure sampling adequacy in order to determine the factorability of the whole matrix. The two tests were Bartlett's test of sphericity and the Kaiser-Meyer-Olkin test. Accordingly, a measure of sampling adequacy was conducted by testing KMO for the set of variables included in the analysis. The test result was 0.706, which, when compared with the minimum requirement of 0.50 as per standard [31,32], implies that the data is appropriate for PCA. The result of Bartlett's test of sphericity was significant ($p<0.005$, $p=0.000$). This, in combination with the fact that the KMO measure was greater than 0.5, means that we can conclude that the data is appropriate for PCA and factor analysis (see Table 1 below).

Based on the results, it is appropriate to proceed with PCA to examine the correlation between unemployment and the other variables, namely total population growth, investment, economic growth and working-age population growth.

Table 1. KMO and Bartlett´s test for data appropriateness

Kaiser-Meyer-Olkin measure of sampling adequacy.		.706
Bartlett's test of sphericity	Approx. Chi-Square	153.308
	Df	10
	Sig.	.000

Source: Authors.

It is also important to check the variability of the data and to know whether there is a strong relationship between the variable and the component in isolation, which can be achieved using a component matrix. Since all the variables in the first component of the correlation matrix are greater than 0.3, we can conclude that the variables are strongly correlated. However, in the second component, only three variables (investment, unemployment and real GDP growth rate) are correlated with the component. Therefore, as can be seen in Table 2 below, total population, working-age population, investment, unemployment and real GDP growth rate strongly correlate in the component. In the case of the second component, the real GDP growth rate strongly correlated with the component, followed by investment and unemployment.

Table 2. Extracted component matrix

	Component	
	1	2
Population total, in millions (mid-year population)	.983	
Working-age population (in millions)	.972	
Investment	.895	.398
Unemployment	-.891	.325
Real GDP growth rate (in %)	.472	-.855

Source: Authors (Extraction Method: Principal Component Analysis).

3.4 Principal Component Analysis

After testing the suitability of the data for PCA, we conducted dimension reduction by ignoring the less significant components based on Kaiser's criterion, or the eigenvalue rule (only factors with an eigenvalue of 1.0 or more are retained for further investigation). Accordingly, as can be seen in Table 3 below, we retained only the first two components for further investigation (eigenvalues were 3.731 and 1.059, respectively). We ignored the remaining three components, for which the eigenvalue was less than 1.0, which means that the component has less information than a single variable. The first two artificial components explain 95.804% of the total variance, which means that we could keep 95.804% of the information after the dimension reduction.

Table 3. Total variance explained by the components

Component	Initial eigenvalues			Extraction sums of squared loadings			Rotation sums of squared loadings		
	Total	% of Variance	Cumulative %	Total	% of Variance	Cumulative %	Total	% of Variance	Cumulative %
1	3.731	74.617	74.617	3.731	74.617	74.617	3.254	65.089	65.089
2	1.059	21.187	95.804	1.059	21.187	95.804	1.536	30.715	95.804
3	.183	3.655	99.459						
4	.021	.419	99.878						
5	.006	.122	100.000						

Source: Authors (Extraction Method: Principal Component Analysis).

In order to decide which components should be retained for further investigation, [33] also suggests finding the point at which the shape of the curve changes direction and becomes horizontal and to then retain all the components above the elbow, or break in the plot, because these components contribute the most to the explanation of the variance in the dataset. Hence, by using the Cattel scree plot test, the first two components explain much more of the variance than the remaining components; we, therefore, extracted two components only for further investigation. Alternatively, the remaining principal components account for a very small proportion of the variability (close to zero) and are probably unimportant. The scree test also supports the decision made to retain the first two components by using the eigenvalue rule as depicted.

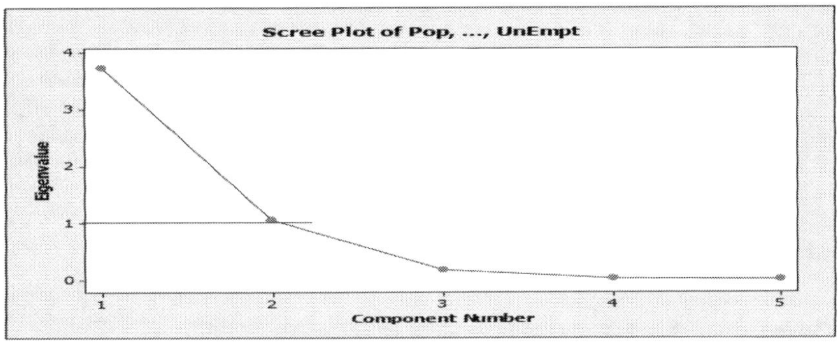

Figure 8. Scree plot for principal component analysist

Source: Authors.

The visualization of the loading plot in Figure 9 shows that the first component of the variables working-age population, total population and investment are positively correlated in the same direction, but correlated in the opposite direction with unemployment. In the second component, the variable real GDP growth rate is a lonely variable, independent of any other variables.

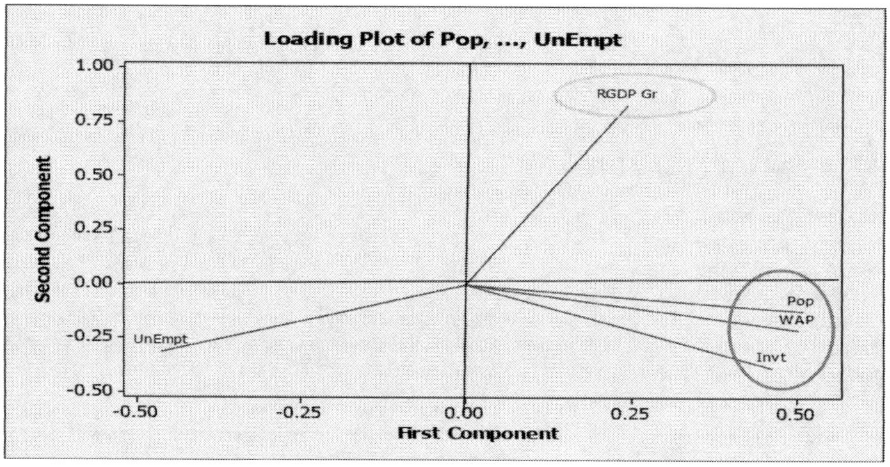

Figure 9. Loading plot for principal component analysis

Source: Authors.

4 Conclusion

At present, unemployment, in particular among those aged 15-24 years, remains a global challenge for the achievement of inclusive economic growth. Different findings reveal that the majority of employed people are in informal employment and many of them live in extreme poverty. This study aimed to examine the correlation between unemployment and selected independent variables based on time series data from 2000-2019. The results showed that unemployment is positively correlated but in the opposite direction to total population, working-age population, investment and economic growth. By looking at trend data, it is clear to see that despite economic growth and investment, unemployment is increasing. Similarly, the increase in the total population and the working-age population also create an increase in unemployment. Policymakers must therefore come up with a strategy that increases economic growth and investment and creates job opportunities, in particular for the young population. Within this context and that of population growth, it is possibly advisable to set a maximum birth limit to reduce the highly increasing population growth rate.

Beyond this, the COVID-19 pandemic is expected to result in companies failing and going into bankruptcy, resulting in reduced economic growth and rising unemployment. This will be exacerbated by falling investment due to the current political instability, ethnic problems and protests, which in all likelihood could lead to social unrest and instability, contributing to a further increase in unemployment. Policymakers, the government in particular, must call for solidarity to combat the pandemic and put forward possible remedies to save the lives of the most vulnerable people during this period since staying at home without access to food is not possible for them. On the other hand, the federal government must give due attention to mitigating the rapid spread of COVID-19 and not being part of creating another pandemic, which may cause further political instability and ethnic problems.

References

1. UN. Secretary-General, Progress towards the Sustainable Development Goals: report of the Secretary-General [online]. Available at: https://digitallibrary.un.org/record/3810131
2. International Labour Organizaton, *Time to Act for SDG 8 : Integrating Decent Work, Sustained Growth and Environmental Integrity* [online]. Available at: https://www.ilo.org/global/publications/books/WCMS_712685/lang--en/index.htm (2019)

3. H. Mokona, K. Yohannes, G. Ayano, Youth unemployment and mental health: Prevalence and associated factors of depression among unemployed young adults in Gedeo zone, Southern Ethiopia. *International Journal of Mental Health Systems*. **14**(1), 1–11 (2020)

4. United Nations, *World Youth Report: Youth and the 2030 Agenda for Sustainable Development*. [online]. Available at: https://www.un.org/development/desa/youth/world-youth-report/wyr2018.html (2018)

5. OECD, *Unemployment rate (indicator)* [online]. Available: https://data.oecd.org/unemp/unemployment-rate.htm (2020)

6. World Economic Forum, *Disrupting Unemployment Business - led Solutions for Action*. [online]. Available at: http://reports.weforum.org/disrupting-unemployment/?doing_wp_cron=1607252530.2552969455718994140625 (2015)

7. International Labour Organization, *World employment social employment social outlook-trends*. [online]. Available at: https://www.ilo.org/global/research/global-reports/weso/2019/lang--en/index.htm (2019)

8. International Labour Organization, *Women and men in the informal economy: A statistical picture (third edition). International Labour Office – Geneva*. [online]. Available at: https://www.ilo.org/global/publications/books/WCMS_626831/lang--en/index.htm (2018)

9. International Labour Organization, *ILO Monitor: COVID-19 and the world of work. Third edition Updated estimates and analysis*. [online]. Available at: https://www.ilo.org/global/topics/coronavirus/impacts-and-responses/WCMS_743146/lang--en/index.htm (2020)

10. International Labour Organization, *World Employment and Social Outlook: Trends 2018 (Geneva)*. [online]. Available at: https://www.ilo.org/wcmsp5/groups/public/---dgreports/---dcomm/---publ/documents/publication/wcms_615594.pdf (2018)

11. International Labour Organization, *Global Employment Trends for Youth 2020: Technology and the Future of Jobs*. [online]. Available at: https://www.ilo.org/global/publications/books/WCMS_737648/lang--en/index.htm (2020)

12. United Nations, *Transforming our world: the 2030 Agenda for Sustainable Development*. [online]. Available at: https://sustainabledevelopment.un.org/post2015/transformingourworld (2015)

13. IMFBlog, *Chart of the Week: Sub-Saharan Africa's Growth—A Tale of Different Experiences* [online]. Available: https://blogs.imf.org/2018/12/13/chart-of-the-week-sub-saharan-africas-growth-a-tale-of-different-experiences/ (2018)

14. World Bank, *Promoting rapid economic growth and improving service delivery* [online]. Available: https://www.worldbank.org/en/country/ethiopia/overview (2019)

15. AfDB, *African Economic Outlook 2019-Macroeconomic Performance and Prospects*. [*online*]. Available at: https://www.icafrica.org/en/knowledge-hub/article/african-economic-outlook-2019-362/ (2019)

16. United Nations, *Transforming Ethiopia together*. [online]. Available at: https://ethiopia.un.org/sites/default/files/2019-08/Ethiopia_UN_book_%202018.pdf (2018)

17. F. Zeru, A. Bitga, and J. Boyson, *Usaid/Ethiopia Cross-Sectoral Youth Assessment Situational Analysis*. [online]. Available at: https://www.youthpower.org/resources/usaidethiopia-cross-sectoral-youth-assessment-situational-analysis-report (2018)

18. M. F. Bashir, B. Ma, L. Shahzad, A brief review of socio-economic and environmental impact of Covid-19. *Air Quality, Atmosphere & Health*. **13**, 1403–1409 (2020)

19. A. Pak, O. A. Adegboye, A. I. Adekunle, K. M. Rahman, E. S. McBryde, D. P. Eisen, Economic Consequences of the COVID-19 Outbreak: the Need for Epidemic Preparedness. *Frontier Public Health*. **8**, 1–4 (2020)

20. IMF, *Federal Democratic Republic of Ethiopia: the 2019 Article IV Consultation and Requests for Three-Year Arrangements Under the Extended Credit Facility and the Extended Fund Facility, IMF Country Report*. [online]. Available at: https://www.imf.org/en/Publications/CR/Issues/2020/01/28/The-Federal-Democratic-Republic-of-Ethiopia-2019-Article-IV-Consultation-and-Requests-for-48987 (2020)

21. UNDP, *Ethiopia National Human Development Report 2018: Industrialization with a Human Face*. [online]. Available at: http://hdr.undp.org/sites/default/files/ethiopia_national_human_development_report_2018.pdf (2018)

22. T. Abebe, Lost futures? Educated youth precarity and protests in the Oromia region, Ethiopia. *Children's Geographies*. **18**(6), 584–600 (2020)

23. A. Sumner, C. Hoy, E. Ortiz-Juarez, Estimates of the impact of COVID-19 on global poverty. *Unu-Wider*. **43**, 1–9, (2020)

24. MOH, EPHI, *COVID-19 pandemic preparedness and response in ethiopia; weekly bulletin*. [online]. Available at: https://www.ephi.gov.et/images/novel_coronavirus/EPHI_PHEOC_COVID-19_Weekly_Bulletin_27_English_11022020.pdf (2020)

25. Z. Shigute, A. D. Mebratie, G. Alemu, A. Bedi, Containing the spread of COVID-19 in Ethiopia. *Journal of Global Health*. **10**(1), 1–4 (2020)

26. V. Erokhin, T. Gao, Impacts of COVID-19 on trade and economic aspects of food security: Evidence from 45 developing countries. *International Journal of Environmental Research and Public Health*. **17**(16), 1–28 (2020)

27. FAO, The state of food security and nutrition in the world 2018: building climate resilience for food security and nutrition. [online]. Available at: http://www.fao.org/3/I9553EN/i9553en.pdf (2018)

28. World Bank Group, *Global Economic Prospects*. [online]. Available at: https://www.worldbank.org/en/publication/global-economic-prospects (2020).

29. M. R. Breines, Ethnicity across regional boundaries: migration and the politics of inequality in Ethiopia. *Journal of Ethnic and Migration Studies*. **46**(15), 3335–3351 (2020)

30. W. Wondimu, B. Girma, Challenges and silver linings of COVID-19 in Ethiopia –short review. *Journal of Multidisciplinary Healthcare*. **13**, 917–922 (2020)

31. H. F. Kaiser, A second generation little jiffy. *Psychometrika*. **35**(4), 401–415 (1970)

32. H. F. Kaiser, An index of factorial simplicity. *Psychometrika*. **39**(1), 31–36 (1974)

33. R. B. Cattell, The Scree Test For The Number Of Factors. *Multivariate Behavioral Research*. **1**(2), 245–276 (1966)

34. Worldometer, *World population*. [online]. Available at: www.worldometers.info (2020)

35. EPHI, *COVID-19*. [online]. Available at: https://COVID19.ephi.gov.et/COVID-19/ (2020)

Influence and Intensity of Exogenous Factors on the Development and Stability of Rural Areas

Petra Pártlová[1,*]

[1]Institute of Technology and Business, Department of Management, Okružní 517/10, 37001 České Budějovice, Czech Republic

Abstract. This contribution focuses on the issue of the influence and intensity of exogenous factors on the development and stability of rural areas. The results are presented at two levels. The first reflects the monitoring of the influence of external factors on the external environment of companies, i.e. those factors which they have limited possibilities to influence, and how to eliminate the potential negative impacts thereof. The second provides information on the results and evaluation of four basic parameters in the field of knowledge skills, namely contenders, competition, coordination and cooperation. The evaluation of the goals set are based on the results of a questionnaire survey, supplemented by controlled interviews with business managers and the managements of local action groups (LAGs) in the Region of South Bohemia. The created database is analysed using the method of content analysis and the simultaneous use of statistical methods. The main outputs of the solution determine the significance and strength of the external environment on companies, which is mostly influenced by competitors, suppliers, customers and business partners. In contrast, universities and research institutes do not have a significant impact on the business environment. Similar results apply to LAGs, where in most cases the LAGs cooperate with each other and are not considered competitors, thereby also reporting good cooperation with towns, micro-regions and the regional authorities of South Bohemia. In contrast, LAGs reported none or poor cooperation with the National Network of LAGs and ministries.
Keywords: external factors; companies; LAG; actors; rural areas

1 Introduction

Rural development policy at the turn of the 20th and 21st centuries was the subject of harsh criticism, which was followed in 2006 by a relatively strong trend towards a new paradigm of rural areas, as incorporated in a new concept by the OECD [1]. The aim of this concept is primarily the growth of rural competitiveness and to look for the local benefits in an area, on the basis of which financial and non-financial support should be directed. A very important factor in this is that new rural development policies should be relevant to the key actors because they bear the responsibility for all the important activities that strengthen development and stability [2-5].

Rural development policy is very complex because it affects many areas of life of the population, is influenced by the historical and cultural contexts of a region, and its implementation depends to a large extent on the activity and cooperation of the local actors [6, 7].

With regards to the issue at hand, it is possible to distinguish three groups of actors in rural areas: a) those who participate and are involved in the process of dealing with rural development (e.g. cooperating municipalities and entrepreneurs); b) actors and entities responsible for a certain activity (e.g. citizens and their associations, visitors, etc.); and c) the affected, i.e. entities in whose area of interest the activity is carried out (e.g. regional authority, administration of a protected landscape area or national park, ministry, etc.) [8, 9].

The important actors are the mayors and the representatives of the municipalities, who by the virtue of their position have and should have open access to all the other actors, whereby the bottom-up principle should be applied [10]. Other important actors are ministries, with direct influence on the municipalities for their delegated competence, and within the given laws. Regional authorities may be responsible for all the functions of rural development, from methodically helping and controlling the activities of municipalities through to active

* Corresponding Author: partlova@mail.vstecb.cz

participation, with the tools they have at their disposal, to reduce or limit the disparities between municipalities, whilst at the same time, developing their activities across a whole region.

Companies are also important actors. The private sector has great potential for development activities. Their cooperation is based on the principle of profitability. They are willing to support a number of development activities, but expect some form of reciprocity [11]. Although the representatives of local institutions (such as schools and libraries) have little power to solve more fundamental problems within municipalities, there are synergies to be found with other actors [9, 12]. Other actors, who are not mentioned often, are political parties and their relatively small influence on the leadership of small municipalities. However, in certain circumstances, political parties hold the keys to accessing state subsidies. Local action groups (LAGs) are also important actors, bringing together representatives of all groups. LAGs, similar to regional authorities, perform all three functions [8, 9]. A well-functioning LAG, which wants to make the most of the potential of its territory, applies mainly the features of networking and cooperation [13-15]; it uses new approaches and procedures in guiding the local development of the territory [16].

The intensity of partnerships is one of the factors influencing the direction and intensity of rural development [17]. Important factors influencing the relations and ties in the region include contenders (actors without interaction), competition (interaction arises), coordination (adaptation of actions to avoid conflicts) and cooperation (mutually beneficial partnership, most important for development) [18]. In terms of rural development, a distinction is made between endogenous and exogenous development factors. Endogenous factors are directly linked to the activities of existing rural users, meaning the development actors [8]. Exogenous factors are the external factors representing a certain framework of factors and instruments that affect rural development, whereby individual actors cannot influence them directly (location of the municipality, natural environment, the operation of other regions on the one hand, legislation, financial systems, central development instruments, grant-supported activities, etc. on the other hand [8, 19, 20-23].

The research that forms the basis of this contribution focuses on two levels in terms of the factors of rural development. The first part focuses on an analysis of how external factors influence business activity in the Region of South Bohemia, as well as on the way the intensity of that influence depends on the close surroundings of a company. The second part focuses on local action groups as important actors in the development of rural areas and on the influence of external factors on their activities in terms of cooperation and competition, as well as on how such factors affect their activities in the search for solutions to rural development problems, the implementation of strategy, and the creation of the conditions for further cooperation and networking.

2 Methodology

The primary data sources are a questionnaire survey and controlled interviews. In the first part of the research, the influence of the external environment on the development of business activities in the Region of South Bohemia is evaluated. For this purpose an analysis was conducted of the responses of 102 companies operating in rural areas in the region who were contacted and asked to identify the external environmental factors that influence other actors and their business activities. The results of the survey were processed using the statistical method in the R programme and presented graphically. The chosen techniques are considered suitable for primary data collection whenever phenomena cannot be monitored directly.

The second part of the research covers local action groups (16) in the Region of South Bohemia. A questionnaire survey and field research were conducted. The structure of the questionnaire was divided according to the four core areas of research, namely contenders, competition, coordination and cooperation. Under contest, there were four questions with three possible answers (yes, no, no interaction). Under competition, there were three partial questions with possible yes and no answers, as well as a detailed justification thereof. Under coordination, issues aimed at coordinating activities within LAGs themselves are addressed, including the involvement of local actors in the LAGs, such as the municipalities themselves and their mayors. Furthermore, the cooperation of the towns, the regional authority of South Bohemia, and the ministries involved in the activities of the LAGs are addressed. Under cooperation, the LAGs cooperation with the for-profit sector, private sector, National Network of LAGs and South Bohemia Network of LAGs are analysed.

After completing the questionnaire, a field survey was carried out in combination with guided interviews with the representatives of all South Bohemia LAGs in order to complete the missing parts of the questionnaire.

The research material was subjected to content analysis. Content analysis is one of many possible types of qualitative data analysis. Within this context, the coding method was used [24]. All analysed material was divided into four areas in connection with the 15 research questions. Each unit created was coded on the basis of the answers (YES, NO, and in some cases NO RELATION). The application of coding is especially useful when the aim of the research is to describe different views on a given research topic [25]. The research results are presented here using the descriptive method. The answers are structured into graphs on the basis of which a detailed analysis and evaluation are carried out.

3 Results and Discussion

In principle, a possible tool for the successful management of business activities in regions is a functioning partnership between the private, public and non-profit sectors with the citizens of the region. In rural areas, the cooperation of all interested legal and natural persons with the public authorities is very important. Such cooperation is representative of the whole community in a municipality or region and its essence is the search for and solution of common problems and the achievement of common goals. Such partnerships bring together and multiply the ideas, energy and resources of stakeholders (local people, administrations, entrepreneurs, non-profit organizations, etc.), whereby the quality of the final product depends on it.

3.1 Analysis of the external business environment

The performed analysis shows that companies and their activities are affected by many actors with varying degrees of intensity (see Table 1 and Figure 1). The influence of competitors, suppliers, customers and business partners were identified as being significant to very significant. The companies operating in rural areas also show that associations (such as LAGs) and financial institutions have a large influence. On the contrary, universities, local government, government, political parties and the media have little to no influence. Research institutes and academies of sciences have the least influence on the activities of the companies.

Table 1. Influence of actors in rural areas on companies

Actors in rural areas	No influence	Little influence	Significant influence	Very significant influence	Total
Competitors	11	21	50	20	102
Suppliers	11	28	48	15	102
Customers	8	15	32	47	102
Distributors	26	29	36	11	102
Business Partners	20	16	41	25	102
Financial institutions	24	30	38	10	102
Universities	58	29	13	2	102
Research institutes, academies of sciences	68	26	8	0	102
Local government and self-government	42	38	15	7	102
Political parties	64	20	11	7	102
Government	59	18	18	7	102
Average	53	30	16	3	102
Associations (LAGs)	24	26	20	32	102

Source: Authors.

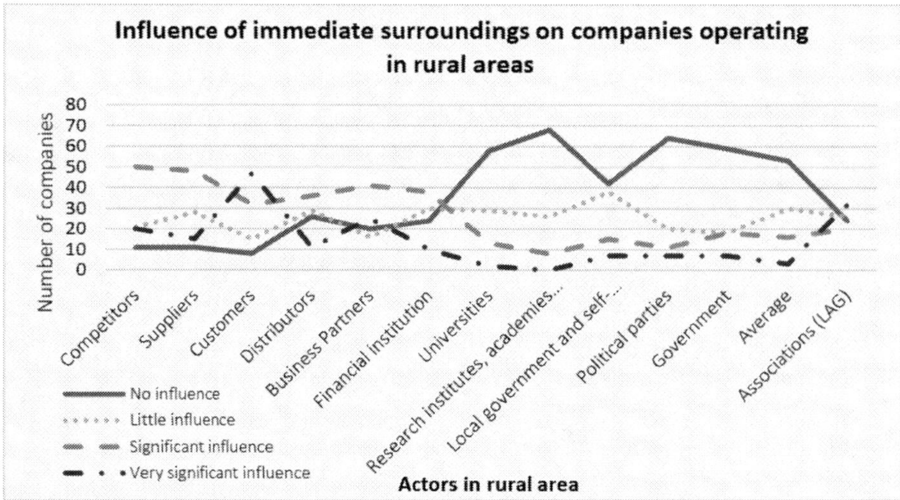

Figure 1. Influence of immediate surroundings on companies operating in rural areas

Source: Authors.

The significance of the intensity of influences on the factors was determined using the R programme.

Calculation:

P value	*2.2E-16*	*x2*	*133.2868*
V	*0.568091803*	*freedom*	*line-1 x column -1*

Statistical methods were used for the analysis, proving that the intensity of the influence depends on the close surroundings of the enterprise (P value = 2.2E-16, $\alpha = 0.05$). The hypothesis was confirmed. Based on the results, V = 0.5681, it is possible to deduce a medium degree of statistical dependence.

Based on the results, it is possible to infer the willingness of companies to cooperate with partners and subjects in the business environment. It is clear that, as far as the other actors in rural areas are concerned, no higher consensus is to be found in the search for such cooperation. Nevertheless, certain forms of partnership are offered, which are cooperation with institutions for business support, cooperation in the field of education, cooperation between companies, with the regions and municipalities.

With regards to the cooperation with institutions and agencies established by ministries (CzechInvest, CzechTrade, ČMZRB, EGAP, chambers of commerce, Society of Science and Technology Parks, unions, associations, regional development agencies, regional advisory and information centres), this is considered to be rather wide ranging. Forms include technical parks, centres, incubators, investment in industrial zones, use of brownfield sites, organizing foreign missions to exhibitions and fairs, etc.

Another possible way is cooperation with research institutions (public and state universities, public research institutions, and private innovation enterprises), which are seen to provide the opportunity to cooperate on research activities, both basic and applied. Innovation vouchers are also a form of cooperation, which is a form of counselling, including expert and support services, especially for SMEs.

A very important form of cooperation, as evidenced by the previous figure, is the cooperation between companies. Companies associate with one another through business networks (networking) and clusters. Such clusters allow companies to react flexibly enough to adapt to the increased speed of technological change and the rapidly growing competitiveness required in global markets.

Cooperation between the region and the business sector is based on a certain form of participation, the main task of which is the creation of conceptual documents, participation in specific development activities or the creation of public alliances (such as public-private partnerships (PPP)). The aim of such cooperation is to support the development of innovation with a suitable infrastructure.

The last type of cooperation that has influenced the development of rural areas in the last 15 years is the active role of LAGs. LASGs have several roles and carry out many activities in rural areas. LAGs are actively involved in improving the cooperation of all entities in municipalities (municipal authorities, entrepreneurs, non-profit organizations, citizens). Even their form predestines them to influence all actors and help them to overcome

 https://doi.org/10.1051/shsconf/20219001014

isolation and offer such activities that lead to mutual cooperation. LAGs also ensure an integrated approach to rural development, using the specificities and assumptions of a particular rural area.

3.2 Analysis of LAG cooperation factors

Contenders

As revealed by the results (see Figure 2), LAGs do not contend with others. On the contrary, most of them cooperate on projects, strategy development, sharing knowledge, based on the principle of collegiality. LAGs are linked by the common nature of their activities.

The same applies for the second question, which deals with LAG cooperation throughout the Czech Republic. The LAGs mostly cooperate and share information about their activities. However, the relations are no longer as intense as in the case of the regional LAG network. They point out that they are mostly united by the Leader method, which is now applied through community-led local development (CCLD).

Another question was related to the competition of foreign LAGs. Most South Bohemian LAGs do not consider their foreign partners to be rivals, or they have no relation with them and have not considered participating in any cross-border cooperation. The reason for this is the allocation of only small financial funds for such projects.

The last question focused on the possible competition between public administrations and LAGs. Most LAGs perceive public administration on the basis of cooperation. They agree that public administrations have financial resources at their disposal, which they might allocate to local action groups for ongoing projects.

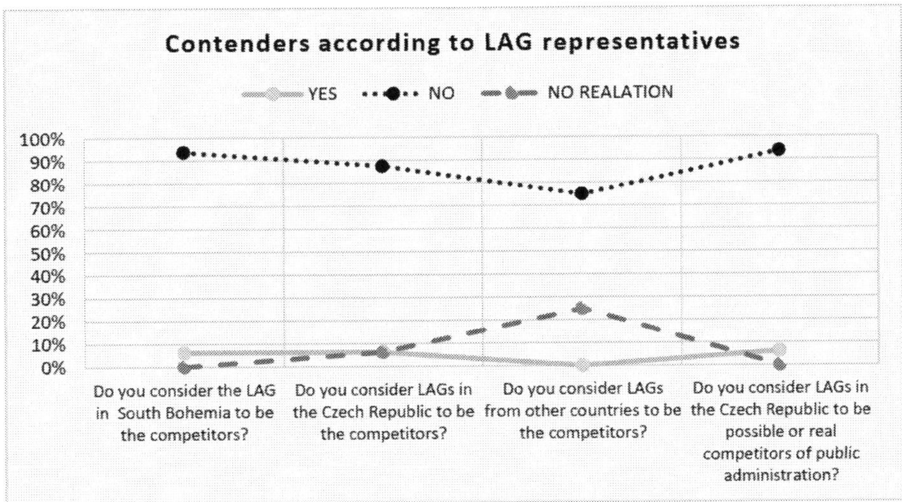

Figure 2. Contenders according to LAG representatives

Source: Authors.

Competition

Under competition, three areas were monitored (see Figure 3). More than two-thirds of the LAGs consider other LAGs to be their allies. They agree that more successful LAGs actually support other local groups and that they create a good image in the Region of South Bohemia. LAGs are not economic entities, so it is much more favourable to use examples of good practice than healthy competition. At the same time, they point out that each LAG has its own territory, its own needs, where it implements projects.

The second question sought to determine whether some form of LAG evaluation should be created. The majority answered that each LAG has its own specifics and that it is very difficult to find the right key with which to evaluate different LAGs. However, many of them added that the evaluation of LAGs would help to point out the less active LAGs and, under certain conditions, would act as a motivational factor. When asked the last question, i.e. whether LAGs should receive financial resources equally, the vast majority of LAGs agreed that it is necessary to respect the territory, its size, population and even the development of the region.

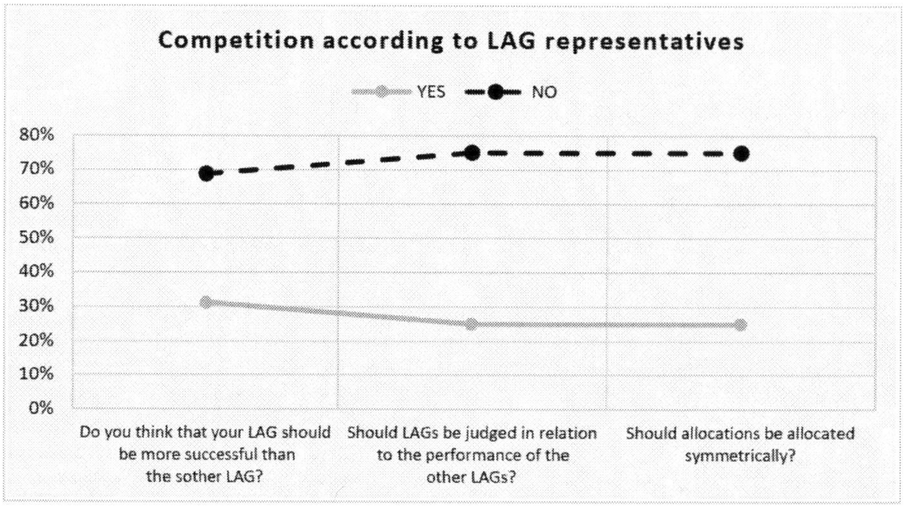

Figure 3. Competition according to LAG representatives

Source: Authors.

Coordination

Under coordination, the questions focused on the cooperation between LAGs and other actors.

With regards to the cooperation between municipalities and LAGs, most municipalities do not cooperate with LAGs. One of the reasons why this is the case is that so far, the municipalities do not have enough information on what the LAGs can offer the municipalities. A change in attitude among municipalities may be instigated over the long-term if LAGs showed more proactive behaviour towards them. Communication with the municipalities takes place mostly through email and telephone communication, and through municipal dailies; personal meetings proved to be the most effective. However, about one-third of the LAGs evaluated the cooperation with municipalities as being very good, and with towns, micro-regions and the regional authority of South Bohemia as good. All the LAGs agreed that the cooperation with the ministry is rather poor. The LAGs would highly appreciate it if the whole system became clearer and simplified (the setup of departments). The results are presented in Figure 4.

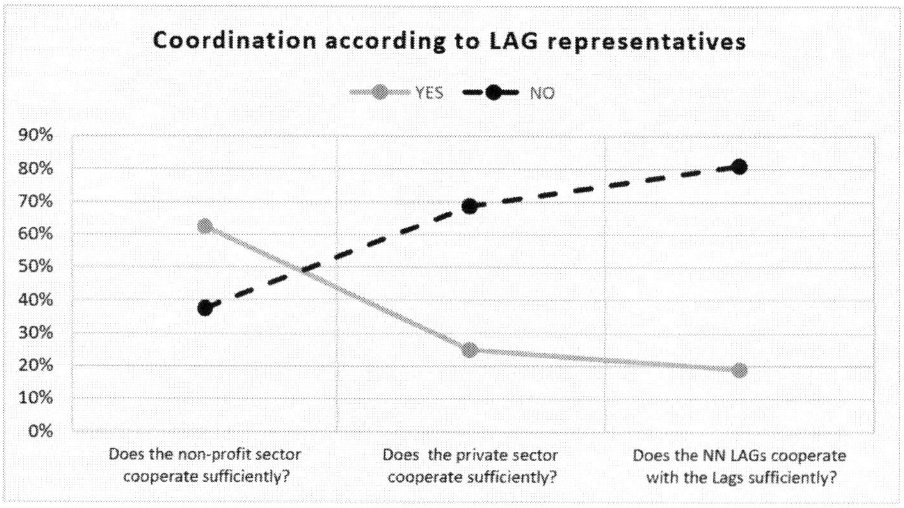

Figure 4. Coordination according to LAG representatives

Source: Authors.

https://doi.org/10.1051/shsconf/20219001014

Cooperation

Under cooperation (see Figure 5), the focus was on partnership with the non-profit sector, private sector and the National Network of LAGs. With regards to the first two, the opinion of the majority is that if the LAG is active and has something to offer, the cooperation works well. The situation has improved significantly, especially in the for-profit sector, with a change in the subsidy policy and the redistribution of the activities into more areas. In the private sector, cooperation with the agricultural sector has been excellent for a long time. With other SMEs the cooperation is worse, but even here, with the new subsidy policy, things are improving. As for the partnership with the NN of LAGs, the opinion of the majority is that such cooperation is limited and at a formal level only, with LAGs seeing little benefit from it. However, they would like to meet more and look for solutions, other activities and tools with which to help and support the development of rural areas.

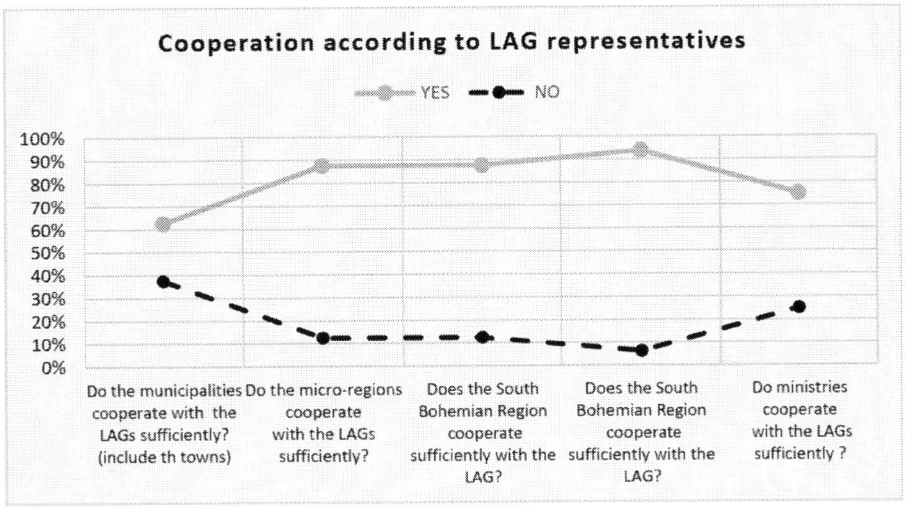

Figure 5. Cooperation according to LAG representatives

Source: Authors.

4 Discussion

Supported by the results, the following can be said.

At present, companies do not utilise all forms of cooperation, as was shown by the analysis of the external environment. The results of the analysis of the actors in rural areas and their influence on the business environment can be summarized as follows:

- The biggest influence on companies are competitors, suppliers, customers and business partners.
- Universities, research institutes and academies of science have the least influence on the activities of companies.
- The activities based on cooperation with LAGs are growing in intensity.

LAGs are advised to focus on the following in the foreseeable future:

- To enter into a wide-ranging discussion on the future of Czech LAGs and their position in society; improve promotion across the country; actively work and develop the Leader platform; simplify the organizational structure of LAGs.
- To involve ministries, public administration bodies in South Bohemia and experts in finding a joint solution for rural areas.
- To map the possibilities of changing the structure for the allocation of funds to LAGs (corresponding income for LAGs), related to the setting of clear rules for the redistribution thereof, the prioritization of activities and the anchoring of LAGs through legislation.

- To prepare for the future, by defining a strategy, and where possible to introduce a single programme designed by LAGs and financed with EU funds. Today it is fragmented under different ministries – Agriculture, Local Development and Labour and Social Affairs (3-4 operational programmes), etc. The current system disproportionately increases the time required for the organization of work, making the activities of LAGs more complicated.
- To inform and advise municipalities, providing support for destination management, cultural and educational events - better and more efficient redistribution of funds according to the specifics and potential of an administered area.

5 Conclusion

Rural areas are essentially anthropologic-ecological systems, which presupposes the harmonization of all three of its most important parts - scientific, economic and social. The interconnection and balance guarantee sustainability. The human factor plays a crucial and, in a way, regulatory role here.

In rural areas, the anthropogenic factor appears in various forms, such as a resident, entrepreneur, vacationer, tourist, conservationist, or as a member of various interest groups, communities and associations.

However, for all forms of activity in rural areas, the following principle applies: man is an integral part of nature and the countryside, neither superior nor subordinate, however unique homo sapiens are, and any wrong decision can disrupt rural equilibrium, which is a condition for its stability and sustainability.

This principle should lead actors operating in rural areas towards mutual cooperation, harmony and, if necessary, active cooperation in eliminating discrepancies.

In this area, social organizations, voluntary associations and citizens' associations have an irreplaceable place, of which LAGs play a crucial role. The citizens who permanently work in such areas know the local conditions and are the bearers of traditions and historical memories. It is therefore a pity that their irreplaceable mission in Czech rural areas is not always appreciated by the governing bodies of the state, and in some cases by the self-governing bodies of municipalities and the entrepreneurs operating in the countryside. As the evidence from the survey conducted in the Region of South Bohemia showed, local action groups cooperate with one another and work together at a very good level. The truth is that this is the result of their informal activities, which is also one of the principles for the proper functioning of LAGs.

The LAGs in South Bohemia would definitely like to be more involved in deciding the direction and development of the Region of South Bohemia. Such involvement is one of their priorities and is perceived by them to be their mission.

Through this contribution, the author aims to open up both public and scientific discussions on the issue, which is important for the stability of Czech rural areas.

Acknowledgement

This research was funded by the Technology Agency of the Czech Republic, ETA programme, Project Reg. No. TL01000349 – Stabilization and development of SMEs in rural areas.

References

1. OECD, *The New Rural Paradigm – Policies and Governance.* Paris: OECD Publishing (2006)
2. P. Rumpel, O. Slach, J. Koutský, *Měkké faktory regionálniho rozvoje* [*Soft factors of regional development*]. Ostrava: Ostravská univerzita (2008)
3. M. Ambrosio-Albalá, J. Bastiaensen. *The new territorial paradigm of rural development: Theoretical foundations from systems and institutional theories*. Antwerp: Institute of development policy and management (2010)
4. J. Falkowski, Political accountability and governance in rural areas: Some evidence from the Pilot Programme LEADER in Poland. *Journal of rural studies*. **32**, 70–79 (2007)
5. M. Pělucha, *Teorie a realita podpory územní soudržnosti v politice rozvoje venkova EU v ČR ve střednědobém horizontu let 2004–2013* [*Theory and reality of support for territorial cohesion in the EU rural*

development policy in the Czech Republic in the medium term of 2004–2013]. Habilitační práce. Praha: NF VŠE (2015)

6. Z. Varvasovszky, R. Brugha, Stakeholder analysis. *Healthpolicy and planning*. **15**(3), 338–45 (2000)
7. R. Perlín, A. Šimčíková, Criteria of a successful rural municipality. *Europa XXI*. **17**, 29–43 (2008)
8. J. Binek, H. Svobodová, J. Holeček, I. Galvasová, K. Chabičovská, *Synergie ve venkovském prostoru. Aktéři a nástroje rozvoje venkova* [*Synergies in rural areas. Actors and tools of rural development*]. Brno: GaREP Publishing (2009)
9. J. Binek, H. Svobodová, J. Holeček, I. Galvasová, J. Martének, *Synergie ve venkovském prostoru. Paradoxy rozvoje venkova* [*Synergies in rural areas. Paradoxes of rural development*]. Brno: GaREP Publishing (2011)
10. J. Bernard, Individuální charakteristiky kandidátů ve volbách do zastupitelstev obcí a jejich vliv na volební výsledky [Individual characteristics of candidates in municipal elections and their influence on election results]. *Sociologický časopis*. **48**(4), 613–640 (2012)
11. E. Hamplová, K. Provazníková, Opinion and Attitudes of Entrepreneurs of Small and Medium-Sized Business in the Czech Republic Conditions. *Procedia Economics and Finance*. **23**, 942–947 (2015)
12. K. Knickel, H. Renting, Methodological and Conceptual Issues in the Study of Multifunctionality and Rural Development. *Sociologia Ruralis*. **40**(4), 512–528 (2000)
13. A. Faloudi, From European Spatial Development to Territorial Cohesion Policy. *Regional Studies*. **40**(6), 667–678 (2006)
14. R. D. Putnam, Bowling Alone: America's Declining Social Capital. *Journal of Democracy*. **6**(1), 65–78 (1995)
15. G. Bosworth, Community-Based Entrepreneurship and Rural Development. International Journal of Entrepreneurial Behavior & Research. *Emerald Group Publishing Limited*. **21**(3), 501–503 (2015)
16. J. Coleman, Social Capital in the Creation of Human Capital. *American Journal of Sociology*. **94**(1), 95-120 (1988)
17. L. Larsson. EU in the village: LEADER II, governance and rural development. In: L. Dargan, M. Shucksmith, LEADER and Innovation. *Sociologia Ruralis*. **48**(3), 274–290 (2002)
18. R. Perlín, Představitelé veřejné správy jako aktéři lokálního rozvoje [Representatives of public administration as actors of local development]. In *Problémy stabilizace venkovského osídlení ČR*. České Budějovice: Jihočeská univerzita v Českých Budějovicích, 133–151 (2000)
19. M. Woods. *Rural Geography. Processes, Responses and Experiences in Rural Restructuring*. London: SAGE (2005)
20. P. Pospěch, M. Delín, T. Doucha, J. Drlík, F. Nohel, D. Spěšná, *Vynalézání venkova v ČR po roce 1989* [*Rural invention in the Czech Republic after 1989*]. Brno: Centrum pro studium demokracie a kultury (2014)
21. M. Hampl, M. Marada, Sociogeografická regionalizace Česka [Sociogeographical regionalization of the Czech Republic]. *Geografie*. **120**(3), 397–421 (2015)
22. K. Hoggart, H. Buller. *Rural development. A Geographical Perspective*. New York: Routledge (2016)
23. V. Skála, Vliv sociálního kapitálu na rozvojový potenciál venkovských obcí [The influence of social capital on the development potential of rural communities]. *Acta Politologica*. **9**(2), 40–56 (2017)
24. G. R. Gibbs, *Analyzing Qualitative Data*. Thousand Oaks: Sage Publications (2007)
25. R. Švaříček, K.Šeďová, *Kvalitativní výzkum v pedagogických vědách* [*Qualitative research in pedagogical sciences*]. Praha: Portál (2007)

Analysis of E-Consumer Behaviour – Selected Findings from an Analysis of Czech E-Shops and their Customers during the Global Pandemic

Frantisek Pollak[1,*], *Michal* Konecny[1]

[1]Institute of Technology and Business, Faculty of Corporate Strategy, Okruzni 517/10, 37001 Ceske Budejovice, Czech Republic

Abstract. The study presented in this contribution deals with the issue of using modern methods of data collection for the purposes of analysing consumer behaviour. Professional literature as well as business practice offer us an enormous number of approaches, traditional or innovative, with which it is possible to obtain and then evaluate data in order to understand specific patterns of consumer behaviour. Following on from our previous research in the field, we have applied a method of collecting and evaluating customers e-data in real time. As a source of data, we used the virtual social network Facebook. As a sample we selected customers of the five largest e-shops operated in the Czech Republic. On a sample of more than one and a half million users we monitored the B2C communication of e-shops and subsequently the C2B Facebook interaction of their customers during the lockdown caused by the COVID-19 pandemic. In the observed period we analysed hundreds of thousands of interactions and then outlined basic trends and specific characteristics of consumer behaviour suitable for further research. The selected findings resulting from the implemented analyses contribute to the creation of a knowledge base of a qualitative nature, which can help to define research goals in the future process of examining the effects of the pandemic on various aspects of business.
Keywords: e-commerce; Facebook; COVID-19; customer interactions; consumer behaviour; lockdown; pandemic

1 Introduction

The application of innovative practices and the management of innovation in general, within the context of the continuous development of ICT, has been an important topic of business practice for several decades [1]. The advantage of using ICT in companies lies mainly in the possibility of more efficient optimization of the supply-customer chain and also in the possibility of a better approach to customers themselves [2]. E-commerce is increasingly becoming an integral part of the core business of most businesses operating in an increasingly globalized market [3]. Under standard conditions, the transition to e-commerce is a continuous process, necessary in the long run, but mostly not yet conditioning the very existence of companies in the short term. Due to the unusual situation in the global market, namely the pandemic caused by the COVID-19 virus, companies suddenly had to reconsider their position on applying e-commerce tools to their portfolios, as countries across the Western market began to close their economies to slow the spread of the pandemic. Analysts estimate that the damage to the global economy may be in the region of 2-4% of world GDP [4]. The disease, which broke out in the Chinese city of Wuhan in December 2019, probably due to the consumption of an unidentified animal [5], shows no signs of slowing down and has claimed more than a million lives to date [6]. The aim of the study is to present partial qualitative results of currently implemented comprehensive research into selected characteristics of consumer behaviour and its changes during the first COVID-19 lockdown. Following on from our previous research in the field, we have applied a method of collecting and evaluating data in real time. On the sample of more than one and a half million Facebook users / customers of the five biggest e-shops on the Czech market, we monitored the B2C communication of the aforementioned e-shops and subsequently the C2B interaction of their user groups during the first lockdown caused by the COVID-19 pandemic. In the observed period, which created an almost model situation for our research, we analysed hundreds of thousands of users' interactions and then outlined specific characteristics of consumer behaviour suitable for further research. The selected findings resulting from the implemented analyses contribute to the creation of a knowledge base of a qualitative nature, which should help to expand general knowledge in the field of innovation in management, crisis communication and marketing approaches.

[*] Corresponding author: frantisek.pollak@mail.vstecb.cz

1.1 Basic definition of key terms and contexts

1.1.1 E-commerce

As far as practice is concerned, there is often confusion in the identification of the terms e-commerce and e-business. One point of view is that the two terms are synonymous because the two terms include business processes that are carried out electronically, in particular via the Internet. Another point of view is that the two concepts are perceived differently because e-commerce is a subset of e-business [7]. According to Turcu and Feraru [8], e-commerce and e-business are somewhat similar and also have some of the same characteristics in that both processes seek to increase profits, reduce costs and improve business processes to generate added value for products and services. In general, it can be stated that e-commerce is part of a broad portfolio of e-business tools. Let us now look in more detail at the issue of e-commerce, specifically its benefits for businesses, consumers and society as a whole.

Benefits of e-commerce for organizations:

- organizations have the opportunity to expand their local markets to national and international markets, which includes minimal capital expenditure, and also have the ability to easily and quickly find a number of consumers, suppliers or the best business partners around the world;
- lower costs for organizations in terms of acquiring, processing, transforming, distributing, storing or analyzing information;
- organizations have the opportunity to reduce costs and inventory within the supply chain;
- organizations have the possibility of lower capital expenditures on business;
- communication costs for organizations are lower due to the use of the Internet, which is cheaper;
- formation of process-organizational systems in organizations;
- other benefits include faster access to information, shorter time, better productivity, improved customer service, and a wide selection of business partners [9].

Benefits of e-commerce for consumers:

- the opportunity to buy and secure transactions at any time, any day and also from anywhere;
- the opportunity to compare prices faster and to buy at lower prices;
- fast delivery;
- much wider choice of sellers who provide goods and services;
- the opportunity to participate in virtual auctions;
- the opportunity to take advantage of discounts due to increased competition;
- the opportunity to obtain important and detailed information about products and services at any time and from any place;
- connection to other consumers, through which they have the opportunity to exchange experiences and opinions about different products or services within virtual communities [9].

Benefits of e-commerce for society:

- the possibility of working within one place or from home, which eliminates the need to commute to work on a daily basis, which also leads to a reduction in traffic on the roads and also to air pollution;
- the possibility to buy larger quantities in case of discounts;
- the possibility of using products or services, hitherto unknown or inaccessible to customers from less developed countries, which gives them a considerable ability to assert themselves, especially the possibility of education or professional growth;
- simpler, faster, qualitatively better and lower-cost provision of public services, irrespective of whether it concerns social services, healthcare services or education [10].

Rajib [11] deals with the main aspects of e-commerce, to a large extent influencing the functioning of e-business, which need to be considered before the e-commerce itself can be created. The first, very important features are flexibility and adaptability. Before an e-shop itself is created, the operator of the online store should be sure that the website of the store will be displayed correctly to each internet user and that modern technological elements will be used in creating the online store. In the event that outdated forms or outdated visualizations are used, the website may not be of interest to

Internet users, which could cause a decrease in traffic to the website. Rajib [11] cites ease and instinctive use as another important element in creating an online store website. He considers it unnecessary to use too many functions, which are complicated, and which are not necessary for the Internet user, respectively does not know how to use them. Of course, the price of creating an online store is also determined by the complexity of the functions. Many beginning e-commerce website owners do not have the capital to consider the more complex features of a website. Within this context, the most sensible approach is to use a smaller budget to start with and to then add additional features to the online store over time. According to the author, the most important feature is security, which is based on the fact that customers are required to provide their personal data when making a purchase. In the event that a customer registers a leak of information to a third party, they will most likely stop using the website of the online store and never purchase anything through it again. Also of importance is a high quality payment system. The online store must be able to accept multiple customer payments at once. The proven and most used payment systems in the world are Internet banking, payments via credit or debit card, or the PayPal system [12-13].

1.1.2 M-commerce

Brzozowska and Bubel [14] argue that e-business is mainly a process of managing an organization with the help of mobile information and communication technologies (ICT), or through their direct use. Prihastomo et al. [15] are of the opinion that today we live in a modern age in which mobile phone technology is already relatively advanced. M-business is created by activities, processes or applications implemented or supported by mobile technologies. Gála [16] highlights several differences between m-commerce and e-commerce. These differences are:

- a mobile device is independent of place, respectively location;
- a mobile device together with the applications on it have a reach anytime and anywhere, unlike e-commerce, which requires an Internet connection from a specific place, i.e. they differ in the scope of reach;
- a mobile device is much cheaper and ultimately provides better accessibility for consumers compared to computers or laptops;
- the ability to operate a mobile device immediately without the need to instal software;
- within m-commerce, immediate identification of the user on the basis of SIM card, which is registered in a specific name or to a specific subject;
- the possibility of one-to-one marketing via mobile devices, which from the merchant's point of view means that they are in direct contact with their customers;
- knowing the immediate location of a mobile device user is appropriate from a marketing point of view, specifically due to the implementation or targeting of its various forms to the user.

A study conducted by the Association for Electronic Commerce (APEK) found that m-commerce is on the rise in the Czech Republic and that young people in particular make purchases via mobile devices. According to this study, at least one purchase was made on a smartphone or tablet by half the people questioned. As part of its current research, APEK asked consumers about their experience with mobile shopping. Jan Vetýška, executive director of APEK, is quoted as saying that "50% of respondents made at least one purchase using a smartphone or tablet". This is 8% more than in 2017. The fact that shopping via mobile technology is most popular with the youngest consumers is not considered a big surprise. Within the age category 15 - 24 years, 74% of respondents bought via smartphone or tablet. Within the age category over 55 years, only 21% of respondents confirmed their experience with this form of shopping. Vetýška is further quoted as saying "shopping via mobile devices is growing steadily. However, it must be said that its popularity in the Czech Republic and Europe in general is still very low compared to, for example, Asia". For historical reasons, the inhabitants of Asian countries often jumped from owning a computer directly to a smartphone, therefore the vast majority of orders in this important region are made using a mobile device. However, Czechs still place the majority of their orders via a computer or laptop. Mobile phones are also used by customers when shopping in brick-and-mortar stores. 39% of respondents confirmed that they occasionally use their smartphone to compare the prices of an online store with those in the brick-and-mortar store they are currently in where they are viewing the product. In exceptional cases, another 25% of respondents consult their smartphone in this way. However, customers behave similarly in relation to e-shops. 22% of respondents, when asked how mobile devices had changed their shopping behaviour, answered that they search for and compare goods in online stores on their mobile phones but choose a computer to process the order. In answer to the same question, 29% of respondents said that they search for information about goods or services more often through these handy devices, and 23% of respondents answered that they search for ratings or reviews from other users more than in the past. As part of the global trend, we can continue to see the growing importance of mobile sales. For this reason, too, e-shops are directing their investments to the development of responsive websites and their own sales applications. Nevertheless, it is not possible in the near future to expect similar dominance in the Czech Republic or even in Europe as can be seen in the eastern markets [17].

According to a study carried out by Worldpay payment technology, m-commerce will overtake computer purchases by 2023. As for German consumers, they are advocates of convenience and this is reflected in the way they do business. Mobile trading is still growing and is becoming the most popular shopping channel. Based on Worldpay payment technology data, we can expect the volume of m-commerce to reach EUR 44.3 billion by 2022, with a 16% annual increase. Based on calculations, Germany aims to become the world's sixth largest e-commerce market by 2022. In its Global Payments Report 2018, Worldpay also cites electronic invoicing as the fastest growing payment method for German consumers. By 2022, the market share of e-invoicing could triple, making it the most popular online payment method. The increase in popularity is also visible in e-wallets (11% annual increase), which puts them ahead of bank transfers, credit and debit cards. With the introduction of PSD2 (Payment Services Directive), the European Union's second directive on payment services in Europe, bank transfers could regain popularity during 2020, depending on how open banking solutions are adopted across the continent. The Worldpay study found that m-commerce accounts for $ 1.14 trillion in global e-commerce sales and that global mobile commerce will grow by 19% by 2023. By 2022, growth is expected in China (€ 1.03 trillion), the US (€ 320 billion), the United Kingdom (€ 100 billion), France (€ 54.5 billion), India (€ 48 billion) and Germany (€ 44 billion), which are among the world's largest m-commerce markets [18]. A study conducted by PayPal, a global market research and consulting firm, found that consumers were more likely to buy smartphones than ever before, but there was still an interest in keeping their mobile business safe. A global study of 22,000 consumers of all ages found that up to 80% of consumers made a smartphone purchase, but in e-commerce, only 63% of organizations are optimized to receive mobile payments. "This is a big gap in businesses that are missing out by not taking advantage of the explosive growth of m-commerce," PayPal said in a statement. "Businesses need to offer a mobile-optimized experience if they are interested in attracting and retaining younger consumers, such as Gen Z and Gen Y." Giving priority to mobile phones could be a priority. The survey found that the main reason for the consumer to refrain from buying through m-commerce was that the trader could not provide them with the method of payment that they preferred. As a result, 21% of respondents stated that they therefore abandoned the purchase. Other key findings include the following:

- 72% of consumers in the US use a smartphone for online payments, but only 57% of organizations say they offer optimization for mobile devices. This is the second lowest value in all the countries surveyed.
- 48% of women prefer to buy via mobile device, for men it is 39%. 19% of women shop via their mobile device at work, compared to 26% for men.
- The supply and demand for m-commerce is in line between traders and consumers, with 1 in 3 (women and men) in both categories stating that they use m-commerce.
- Facebook, Instagram and WhatsApp are among the most widely used m-commerce platforms worldwide (35% of respondents are Facebook-owned users).
- 87% of brick-and-mortar merchants received a contactless payment, with 50% of these payments made via smartphone.
- India is a leader in the use of m-commerce, with 70% of consumers preferring m-commerce purchases and 81% of retailers optimized for m-commerce.
- With 83%, Italy is the top European market for consumers shopping online via smartphone, with 65% of Italian organizations claiming they have a mobile-optimized website or application.

In terms of trust and security, 51% of respondents worldwide are less likely to buy via mobile technology due to concerns about their security. In some countries, the percentage is lower. An example is Japan, where only 28% of respondents said they were less likely to buy via mobile technology due to security concerns. The biggest security concerns for m-commerce are in the UK (64%), Australia (63%) and the US (58%). This is then passed on to retailers, with 20% of them citing consumer safety concerns as a reason for their reluctance to optimize for m-commerce. In a separate survey of holiday purchases in the US and Canada, AdColony [19] found that a mobile phone is considered as secure as a desktop computer. As part of AdColony's results regarding the purchase of gifts through m-commerce in cooperation with PayPal, focusing on the overall participation in m-commerce, it was found that 81% of respondents bought gifts via mobile phone. 45% of respondents said that they shop regularly via smartphone.

Other findings from the AdColony [19] survey:

- consumers bought various types of goods and services through m-commerce, whether they were books, event tickets or travel. Through m-commerce, the most purchased were clothes (43%) or household goods (20%);
- more than half of the respondents stated that they would never buy anything through a mobile device if they bought directly from advertising. However, if the product were relevant to them, 71% of respondents said they would buy the product;
- if the consumer visits the store, they will not make an online purchase unless they come across a cheaper price or if the store does not have the given size or colour of the item;

 https://doi.org/10.1051/shsconf/20219001015

- the importance of mobile devices when shopping was mentioned by 61% of respondents;
- mobile phones are used by consumers in various ways, but they are mostly used to find out competitors prices (55%) or to obtain discounts (46%).

In this first part of the study, we presented the results of desk research on the issue of e-commerce in order to approach the knowledge base of the analysed issues. Having done so, it is appropriate to undertake a deeper examination of the selected contexts.

2 Methodology

The aim of the study is to present partial qualitative results of currently implemented comprehensive research into selected characteristics of consumer behaviour and its changes during the first COVID-19 lockdown. This follows up on the methodology that we designed and subsequently implemented as part of research into the user activity of customers of SMEs [21]. The research was carried out in 2015. For the purpose of describing changes in consumer behaviour trends, it is used as a reference for the presented analysis.

The first step in the methodology used is to define the object of research, in this case the five largest e-commerce companies operating on the Czech market according Journal E-Commerce Bridge 2019 rankings [20]. The journal ranked the subjects on the basis of traffic on their websites. In the case of the e-shops, traffic together with turnover is the most relevant indicator of activity.

From the point of view of the research file, we focused on Czech customers who buy through e-shops. As for the research sample, it consisted of customers who are also fans of the official profiles of the selected e-commerce companies on the social network Facebook. In terms of numbers, it concerned a sample of 1,594,180 users.

From the point of the observed period, we collected data during the first COVID-19 lockdown from 12th March to 17th May 2020. Due to the fact that in the observed period e-shops represented one of the few or only alternatives for the purchase of goods and services, as most traditional infrastructure was closed, it can be said that model conditions for complex research were created.

From the point of view of the monitored parameters, the main focus was monitoring the timing and the number of posts, subsequently the number and types of interactions in which the users responded to the posts. In other words, we monitored the B2C Facebook e-marketing communication activities of the selected subjects and the C2B feedback (Like, Comment and/or Share) in predefined time frames. For the purpose of data unification and with the aim of in-depth analysis, a time frame of 24 hours was set for subtracting the frequency of interactions per published post on the subject's profile. After this time, we subtracted the final number for each of the observed posts interactions. It is known from empirical practice that the dominant number of interactions is recorded immediately after the publication of the post. Subsequently, post durability is reduced, as it is replaced or literally displaced by new contributions and content. The findings are therefore formulated as the number of interactions in relation to posts published on a particular day in a model week, although a particular interaction may have occurred within 24 hours of publication. As mentioned above, the dominant number of interactions occurs immediately; in any case, it is possible to view this methodological sequence as a partial limitation of the study. Using available resources and taking into account the quantity of data, we considered this limitation acceptable.

In this phase of the research, the data were evaluated by descriptive statistics, the basic trends visualized, compared with the reference values of research carried out in the past, and subsequently discussed in order to identify the basic qualitative findings for subsequent in-depth research. It is necessary to note that for the purposes of the research, the subjects were not compared with each other, as the aim of the research was not to optimize the processes of a particular company, but to design industry-wide indicators through relevant companies.

During the observed period from 12th March to 17th May 2020, we analysed hundreds of thousands of users' interactions and then outlined specific characteristics of consumer behaviour suitable for further research. The research provided comprehensive findings, of which we selected specific ones resulting from the implemented analyses, in order to contribute to the creation of a knowledge base of a qualitative nature, which should help to expand knowledge in the field of innovation in management, crisis communication and marketing approaches.

3 Results and Discussion

The specific situation created a model environment for examining the specifics of consumer behaviour. As previously stated in the introduction, e-commerce brings companies a wide range of advantages over traditional forms of trading. This includes, among other things, the advantage of feedback in real time. Academics and practitioners alike are being more and more confronted with the fact that the value of a company no longer lies in its economic indicators, but in its benefits for society. Customers transform themselves from nameless consumers into active followers, into communities that have their own dynamics, express assertive feedback, and largely determine the value of an organization through their attitudes, especially if it concerns a publicly traded organization. Examples of such organizations are the showcases of Western economies such as Apple or Tesla. At this point, the question has to be asked, how does the collapse of the traditional consumer supply chain, forced by the lockdown of the economy, affect the interaction of these active and assertive communities?

In the following table we present the basic indicators of the activity of the monitored companies, according to the Facebook profiles of the five largest operators of e-commerce services in the Czech Republic [20], namely: Alza.cz, Mall.cz, CZC.cz, Aukro.cz and Lidl ČR on the basis of the accumulated number of posts during the monitored period. These activities are followed up by the basic numbers of user interactions in relation to those activities, as calculated through descriptive statistics. The results are as follows:

Table 1. Indicators of overall activity

	Number of posts (n)	Number of likes (n)	Number of comments (n)	Number of shares (n)	Number of days. (n)
SUM	689	349,315	399,243	72,904	67
Daily Max	18	28,082	96,261	11,559	67
Daily Min	4	391	34	18	67
Daily Average	10	5,214	5,959	1,088	67
Daily Modus	11	-	-	42	67
Daily Median	11	3,556	924	325	67
Total fans 1,594,180					

Source: Authors.

During the monitored period of 67 days, more than 800,000 user interactions were recorded for a total of 689 posts/contributions published on the profiles of subjects. If the total number of fans of the subjects is taken into account, it can be said that the groups are relatively active.

During the monitored period, a total of almost 30,000 likes, 100,000 comments and more than 11,500 shares were cumulatively recorded as a maximum value in one day.

These are relatively high numbers, in particular when taking into consideration that the proportions correspond approximately to the nature and value of the type of interaction as we know them from previous reference research carried out in the past [21]. The following visualizations bring into picture the interactions during the reference model week and model week during the COVID-19 lockdown:

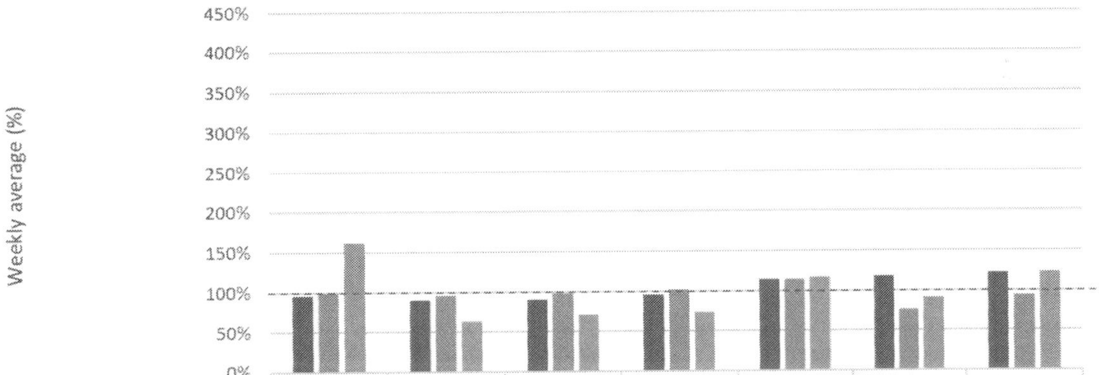

	Monday	Tuesday	Wednesday	Thursday	Friday	Saturday	Sunday
ø Number of likes (%)	96%	90%	90%	96%	114%	118%	122%
ø Number of comments (%)	99%	96%	98%	102%	114%	76%	94%
ø Number of shares (%)	162%	63%	71%	74%	117%	91%	123%

Figure 1. Interactions during the reference model week

Source: [21].

Legend: Number of posts was not monitored during the reference period

As can be seen in Figure 1, in the previous research we identified that, from the point of view of the model week, user interactions were distributed relatively evenly. From the point of view of the maximum interaction "share", we recorded the value of maximum relative abundance on the first day of the working week. In terms of "like" and "comment" interactions, the values during the week oscillated around the average. Above-average values were subsequently recorded at the end of the working week and during the weekend. With a certain degree of abstraction, it can be stated that, under normal conditions, the activity of users at the beginning and end of the working week is partly related to the gradual increase or decrease in work commitment. Subsequently, during the weekend, users focus their online activities on B2C communication in its simplest form, specifically by liking marketing communications on the fan pages of the companies.

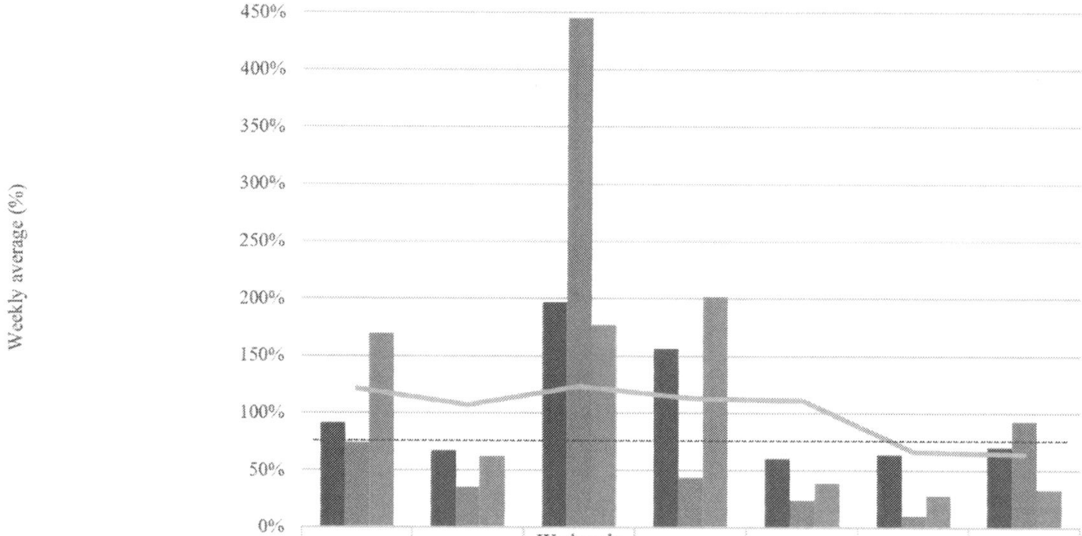

	Monday	Tuesday	Wednesda y	Thursday	Friday	Saturday	Sunday
ø Number of likes (%)	91%	67%	197%	156%	60%	64%	70%
ø Number of comments (%)	74%	35%	445%	44%	24%	11%	93%
ø Number of shares (%)	170%	62%	177%	202%	39%	28%	33%
ø Number of posts (%)	121%	107%	123%	113%	111%	66%	64%

Figure 2. Interactions during the model week - COVID-19 lockdown

Source: Authors.

Figure 2 shows selected characteristics of consumer behaviour in the form of user group interactions in relation to the marketing communications of the monitored e-shops. The recorded data was classified during the observed period on the basis of the days of the week, in order to model the cumulative relative numbers of the individual groups of user interactions over time. On the x-axis are both B2C communications in the form of the number of contributions/posts of the companies, as well as the feedback in the form of interactions to these posts, i.e. the C2B communications of customers with the e-shops. On the y-axis is an indicator in the form of a weekly average. It presents the arithmetic mean of the individual cumulative frequencies of that interaction during the observed period. For better orientation, the weekly average of interactions is marked with a dashed blue line.

As previously mentioned, these are the partial results of a comprehensive study. As a result, in the discussion of the findings, due to the scope limitations of the contribution, the focus is only on the selected characteristics. A closer examination of how the sample behaved during lockdown shows that on the first working day of the week, a relatively major increase in the number of marketing communications from the companies was recorded. In terms of numbers, this is a 21% increase compared to the measured average. In terms of customer interactions, the average relatively cumulative number of interactions in the form of likes and comments is only below average. On the other hand, interactions in the form of shares, which on the basis of our research are evaluated as the most valuable, achieve a value significantly above-average. Specifically, this is the third highest value indicated during the model week. As it is a specific interaction traditionally caused by the marketing activity of a company, it can be partly attributed to the increased Monday activity of the companies, which keep their customers engaged in the form of various competitions and games. At the same time, it is known that this user activity is not enforceable in terms of expected numbers. Therefore, its abundance can be affected by both character and timing. It can be deduced from the data that users are looking for a way to monetize their "efforts" after the weekend. Simple interactions in the form of comments and likes are likely to be replaced by more targeted online activity. How this is reflected in their work commitment and performance is questionable. The second working day of the week has the character of below-average indicators on the part of both companies and their customers. Each of the monitored parameters gradually, in some cases significantly, decreases. It is possible to assume users increased focus on their work and activities outside the online environment. However, to confirm this it is necessary to examine user interactions in terms of hours within the model day. The third working day of the model week during the lockdown, is characterized by a large increase in user interactions to only an above-average number of posts on the profiles of the monitored e-shops. The indicator "comments" achieves a value significantly above-average. It is possible to assume that users use this form of interaction for the purpose of socialization in the middle of the working week. It is also possible to assume that this indicator has a cumulative nature from the point of view of trends, when in the days preceding and following the monitored parameter, the monitored period is cut. Despite the absence of more complex data, it is therefore possible to indicate the trend of so-called Wednesday comments. We see a similar, although not so significant, trend in the case of interactions in the form of likes. Thursday is marked by the second highest average activity of the companies and well above-average activity of users in the form of likes and shares. The sharing parameter reaches its weekly maximum, where it exceeds the weekly average by more than two times. After Wednesday's maximum, the comment parameter records only trace values at the level of 44% of the weekly average. The sharing parameter reaches its weekly maximum. The question of labour productivity during the model lockdown week therefore opens up again, which is something we will address as part of our continuous research into the issue in the near future. On Friday, user interactions fall sharply for all the monitored parameters, whereby only the likes parameter reaches a level higher than 50% of its weekly average. At the same time, from the point of view of positions with the character of marketing communications published on the profiles of the monitored companies, this is an above-average saturated day. Due to other measured values, it is possible to classify this day in terms of "weekend" interactions, whereby users are most likely to dedicate their time to life outside the online world. Respectively, their stay in the online space is greatly reduced. The mean for the monitored indicators is below-average for Saturday and Sunday. Saturday in particular is marked as an “offline” day, where both the companies and their customers have their attention drawn in a different direction. From the point of view of the values measured before and during the pandemic, it is possible to express only statements in the form of research questions. Specifically, how exactly did the pandemic affect e-consumer behaviour? How negatively did the lockdown affect labour productivity? Were users distracted by activities unrelated to their work during home office? If so, to what extent? Our past research [21] indicates certain patterns of behaviour, but in the case of the presented research, the measured values for almost all parameters significantly exceed the original values, even though they basically confirm the trends we indicated from the pre-crisis period. The weekend is preferred for activities outside the online world, while the working week is the preferred time for various aspects and activities of e-consumer behaviour. The individual days of the week are significantly specific in terms of the expected interactions.

4 Conclusion

The global pandemic has severely disrupted traditional supplier-customer ties. During the 67 days of first lockdown caused by the new coronavirus, the market's attention shifted from one day to the next from offline to online. With the closure of brick-and-mortar stores (excluding stores selling daily consumer products and medicines), both providers and consumers of goods and services were forced to significantly modify their usual patterns of behaviour. It is a big surprise that the supply chain did not experience any significant problems within our sample. The best players on the market were able to adapt to the new conditions relatively quickly, which customers appreciated, and which was reflected in the considerable number of positive interactions. Consumer behaviour, and the specifics thereof, have also adapted to the situation. Past practice and empirical studies [22-24] have shown that some behavioural adaptations were temporary and consumer behaviour has reverted to its original state, but that other behavioural adaptations have become the new norm. It is therefore necessary to prepare for this challenge, to apply innovative procedures and methods to the usual routines and to confront this non-standard challenge. Evolutionary theory teaches us that the most adaptable, not necessarily the best will survive. As stated at the beginning of the study, the observed period created a model situation for our research. We analysed hundreds of thousands of users' interactions and then outlined specific characteristics of e-consumer behaviour suitable for further research. The selected findings resulting from the implemented analyses contribute to the creation of a knowledge base of a qualitative nature, which should help to expand knowledge in the field of innovation in management, crisis communication and marketing approaches. From the point of view of the limitations of the presented study, it is necessary to point out the fact that despite the significantly representative sample, the data presented in this partial study of the currently implemented comprehensive research are only of a qualitative nature. The findings and trends indicated and discussed in the study are based mainly on abstraction because there is still no relevant data that we can use as basic values for the pre- and post-crisis period. As part of the continuous research, we want to focus on monitoring trends during the second wave of COVID-19 and related measures with a lockdown nature, as well as analysing data in the coming period of economic recovery after the pandemic. Only then will it be possible to evaluate the research in terms of quantitative data, and thus formulate relevant findings for science and practice. In time, the ambition of the authors is to contribute to the development of a qualitative knowledge base for continuous research in the field.

Acknowledgement

This research was funded by the Technology Agency of the Czech Republic, Programme of ETA, Project Reg. No. TL 0200215- Digital Transformation for Business Model Innovations in Small and Medium Sized Enterprises in the Czech Republic.

References

1. D. Chaffey, *Digital marketing*. London: Pearson (2015)
2. T. Mazzarov, *SMEs engagement with e-commerce, e-business and e-marketing* [online]. Available at: https://doi.org/10.1080/13215906.2015.1018400 (2015)
3. R. Madleňák, L. Madleňáková, *Základy elektronického obchodovania* [*Basics of e-commerce*]. Žilina: EDIS-vydavateľstvo ŽU (2016)
4. S. Luo, K. P. Tsang, *How Much of China and World GDP Has The Coronavirus Reduced*? [online]. Available at: https://www.coronavirusandtheeconomy.com/ongoing-research/how-much-china-and-world-gdp-has-coronavirus-reduced (2020)
5. Z. Wu, J. M. McGoogan, Characteristics of and important lessons from the coronavirus disease 2019 (COVID-19) outbreak in China: summary of a report of 72 314 cases from the Chinese Center for Disease Control and Prevention. *Jama*. **323**(13), 1239–1242 (2020)
6. World Health Organization, *WHO Coronavirus Disease (COVID-19) Dashboard* [online]. Available at: https://covid19.who.int (2020)
7. A. Khurana, *Ebusiness vs. Ecommerce – Understanding the Difference: Ebusiness vs. Ecommerce* [online]. Available at: https://www.thebalance.com/ebusiness-vs-ecommerce-1141573 (2016)
8. O. L. Turcu, A. Feraru, *E-business – A Contemporary Realistic Form Of Successful Business*. Economics Edition, "Vasile Alecsandri" University of Bacau, Faculty of Economic Sciences (2013)
9. D. Benešová, *E-business v službách* [E-business in services]. Bratislava: Vydavateľstvo Ekonóm (2010)
10. M. Bodiš, Procesy elektronického obchodu [E-commerce processes]. *Trendy v podnikaní.* **4**(2), 4–10 (2014)
11. R. Rajib, *Learning more ecommerce and online business*. Morrisville: Lulu Press, Inc. (2016)
12. R. Čiarnienė, G. Stankevičiutė, Theoretical Framework of E-Business Competitiveness. Kaunas. *Procedia – Social and Behavioral Sciences*. **213**, 734–739 (2015)
13. K. Bystrická, J. Jesenič, M. Tužinský, *Internetový obchod* [*Eshop*]. 2nd ed. Bratislava: Wolters Kluwer (2016)
14. A. Brzozowska, D. Bubel, E-business as a new trend in the economy. International Conference of Communication, Management and Information Technology. *Procedia Computer Science*. **65**, 1095–1104 (2015)
15. Y. Prihastomo, A. N. Hidayanto, H. Prabowo, M. Widjaja, The Key Success Factors in E-marketplace Implementation: A Systematic Literature Review. *2018 International Conference on Information Management and Technology (ICIMTech)* (2018)
16. L. Gála, J. Pour, P. Toman, *Podniková informatika: počítačové aplikace v podnikové a mezipodnikové praxi, technologie informačních systémů, řízení a rozvoj podnikové informatiky* [*Business informatics: computer applications in business and intercompany practice, information systems technology, management and development of business informatics*]. Praha: Grada Publishing, a.s., (2006)
17. P. Michl, *Studie APEK: M-commerce v Česku roste, přes mobil nakupují nejčastěji mladí* [*APEK study: M-commerce is growing in the Czech Republic, young people most often shop via mobile phones*] [online]. Available at: https://www.focus-age.cz/m-journal/aktuality/studie-apek--m-commerce-v-cesku-roste--pres-mobil-nakupuji-nejcasteji-mladi__s288x14062.html (2018)
18. S. Steiger, *Studie: M-Commerce überholt Desktop-Shopping bis 2023* [Study: M-Commerce will overtake desktop shopping by 2023]. [online]. Available at: https://www.ibusiness.de/akzuell/db/248674steiger_2.html (2019)
19. D. P. Clement, *Study Finds 80% of Consumers Shop on Smartphones, M-Commerce Gains* [online]. Available at: https://multichannelmerchant.com/ecommerce/study-finds-80-consumers-shop-smartphones/ (2019)
20. Ecommerce Bridge, *The largest e-shops in Slovakia and in the world in 2019* [online]. Available at: https://www.ecommercebridge.sk/najvacsie-e-shopy/ (2020)
21. F. Pollák, P. Dorčák, The effective use of facebook by small and medium-sized enterprises operating in Slovakia. *Market*. **28**(1), 79–91 (2016)
22. P. Dorčák, P. Markovič, F. Pollák, Multifactor analysis of online reputation as a tool for enhancing competitiveness of subjects from automotive industry. *Journal of Economics*. **65**(2), 173–186 (2017)
23. F. Pollák, P. Dorčák, P. Markovič, Reputation Management, *Promotion and Marketing Communications*, Umut Ayman and Anıl Kemal Kaya, London: IntechOpen. [online]. Available from: https://www.intechopen.com/books/promotion-and-marketing-communications/reputation-management (2019)

24. J. Straková, P. Pártlová, J. Váchal, M. Vochozka, Excellent top manager system (ETMS) - Quality management tool tertiary education. *Proceedings of the 29th International Business Information Management Association Conference - Education Excellence and Innovation Management through Vision 2020: From Regional Development Sustainability to Global Economic Growth*. 1844–1852 (2017)

Innovations in the Management of E-Commerce Entities Operating on the Slovak Market – Analysis of Customer Interactions during the COVID-19 Pandemic

Frantisek Pollak[1,*], *Bozka* Malinak[2], *Peter* Markovic[3], *Robert* Vilagi[3]

[1]Institute of Technology and Business, Faculty of Corporate Strategy, Okružní 517/10, 37001 České Budějovice, Czech Republic
[2]Acuity Engineering & Consulting Service Ltd, 10327 122 St NW, T5N1M1 Edmonton AB, Canada
[3]University of Economics in Bratislava, Faculty of Business Management, Department of Business Finance, Dolnozemská cesta 1/b, 852 35 Bratislava, Slovakia

Abstract. This paper deals with the issue of innovation in e-commerce, specifically with the application of innovative methods of evaluating customer interactions in the environment of the virtual social network Facebook. The first part presents the results of desk research on the issue of the utilization of virtual social media Facebook in e-commerce in order to approach the knowledge base of the analysed issues. The aim of the second part of the study presented in the paper was to describe and evaluate the activities of customer groups of the five largest e-commerce entities operating in the Slovak virtual market. The data needed for the analysis were collected during the state of emergency declared in connection with the COVID-19 pandemic. From the point of view of findings, the initial trends in the customers interactions were declining online activity over the weekend and an increase in online activity at the beginning of the working week. In general, it can be stated that data were analysed in a highly non-standard market situation. With respect to the social media e-communication of selected e-commerce entities, it was possible to outline specific patterns of consumer behaviour. As a result, it was possible to create a qualitative knowledge base for further research into the impact of the COVID-19 pandemic on national and international economies, for both economics and management sciences.
Keywords: facebook; e-commerce; COVID-19; customer interactions; customer behaviour; lockdown; Slovakia; pandemic

1 Introduction

For almost two decades, literature [1-5] has provided estimates and predictions of the possible impact of innovations, especially e-commerce on the global market. We are confronted with concepts where traditional stores turn into a form of showroom, respectively even further in the predictions, that sales will exceed those in traditional bricks and mortar stores [6]. The customer therefore gets the space to create a product according to their ideas and according to their needs. This method gives the customer a sense of product and even brand development. Businesses therefore obtain data for the needs of targeted communication with their target groups. Based on customer behaviour, businesses then analyse customers habits, ways of preferred presentation and taste. Entrepreneurship through customer care is part of modern business strategy textbooks, where online is no longer only recommended, but rather required [7]. At the end of 2019, it was inconceivable that bricks and mortar stores could be temporarily closed by the state. The unprecedented situation of the first half of 2020 literally cleared the market in an unthinkable way. As a result, small and medium-sized companies have been faced with the difficult task of finding new ways of communicating with customers, as well as new ways of selling goods and providing services. Online business today does not fill the position of an ancillary service, but under the pressure of circumstances it has become a necessary part of the tools for the basic survival of companies across the economy. Traditional companies built on history-proven foundations and forms of sales no longer have a choice. The situation surrounding the COVID-19 pandemic has also alerted traditionalists and staunch opponents of social networks. However, it should be borne in mind that this situation is not the result of market formation or the global market economy, but an externality, the black swam that only few risk managers anticipated. From a financial point of view, the term reserve fund has taken first place in the value ranking of corporate strategy. Once a term, envisaged only by multinational companies, it has become an essential part of the financial management of small and medium-sized companies. Creating a financial reserve to cover short-term outages is also not the answer. The pandemic has created a new chapter in global risk management that is moving into the microeconomy in response to customer retention, taking into account real-time government actions. The level of preparedness of companies for unexpected circumstances will be verified only by the fact itself. If we look at the repeatability of similar externalities, we find that we have a decimal sinusoid here. This fact

* Corresponding author: frantisek.pollak@mail.vstecb.cz

creates the possibility of a model of preparation for a given situation, although it should be noted that each pandemic brings new inputs to economic analysis. At present, consumers are placing greater emphasis on online communication at the level of their own comfort, whereby interactions result only from everyday needs. However, in a time of pandemic, companies are literally forced to take measures that do not come as market impulses, but rather as state regulations. All of these acts in business management function as a catalyst for innovation. Within this context, even the concept of corporate assets has become multidimensional. As part of our study, we will look at the issue of customer groups and their interactions, as a generator of additional value and competitiveness. The aim of this work is to use innovative approaches to examine and describe selected characteristics of customer behaviour in a sharply non-standard environment. The first part presents the results of desk research on the issue of the utilization of virtual social media Facebook in e-commerce in order to approach the knowledge base of the analysed issues. The aim of the second part of the study presented in this paper is to describe and evaluate the activities of customer groups of the five largest e-commerce entities operating in the Slovak virtual market during the state of emergency declared in the Slovak Republic from March to June 2020. The study itself created the opportunity to identify possible patterns of behaviour based on a sample of almost half a million customers. By seeking to create qualitative knowledge for future in-depth research into the impact of the pandemic on both national and transnational economies, the authors hope to contribute to a shift in knowledge in both economic and managerial sciences.

1.1 Basic definition of key concepts

1.1.1 Concept of social networks

At the end of the 1970s, researchers Holland and Leinhardt [8] described the starting points for social network research on the basis of individual links between various actors. Terms such as social and opinion clusters and equilibrating processes in social networks defined the basic preconditions for in-depth research Australian professor Garry Robins [9] noted that social network research is conceptually different due to its unique theoretical imperative rather than its collection of methods. It is possible to fully identify with this statement, especially if we consider the claim that network structure affects collective outcomes, and locations within networks affect actor outcomes. American researchers [10] subsequently define virtual social networks as web services, which enable individuals to:

- create a public profile within a bounded system;
- articulate a list of other users with whom they share a connection;
- view and traverse their list of connections within the system.

The communication element is highlighted in Macek's definition [11], according to whom virtual social networks are egocentric communication platforms, in which user profiles through which users can view profiles of other users and communicate with them, take the central position. Virtual social networks are based on a combination of different methods of communication, where sharing is the most important one. It is a public space, which allows users to fulfil social norms, express their opinions publicly, learn from reactions of others and give events and statements the value of reality by confirming them as witnesses [12]. The concept of a social network can be viewed from two perspectives. The first perspective is a part of sociology, the second is as a part of the Internet. From a sociological perspective, a social network using a sociological dictionary is defined as follows: A social network is a set of entities connected by exchange relationships. Entities form the node points of the network, while relationships are expressed by connecting lines between these points. A social network can therefore be understood as interconnected groups of people integrating and influencing each other, while the sphere of their interactions includes varied content based on common experiences, interests, religion or racial affiliation [13]. From the perspective of the Internet, a social network is defined as an Internet service that allows registered users to create and edit a personal/corporate public/non-public profile and to use it to communicate with other users, share various kinds of information, photos, videos, chat, and perform other activities [14-15]. Virtual social networks go far beyond technology and media. It is one of the most prominent sociocultural phenomena of this decade [16]. By developing new types of everyday interactions, which implies the possibility of new types of relationships, social websites like Facebook, MySpace and LinkedIn have fundamentally changed our lives and work style and the way we connect with others [17-18]. As individuals, we have two sources of personal competitive advantage: human capital and social capital. Human capital includes talent, intellect, charisma and formal authority, which is important for success, but cannot often be controlled directly. On the other hand, social capital is derived from our relationships. Research shows [10] that due to the transfer of their network to the web, people are more capable and more efficient in assembling and using social capital. Consciously or unconsciously, they use Facebook and LinkedIn, for example, as a tool for maximizing their social capital from relationships:

- private information – frequent, informal communication which happens on social networks, contains private information. The emotional relationship between individuals on a social network is transferred to their offline relationships and therefore increases the likelihood of information exchange;

- varying offer of skills - HR managers, recruiters and others can simply access LinkedIn or Facebook to search the profiles of and view the activities of those members who have the corresponding skills required and contact them directly;
- energy and attention of others - instead of filling their own network with spam in the form of bulk mail, social network members can passively offer opportunities on their profiles in the form of status and therefore have the approaching parties articulate their interest themselves.

1.1.2 Concept of e-communication

Virtual social networks are based on a combination of different methods of communication, where sharing is the most important one. It is a public space, which allows users to fulfil social norms, express their opinions publicly, learn from the reactions of others and give events and statements the value of reality by confirming them as witnesses [10]. The most important advantage of online social networks is social capital, which is a result of the formation and maintenance of interpersonal relationships [19]. From another perspective, the main reason for using online social networks is keeping in touch with friends and acquaintances. This is also true of our memories. Four of the most common reasons for users are contact with friends, planning with friends, organizing events and getting advice and recommendations. Companies and other institutions may use online social networks to, for example, inform about interesting events because sending invitations to interesting events is one of their basic functionalities [20]. Social networks are also an effective tool for public relations. If users are offered an interesting topic, they can very efficiently spread all types of petitions and similar initiatives through social networks. Social networks also enables companies to bring brands closer to their customers, increase their popularity and strengthen their image [21]. The functionality of social networks enables users to add their favourite brands to their personal profiles. Foreign research suggests that fans of brands on online social networks are much more likely to buy a brand or recommended it to their friends [22-23]. Marketers need to be where their customers and potential customers are, which is, more and more frequently, on social networking sites [24]. Sites such as Facebook have hundreds of millions of active users. A billion minutes are spent on Facebook every day. Social networks are a rapidly growing global phenomenon that extends across all continents [17].

2 Methodology

As mentioned in the introduction, e-commerce is increasingly becoming an integral part of the core business of most businesses operating in an increasingly globalized market. Under normal circumstances, the transition to e-commerce is a continuous process, which is necessary in the long run, but which is not yet conditioning the very existence of companies in the short term. However, what if a large part of the traditional B2C market suddenly collapses? How will the absence or a significant reduction in the supply of bricks and mortar stores complicate the already extremely difficult situation of managing an online reputation? On 15th March 2020, a state of emergency was declared in the Slovak Republic in connection with a pandemic caused by a new coronavirus. A lockdown saw bricks and mortar stores close and a significant slowdown in the economy. This hitherto unthinkable state created specific conditions in which it was possible to observe on a daily basis highly non-standard situations, which created an environment for research in almost all areas of scientific research.

The aim of this work is to use innovative approaches to examine and describe selected characteristics of customer behaviour in a strictly non-standard environment. The first part presents the results of desk research on the issue of the utilization of virtual social media Facebook in e-commerce in order to approach the knowledge base of the analysed issues. The aim of the second part of the study presented in this paper is to describe and evaluate the activities of customer groups of the five largest e-commerce entities operating in the Slovak virtual market [25] during the state of emergency declared in the Slovak Republic from March to June 2020.

In the period from 15th March to 13th June 2020, the interactions of customer groups were systematically recorded on a daily basis in the environment of the virtual social network Facebook.

The customer groups in question were fans of the profiles of the five largest e-commerce operators in the Slovak market.

For the purpose of the analysis, the interactions of these customer groups are represented by the interactions of the individuals to contributions that have a marketing communication nature, and which the subject / profile manager of the subject published during the observed period on the official profile of the monitored company.

The reactions took the following forms:

- Like;
- Comment;
- Share.

The collected data were classified, evaluated by means of descriptive statistics and then graphically interpreted in order to illustrate the key findings. The analysis process as well as the results thereof are subsequently discussed.

The study itself created the opportunity to identify possible patterns of behaviour based on a sample of almost half a million customers of the five largest e-commerce operators in the Slovak Republic. By seeking to create qualitative knowledge for future in-depth research into the impact of the pandemic on both national and transnational economies, the authors hope to contribute to a shift in knowledge in both economic and managerial sciences.

3 Results and Discussion

The specific situation created a model environment for examining the specifics of customer behaviour. As we stated in the introduction of this article, social media communication, as an integral part of e-commerce, gives companies a wide range of advantages over traditional forms of trading. This includes, among other things, the advantage of real-time feedback. Academics and practitioners alike are increasingly being confronted by the fact that the value of a company no longer lies in its economic indicators, but in its benefits to society.

Customers have transformed themselves from nameless consumers into active followers, into communities that have their own dynamics, express assertive feedback, and largely determine the value of an organization through their attitudes, especially if it is a publicly traded organization. Examples are the showcases of western economies such as Microsoft or Google. At this point, we must ask the question, how does the collapse of the traditional consumer supply chain, forced by the lockdown of the economy, affect the interactions of these active and assertive communities?

In the following table we present the basic indicators of the activities of the monitored entities on the basis of the accumulated number of posts on the Facebook profiles of the five largest operators of e-commerce services in the Slovak Republic, namely: Alza.sk, Mall.sk, nay.sk, Martinus.sk and Bonprix SR. In addition, the basic numbers of user interactions in relation to those activities were calculated using descriptive statistics. The results are as follows:

Table 1. Indicators of overall activity

	Number of posts (n)	Number of likes (n)	Number of comments (n)	Number of shares (n)	Number of days. (n)
SUM	668	171,138	59,684	8,795	91
Daily Max	14	6,497	5,854	1,152	91
Daily Min	3	52	0	3	91
Daily Average	7	1,881	656	97	91
Daily Modus	9	910	488	6	91
Daily Median	7	1,016	358	29	91
Total fans approx. 500,000*					

Legend: Bonprix had a geographically allocated profile which displayed only the total number of fans globally for all geographic locations together, i.e. more than six million fans. For our purposes, we estimated the number of fans proportionally and added this to the measured number of fan groups of the other monitored entities. The total number is therefore only an approximate estimated value.

Source: Authors.

In the monitored period of 91 days, we recorded almost a quarter of a million user interactions for a total of 668 posts/contributions published on the profiles of the subjects. If the total number of fans of the subjects is taken into account, it is possible to state that the user groups are relatively active. During the period under review, a total of almost 6,500 likes, 6,000 comments and more than 1,100 hundred shares were cumulatively recorded as a maximum value in one day. These are moderately high numbers when compared to the nature and value of the type of interactions identified in reference research carried out in the past [19].

A more detailed look at the numbers reveals that even among the best, there are still entities that do not communicate with their customer groups on a regular daily basis, while regular and targeted communication is essential for both creating and subsequently building relationships in the online environment. At the same time, it is possible to identify that from the point of view of user interactions there is room for improvement. In the case of the more active e-commerce entities, user responses were initiated by regular activities of an active marketing communication nature, such as competitions.

On the other hand, a significant percentage of posts received only few or no responses from customer groups. It should be noted that marketing activities, unless it is an activity of a social marketing nature, which do not instigate a customer response, are either incorrectly implemented or insufficiently prepared. This represents a wasted opportunity and an inefficient use of resources. Especially in turbulent times, it is essential to maximize efforts towards efficiency, as it is then that active and efficient competitors can deprive others of long-established market advantage. The following figure illustrates the user interactions during a model week:

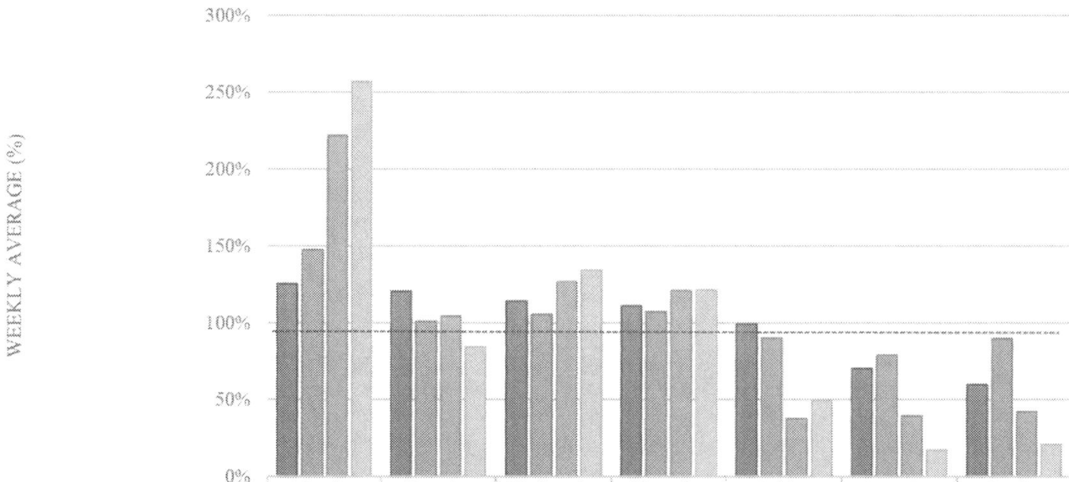

	Monday	Tuesday	Wednesday	Thursday	Friday	Saturday	Sunday
ø Number of posts (%)	125%	121%	114%	111%	99%	70%	60%
ø Number of likes (%)	148%	101%	105%	107%	90%	79%	90%
ø Number of comments (%)	222%	105%	127%	121%	38%	39%	42%
ø Number of shares (%)	257%	84%	134%	121%	50%	17%	21%

Figure 1. Interactions during a model week

Source: Authors.

The visualization presents user interactions from the perspective of a model week. The x-axis shows the types of the monitored interactions according to the day of the week. The y-axis shows the average cumulative frequencies for the observed interactions, whereby the dashed blue line indicates the level of the mean values with respect to the daily average calculated and presented in Table 1.

As can be seen from the graph of the model week, in terms of expected interactions to posts, the most numerous are on Mondays, which exceeds the daily averages for each of the monitored parameters. Of particular interest is the parameter "shares", which reaches 257% of the average weekly value. From the point of view of the parameter "shares", the least numerous is Saturday, which reaches just 17% of the average daily value of the parameter. In other words, from the point of view of the average relative frequency, shares on interactions posted on Saturday are 83% below the average of a normal week. From the point of view of the no less valuable indicator of interactions "comments", Monday is once again the best. In terms of average relative numbers, this is a little more than twice as powerful as the average daily value of the model week. In terms of ranking, Monday is followed by Wednesday and Thursday. This indicator has been identified in reference research in the past, so this is a recurring trend [19]. From the point of view of the "like" interaction, the most numerous Monday is followed by Thursday and Wednesday. In general, it is possible to observe the trend of a gradual decline in interactions at the end of the week, which can be described as positive with a significant dose of induction. The reason for this is that it may indicate that even during a pandemic, the model week retains the characteristics of the pre-pandemic period, whereby the arrival of the weekend sees the activity of users on social networks gradually decrease. This is especially visible when comparing Mondays and weekends, where the difference is enormous. As previously mentioned, this trend is also evident in previous research.

It is the shared opinion of the authors that there is a conflict with regards to work efficiency and time management at the beginning of the new week. Nevertheless, the model week in the pandemic and lockdown retains the parameters of the model week before the pandemic, which from the point of view of user preferences and habits can be evaluated as a positive phenomenon. At the same time, as can be seen in Figure 1, the activity of the subjects also declines with the arrival of the weekend, with the subjects conducting little more than half of the activity in terms of the number of posts compared to the weekly average.

Based on the findings presented, it can be stated that the values measured create a qualitative knowledge base for deeper quantitative research. The trends indicated in the discussion correlate with the trends identified in past research. The validity of the correlation will be tested in follow-up research.

4 Conclusion

The specific conditions under which we carried out our analysis created almost a model situation for the fulfilment of the research aims. The investigation of the interactions of selected e-commerce entities provides opportunities for future comprehensive quantitative research. The e-commerce market has passed a stress test of unprecedented proportions. Almost overnight, this market had to replace the entire supply of traditional stores, which were forced to close their operations. The enormous pressure on the supply chain tested the stability of the system, whereby problems of an operational nature, but with strategic implications, were solved in real time. The supply and distribution of goods to customers, as well as customer service, are all physical aspects of trading on the Internet. From the point of view of the sample examined, it can be concluded that the processes were handled well. In general, it is possible to observe the trend of a gradual decline of interactions at the end of the week, which can be described as positive with a significant dose of induction. The reason for this is that it may indicate that even during a pandemic, the model week retains the characteristics of the pre-pandemic period, whereby the arrival of the weekend sees the activity of users on social networks gradually decrease. In the monitored period of 91 days, we recorded more than a quarter a million user interactions for a total of 668 posts published on the profiles of the subjects. If the total number of fans of the subjects is taken into account, it is possible to state that the user groups are relatively active. Active feedback not only brings opportunities in the form of real-time user information, but also a significant threat in the form of reputational risk [26-29], which can easily damage a brand. This is particularly true in the case of uncontrollable variables created by an atypical market situation. In this case, the managements of the e-commerce entities examined in our study can be rated positively. During the period under review, no significant crisis situation was noted in terms of user feedback in connection with the pandemic and the market situation. From the point of view of recorded interactions, it is possible to state that despite the stutters at the beginning of the lockdown, with increasing days, the processes were optimized in the desired direction. Positive interactions were maintained by the subjects in the form of competitions, whereas negative interactions mostly related to the longer delivery times for orders. An interesting innovation in marketing communication was applied by the Martinus internet bookstore, which initiated a relatively large group of followers to actively comment on their contributions through a game involving a superhero. We rate this more than positively, as we believe that innovation is the main driver of modern marketing management [30]. This also supports what Petr Drucker claims, that entrepreneurship has only two basic functions: marketing and innovation. Marketing and innovation produce results, everything else is a cost. At the end of this study, it can be stated that through the analysis of the data of the interactions of customer groups in a highly non-standard situation, it was possible to outline specific patterns of consumer behaviour, thereby creating a qualitative knowledge base for further research into the impact of the COVID-19 pandemic on national and international economies, for both economics and management sciences. With regards to the limitations of the research, the relatively limited sample may be an issue from the point of view of the object of research, i.e. the five largest subjects of e-commerce in the Slovak market. It is also considered necessary to carry out a further analysis in order to compare the period monitored with a similar period a year after the end of the pandemic.

Acknowledgement

This article is one of the partial outputs of the currently solved research grant VEGA no. 1/0240/20

References

1. R. Delina, M. Tkáč, Trust Building Mechanisms for Electronic Business Networks and Their Relation to eSkills, World Academy of Science. *Engineering and Technology*. **6**(71), 380–390 (2010)
2. M. Maryška, P. Doucek, R. Kunstova, The Importance of ICT Sector and ICT university education for the economic development. *3rd international conference on new horizons in education – INTE 2012*. **55**, 1060–1068 (2012)
3. F. Reichheld, *The Ultimate Question 2.0 (Revised and Expanded Edition): How Net Promoter Companies Thrive in a Customer-Driven World*. Brighton: Harvard Business Review Press (2011)
4. M. A. Stec, P. B. Filip, M. Grzebyk, A. Pierscieniak, Socio-economic development in EU member states - concept and classifications. *Engineering Economics*. **25**(5), (2014)
5. J. Straková, I. Rajiani, P. Pártlová, J. Váchal, J. Dobrovič, Use of the value chain in the process of generating a sustainable business strategy on the example of manufacturing and industrial enterprisescompanies in the Czech Republic. *Sustainability (Switzerland)*. **12**(4), 1520 (2020)
6. M. Fertik, D. Thomson, *Wild West 2.0: How to Protect and Restore Your Reputation on the Untamed Social Frontier*. New York: Amacom (2010)
7. J. Lee, S.-J. Kim, I. Kwon, Corporate social responsibility as a strategic means to attract foreign investment: Evidence from Korea. *Sustainability*. **9**(11), 2121 (2017)
8. P. W. Holland, S. Leinhardt, *Perspectives on Social Network Research*. Cambridge: Academic Press (1979)

https://doi.org/10.1051/shsconf/20219001016

9. G. L. Robins, *Doing Social Network Research: Network-based Research Design for Social Scientists*. London: SAGE Publications Ltd. (2015)

10. M. Boyd, N. Ellison. Social Network Sites, Definition, History, and Scholarship. *Journal of Computer - Mediated Communication*. **13**(1) (2007)

11. J. Macek, *Úvod do nových médií [Introduction to New Media]*. Ostrava, Ostrava: Ostravská univerzita v Ostravě (2011)

12. V. Bednář, V. *Internetová publicistika [Internet Journalism]*. Praha: Grand Publishing (2011)

13. M. Petrusek, A. Vodáková, *Velký sociologický slovník [Large sociological dictionary]*. Praha: Karolinum (1996)

14. H. Kulhánková, J. Čamek, *Fenomén facebook [Facebook Phenomenon]*. Kladno, Kladno: Jakub Čamek (2010)

15. B. Wierzbinski, B. Communication and information sharing in the process of creating market advantage among small and medium enterprises collaboration. *Hradec Economic Days 2014*. Hradec Králové: Gaudeamus, 437–445 (2014)

16. N. T. Saruc, P. Dorčák, F. Pollák, E-business and its Application in Conditions of Central European Market. *QIP journal*. **17**(1), (2013)

17. C. Shih, *Vydělávajte na Facebooku: Jak využít sociální sítě k oslovení nových zákazníků, vytvoření lepších produktů a zvýšení prodejů [Make money on Facebook: How to use social networks to reach new customers, create better products and increase sales]*. Brno: Computer Press (2010)

18. M. Zeisser, *Unlocking the Exclusive Potential of Social Networks* [online]. Available at: http://www.mckinseyquarterly.com/Unlocking_the_elusive_potential_of_social_networks_2623 (2013)

19. F. Pollák, P. Dorčák, The effective use of facebook by small and medium-sized enterprises operating in Slovakia. *Market - Tržište*. **28**(1), 79–91(2016)

20. M. Karlíček, P. Král, *Marketingová komunikace: jak komunikovat na našem trhu [Marketing communication: how to communicate in our market]*. Praha: Grada Publishing (2011)

21. J. Soviar, Simplification of Marketing Scheme for Business Start-Ups. *Communications - Scientific Letters of the University of Zilina*, **13**(4), 55–57 (2011)

22. M. Zamazalová, *Marketing obchodní firmy* [Business marketing]. Prague: Grada Publishing (2009)

23. R. Delina, Transparency in Electronic Business Negotiations - Evidence Based Aanalysis. *Quality Innovation Prosperity*. **8**(2), 79–89 (2014)

24. F. Sudzina, P. Kmec, The technological paradox and evaluation of the benefits of informatization. *Journal of Economics*. **54**(3), 281–293 (2006)

25. Ecommerce Bridge. *The largest e-shops in Slovakia and in the world in 2019* [online]. Available at: https://www.ecommercebridge.sk/najvacsie-e-shopy (2020)

26. P. Dorčák, P. Markovič, F. Pollák, Multifactor analysis of online reputation as a tool for enhancing competitiveness of subjects from automotive industry. *Journal of Economics*. **65**(2), 173–186 (2017)

27. F. Pollák, P. Dorčák, P. Markovič. *Reputation Management, Promotion and Marketing Communications*, U. Ayman, A. K. Kaya, London: IntechOpen. [online]. Available at: https://www.intechopen.com/books/promotion-and-marketing-communications/reputation-management (2019)

28. J. Vodak, M. Kubina, J. Soviar, D. Zraková, Online Reputation in Automotive. *18th International Scientific Conference on LOGI. Ceske Budejovice: MATEC Web of Conferences*. **134**(1), (2017).

29. D. Zrakova, M. Kubina, G. Koman, Influence of information-communication system to reputation management of a company. *12th International Scientific Conference of Young Scientists on Sustainable, Modern and Safe Transport. Procedia Engineering*. **192**, 1000–1005 (2017)

30. J. Straková, P. Pártlová, J. Váchal, M. Vochozka, Excellent top manager system (ETMS) - Quality management tool tertiary education. *Proceedings of the 29th International Business Information Management Association Conference - Education Excellence and Innovation Management through Vision 2020: From Regional Development Sustainability to Global Economic Growth*. 1844–1852 (2017)

Small and Medium-Sized Enterprises in Rural Areas of the Czech Republic and Their Approach to Management Auditing

Daniel Raušer [1,*]

[1]Institute of Technology and Business, Faculty of Corporate Strategy, Nemanická 436/7, 37010 České Budějovice, Czech Republic

Abstract. Regarding small and medium-sized enterprises (SMEs) and their conditions in the present business environment, it is determinative for them to be prosperous and ambitious. In order to achieve their defined goals, such enterprises are advised to conduct periodical marketing environmental analyses and various audits. However, an audit process that examines overall enterprise management and measures its performance, i.e. management auditing, tends to be overlooked at times or mistaken for other activities. Related research involving Czech SMEs located in rural areas and operating in diverse industries discovered that management audit process capabilities, specifically in terms of organization and resources, are not fully utilized resulting in the enterprises' lower competitiveness. In connection with broader assessment of enterprise management systems, the objective of this paper consists in introducing a particular sub-model titled "Comprehensive and Individual Resource Analysis", which finalizes the previously developed and published "Two-Stage 7S Evaluation Model" [1]. Implementing the eventually completed model of management auditing on a regular basis would allow the SMEs to remove faults (and raise efficiencies) in their management processes, and enhance their enterprise value in the domestic and international markets.
Keywords: management auditing; small and medium-sized enterprises; rural areas; comprehensive and individual resource analysis; management system evaluation

1 Introduction

Nowadays, perhaps all business enterprises, especially small and medium-sized enterprises (SMEs) based outside urban areas, aim for sustainability and profitability supplemented by taking competitive advantages over other enterprises. The most significant aspect of such advantages may be seen, for instance, in thorough knowledge of enterprise environment and customer needs and wishes [2], an efficient use of enterprise management systems [3], internal stability [4], a quality workforce [5], or quality marketing and advertising [6]. Also, enterprises should make regular analyses of their internal environment and that is where management auditing may be used as a pivotal instrument. Furthermore, the main effort of enterprises (including the above SMEs) should reside in continuous improvement of their management systems. That implies recognizing the systems' current state (with their positives and negatives) and possibly designing new or modified systems, which may be viewed as the primary content of the management audit [7]. As well as that, it should be an independent and objective review that assists enterprises in reaching their targets by presenting a systematic, methodical approach that would inspect and help to boost the efficiency of management processes. Moreover, management auditing contributes to discovering problematic area(s) within enterprises, reducing their effective management [8], and supplies enterprise managers with information on the style, quality and manner of their management [9]. However, a number of research papers have shown that mostly SMEs have not been using management auditing in the same frequency and intensity, as opposed to other audits (e.g. environmental, financial, personnel, sales, tax audits, etc.), and the content and purpose of management audits have often been mistaken. Moorthy et al. [10], Botez [11], Aschauer et all. [12] and Ma´ayan, Carmeli [13] all state that management audits tend to be either omitted or replaced with internal audits in relation to audited areas (such as management, production, marketing and finance). Hrušková [14] claims that management auditing is seldom performed by SMEs, and is even being confused with marketing auditing due to the use of identical analyses (e.g. SWOT analysis, the IFE matrix and the EFE matrix) in order to assess strengths and weaknesses. Vágner [15] points to some confusion between management auditing and internal controlling given the similarity of provided services (consulting, information and advisory activities or methodologies on suitable enterprise management). Nevertheless, management as such, should indeed be regarded as the essential area in terms of management auditing. Depending on their discretion, however, managers or auditors may then concentrate on other areas, e.g. human resources, marketing, production, finances and IT systems, as advised by Wheelen, Hunger [4]. In the opinion of Majernik et al. [16], there should also be a focus on quality and sales, with special attention being drawn to risk management as

* Corresponding author: raudan@mail.vstecb.cz

part of the management orientation. The same viewpoint is held by Stojanovic et al. [17] and Rivera et al. [18]. Other authors, e.g. Ujwary-Gil [19], Pascual, Kumar [20] note that certain efficiency of used resources is to be ultimately considered as well.

2 Methodology

An electronic questionnaire survey, structured interviews and model development were used in this paper. The survey (and the related data processing) was carried out between September 2017 and December 2019 and involved only SMEs that are situated in rural areas of the Czech Republic and operate in various industries (e.g. agriculture, automotive, building, engineering, hospitality, textiles, etc.). At the author's request, an alphabetical list of such SMEs officially registered in each of the country's 14 administrative regions had been sent to him from the Association of Small and Medium-Sized Enterprises and Crafts CZ. Since the vast majority of micro-enterprises (with up to 10 employees) do not conduct management audits, as has been confirmed by previous research [1], they were automatically excluded from the survey. Subsequently, enterprises, whose contact details could be found online, were approached via email and members of their management were asked to fill in an electronic questionnaire, comprising three questions and three follow-up questions, and submit their answers afterwards. The aim was to find out whether: a) the SMEs perform management audits (in their true sense, i.e. assessment of the current state of enterprise management systems) and/or other audits (and if they do, then which?); b) the SMEs develop and/or employ any aids (analyses, models, etc.) for assessing their management systems and/or available resources (and if they do, then which?); c) the SMEs consider themselves to be competitive (and if they do, then whether and how do they succeed in keeping up their competitive strategy or advantages?). Next, the survey was supplemented by structured interviews with a group of 12 auditing professionals (with 10+ years of experience) that the author has collaborated with before. The purpose was to learn about their supposed knowledge of management auditing and potentially gather some more valuable facts or details about the issue. As for model development, the primary objective was to extend the already compiled "Two-Stage 7S Evaluation Model" [1] by proposing an additional sub-model (titled "Comprehensive and Individual Resource Analysis"), which would constitute the third stage in management audit implementation and would be a further action connected with two preceding sub-models. To be brief, principles of the first stage are included in the "7S Adaption" sub-model (as based on the 7S McKinsey Framework), i.e. a proposed analysis of enterprise management from the standpoint of efficiency and effectiveness. Here, managers or auditors are able to detect strengths/weaknesses in view of the given enterprise management. That is followed by principles of the second stage incorporated in the "Application IFE - 7S Matrix" sub-model (as based on the IFE Matrix), i.e. an expanded enterprise management analysis linked to assessing specific evaluation factors or their indicators with regard to internal stability within the entire enterprise management. Here, it may be determined whether strengths prevail over weaknesses or vice versa. Additionally, since resources are integral to enterprises and their operations and may facilitate an improvement in their market competitiveness, the third stage principle (reflected in the intended sub-model) is a further analysis (evaluation) of enterprises in relation to their resource utilization systems, hence being the extension of the above Evaluation Model (and specifically referring to the following characteristics: "Factors – Systems – Evaluation Indicators – System of using company (enterprise) resources – Evaluation Questions – Does the enterprise use an optimal combination of its resources?"). Having been completed, the entire model could possibly be presented as a "Three-Stage Evaluation Model of Enterprise Management", and on its examining, managers themselves may ultimately decide on the model's applicability and appropriateness for the respective enterprises and their related activities.

3 Results and Discussion

The number of approached SMEs in rural areas of the Czech Republic eventually totalled 1066. Surprisingly, and contrary to the author's expectations, a considerable number (862, i.e. 80,86%) of them participated in the survey by responding to the above questions. Time-consuming data processing was therefore required, yet it conduced to generate a fairly representative sample being of a highly informative nature, where the highest response (return) rate was reported for SMEs in the Prague Region (17,2%), the Liberec Region (14,8%) and the Karlovy Vary Region (13,6%), whereas the lowest response (return) rate was reported for SMEs in the Ústí Region (4,3%), the Moravia-Silesia Region (2,7%) and the Vysočina Region (1,5%). Considering the actual results, the survey found that only 126 (14,6%) of the SMEs (hereinafter referred to as "enterprises") carry out management audits as such. The survey, as well as the interviews, also proved that 394 (i.e. 45,7%) out of 862 enterprise managers, and 4 out of 12 auditors respectively, do not frequently distinguish management audits from several other audits. To further add to this, the managers perceive the management audit's concept as that of a personnel audit or an audit of managerial characteristics and skills, or even exchange it for an internal audit, whereas the auditors proffer the management audit as a personnel audit when dealing with different enterprises. Moreover, despite the fact that 342 enterprises (i.e. 39,6%) do have analyses of the current state of their management systems prepared and implemented, the results are not summarized, but are assessed rather individually, which does not obviously provide the enterprises with comprehensive evaluations. Specifically, the analyses include SWOT analysis, balanced scorecards, process analysis and internal benchmarking. In spite of that, quite a large number of the managers (449, i.e. 52%) stated that every so often the enterprises managed by them struggled to maintain their competitiveness, while not completely using their available resources, which affects the functioning and market position of the enterprises. Lastly, and perhaps surprisingly, the survey clearly indicated that 96 managers (i.e. 28%) learnt about the management audit for the first time only from the paper's author, and were not even aware of its content or purpose. Conversely, the highest number of performed enterprise audits involved mandatory audits, particularly tax audits, followed by financial, personnel, cost, quality and internal audits. Aside from that, it should be added that the data collected were contributory factors for the author in his preparing the considered resource sub-model (and finishing the full model) and verified the validity of his choice of their development.

As regards the survey results together with findings from the interviews, it is recommended that (not only) the above enterprises conduct management auditing on a regular basis. That may be effected by the aforementioned "Two-Stage 7S Evaluation Model" that has been generated as the main framework in terms of using a suitable systemic approach as well as a standardized procedure, thereby allowing management audit implementation. However, any attempt to do so should largely depend on the willingness of the enterprises, or members of their management, and also on making a goal-directed effort and taking a problem-solving perspective to try and tackle the current situation in a particular enterprise. First, in order for the model to be genuinely applicable, and to fulfill the essence of management auditing (i.e. to evaluate the present state of enterprise management systems) as well as the systemic approach principle, the author believes that basic managerial questions presented in Table 1 are to be acknowledged, answered and discussed by enterprise managers who should realize which areas need to be analyzed as part of management auditing. Additionally, answering the questions is the first step (and simultaneously the basis) of the entire model and should be helpful in guiding the managers (and individual enterprises) to correct answers or solutions. Apart from that, the managerial questions may serve as precautions not to leave out any factors in view of sustainability and market success (for more details, see Píchová, Raušer, [1]).

Table 1. Managerial questions and factors

Managerial Questions	**Factors based on the 7S Model**
What does the enterprise want to achieve and how does it want to achieve it?	Strategy
What staff and composition are needed in the enterprise?	Staff
What abilities and skills does every employee need to have?	Skills
What will be the hierarchical structure of selected employees?	Structure
Which style of management should be used in the enterprise?	Style
What systems are necessary to be used in the enterprise?	Systems
Are all employees well aware of the enterprise events and happenings?	Shared values

Source: Author.

Then, having answered the questions and evaluated the first stage as well as the second stage results, managers may consequently advance to the third stage, i.e. an analysis (or assessment) of enterprise resources (tangible, intangible, financial, human). That is embodied by the following resource sub-model, i.e. the "Comprehensive and Individual Resource Analysis", whose first section includes comprehensive assessment of resources (Table 2), whilst the second section concerns individual assessment of resources (Table 4), with both sections holding the same importance.

Table 2. Comprehensive assessment of resources

Category	**Category Significance: expressed by 1, 2 or 3**	**Evaluation Indicators (Relevance): expressed by *, +, x, - or 0**			
		Tangible	**Intangible**	**Financial**	**Human**
Resource Originality					
Resource Availability					
Resource Sustainability					
Resource Flexibility					
Resource Substitutability					
Resource Importance					
Comprehensive Assessment	- - -	Aggregate Score	Aggregate Score	Aggregate Score	Aggregate Score

Note 1: Category Significance is evaluated as follows:
1 = high significance; 2 = average significance; 3 = low significance.

Note 2: Evaluation Indicators are assigned according to their relevance, where:
* denotes very high originality, very easy availability, very simple sustainability, very high flexibility, very easy substitutability, very high importance;
+ denotes high originality, easy availability, simple sustainability, high flexibility, easy substitutability, high importance;
x denotes low originality, difficult availability, difficult sustainability, poor flexibility, difficult substitutability, low importance;
- denotes very low originality, very difficult availability, very difficult sustainability, very poor flexibility, very difficult substitutability, very low importance;
0 denotes zero (i.e. no) originality, availability, sustainability, flexibility, substitutability, importance.

Source: Author.

The table demonstrates that the comprehensive section encompasses evaluations of resources regarding their originality, availability, sustainability, flexibility, substitutability and importance (the categories generated by the author). Initially, each category is evaluated in view of its significance for a certain enterprise (see Note 1), which is followed by assessment of given resources in relation to the evaluation indicators (see Note 2). Eventually, there is a summary of results respecting the following simple formula: *Significance x Relevance*

As a result, the above assessment may therefore determine whether resources the enterprise makes use of appear to be relevant or significant for it or not. From such a viewpoint, the author outlines the classification of enterprise resources into three types (classes) depending on their total number of gained points (or scores), as illustrated in Table 3.

Table 3. Scores of comprehensive assessment of resources

Significance		Relevance		Category	Comprehensive Assessment			
					Tally of Aggregate Scores			
Evaluation	Points	Symbols	Points		Tangible	Intangible	Financial	Human
1	3	*	4	Originality				
2	2	+	3	Availability				
3	1	x	2	Sustainability				
		-	1	Flexibility				
		0	0	Substitutability				
				Resource Importance				
Classification of Resources to Classes I. – III.								
Eventual Comprehensive Assessment								
Resource Classification Classes				Categorization of Classes	Total Points			
I.				Crucial Resource	72 - 49			
II.				Influential Resource	48 - 25			
III.				Substitutable Resource	24 - 0			

Source: Author.

To be more specific, the first type, i.e. the crucial resource class, represents a resource (resources) vital for continuous operations of the enterprise. The resource(s) are utilized in an adequate amount and form, cannot be substituted by any other resource(s), and manager(s) of the enterprise are strongly recommended to retain the resource(s) in the long term. The second type, i.e. the influential resource class, contains a resource (resources) which have a great effect on running the enterprise, but which are not extremely important. The resource(s) are utilized only to a certain extent, cannot always be fully involved in enterprise activities, and manager(s) of the enterprise are advised to consider either maintaining the resource(s) in their existing amount and form, or decreasing their amount, or even altering their present form. Lastly, the third type, i.e. the substitutable resource class, includes a resource (resources) that are likely to be useful for the enterprise, yet they may not be so necessary and hence could be replaced. Furthermore, since assessing resources in terms of their relevance and significance is anticipated to be subjective from the manager's/managers' perspective, the author suggests a two-step evaluation in that respect. The first step could be seen as assessment of a comprehensive resource analysis performed individually by each manager of the enterprise (and auditors), whereas the second step could reside in assessing the comprehensive analysis carried out in mutual brainstorming.

Finally, when taking account of the second section, i.e. individual assessment of resources, the author proposes that related parameters stated in Table 4 are allowed for tangible, intangible, financial and human resources within the enterprise. The parameters only exemplify perhaps the most elementary features while conducting the analysis, and may be naturally and gradually amended in accordance with conditions or demands given by managers and/or auditors. However, it should be generally accepted for the resources and their corresponding parameters to be observed and evaluated throughout the entire process.

Table 4: Individual assessment of resources

Resources	**Parameters**
Tangible	Age of assets and machinery; Flexibility of assets and machinery; Technical conditions of assets and machinery; Reliability of machinery; Production capacity of machinery; Material provision.
Intangible	Innovative development – patent or trademark development; The level of generated know-how; The level of image among customers and the general public; The level of information security; The level of software security used.
Financial	Balance of capital structure used – in terms of ownership and time; Trends – profit, cash flow, receivables and payables; Trends – ratios of financial analysis and indicators based on cash-flow, used methods/styles of financing.
Human	Number of employees – more detailed allocations to workers, administrative staff and managers; Employee training and qualifications, experience, fluctuation, motivation preferences; Trust in employees; The level of delegating authority to subordinate workers.
Overall Assessment	a) Assessment of Composition and Structure of Resources; b) Assessment of Balance and Combination of Resources; c) Assessment of Intensity of Using Resources.

Source: Author.

In addition to that, it should be noted that (during the entire analysis) managers/auditors ought to be aware of all resources introduced into production as inputs and focus on their appropriate allocation. Concentrating only on resources the enterprise possesses (which may referred to as resources in the narrow sense) would not be desirable. Thus, similar attention should also be drawn to resources that cannot be possessed (e.g. relations with employees, customers, suppliers, creditors or business partners). They may be regarded as resources in the broader sense since they can become necessary and significant elements within the scope of the enterprise´s advantages over competitors provided that such resources are efficiently utilized and mutually complemented, which should be under observation as well.

3 Conclusion

Business enterprise planning should be directed towards sustainable development conditioned, inter alia, by internal stability and efficient use of enterprise management systems. In order to determine their current form, management auditing may serve as a valuable assessment indicator. However, as emerged from the extensive electronic inquiry related to small and medium-sized enterprises (SMEs) in rural areas of the Czech Republic, such an approach to auditing either fails to be noticed or is substituted for divergent types of audits. It also transpired that accessible resources used within the SMEs do not reach their full potential, which interferes with enterprise operations and overall performance. In their effort to rectify the issues, (not only) the SMEs may implement management auditing according to the systemic procedure specified in this paper, where the "Comprehensive and Individual Resource Analysis" sub-model was put forward. It complements the formerly introduced "Two-Stage 7S Evaluation Model" by representing the third stage, and therefore meets the objective of this paper. Although each stage may be performed separately and more frequently, the completed model enables regular application with regard to assessing the current state of enterprise management, helping to increase competitiveness of enterprises as well as their position in the domestic and international markets, and raising awareness of management auditing among managers and other professionally engaged individuals.

References

1. R. Píchová, D. Raušer, Management audit as a tool for managing small and medium-sized enterprises. *Proceedings of the 7th International Conference on Management on "Managerial trends in the development of enterprises in globalization era"*. (2017)
2. M. E. Porter, *Competitive advantage: Creating and Sustaining Superior Performance*. New York: Free Press (1985)
3. P. F. Drucker, *Management challenges for the 21st century*. New York: Harper Business (1999)
4. T. L. Wheelen, D. J. Hunger, *Strategic management and business policy: concepts and cases*. 13th ed. Upper Saddle River: Prentice Hal (2012)
5. M. Armstrong, *Armstrong's handbook of strategic human resource management*. 6th ed. London: Kogan Page (2016)
6. P. T. Kotler, K. L Keller, *Marketing Management*. 15th ed. Boston: Pearson (2016)
7. P. K. H. Spencer, *Internal Auditor at Work*. Hoboken: John Wiley (2004)
8. J. Truneček, *Interní manažerský audit [*Internal management audit]. Praha: Professional Publishing (2004)

9. A. Chambers, G. Rand, *The Operational Auditing Handbook: Auditing Business and IT Processes*. 2nd ed. Hoboken: John Wiley (2010)

10. M. K. Moorthy, et al., The impact of information technology on internal auditing. *African Journal of Business Management*. **5**, 3523–3539 (2011)

11. D. Botez, Internal Audit and Management Entity. *Procedia Economics and Finance*. **3**, 1156–1160 (2012)

12. E. Aschauer, A. Moro, M. Massaro, The auditor as a change agent for SMEs: the role of confidence, trust and identification. *Review of Managerial Science*. **95**, 339–360 (2015)

13. Y. Ma´ayan, A. Carmeli, Internal audits as a source of ethical behaviour, efficiency and effectiveness in work units. *Journal of Business Ethics*. **137**, 347–363 (2015)

14. D. Hrušková, Marketing audit and its impact on business performance. *Conference on Current Problems of the Corporate Sector*. 219–223 (2015)

15. L. Vágner, Controlling as an international economic tool. *Proceedings of the 15th International Scientific Conference on Globalization and its Socio-Economic Consequences*. 829–833 (2015)

16. M. Majerník, N. Daneshjo, R. K. Stofkova, G. Sanciova, Integrated auditing of enterprise management systems in a globalized environment. *Proceedings of the 16th International Scientific Conference on Globalization and its Socio-Economic Consequences*. 1261–1269 (2016)

17. T. Stojanovic, M. Andric, Internal auditing and risk management in corporations. *Strategic management*, **21**, 31–42 (2016)

18. D. E. Rivera, A. S. Villar, M. R. M. Pino, Towards the Concept of Risk Approach in the audit integrated management system. *Quality - Access to Success*, **17**, 41–47 (2016)

19. A. Ujwary-Gil, The concept and stages of the audit of intangible resources: A network approach. *Proceedings of the 17th European Conference on Knowledge Management*. 916–925 (2016)

20. D. G. Pascual, U. Kumar, *Maintenance Audits Handbook*. Boca Raton: CRC Press (2016)

The Effect of Entrepreneurial Marketing on the Performance of Small and Medium-sized Enterprises

Gemechu Abdissa Shuremo[1,*], *Csaba* Bálint Illés[1], *Anna* Törőné Dunay[1]

[1]Faculty of Economics and Social Sciences, Szent István University 1 Páter K. Str., Gödöllő, 2100, Hungar

Abstract: Entrepreneurial marketing is a non-conventional marketing technique that provides business owners a method to successfully operate their business with limited resources. The main objective of this study is to therefore investigate whether entrepreneurial marketing can affect the performance of small and medium-sized enterprises operating in Ethiopia, specifically in the city of Jimma. A purposive sampling technique was employed to gather data from 100 business owners that have been on market for more than 7 years. The data were analysed using Minitab and SPSS statistical software packages. The study showed that some elements of entrepreneurial marketing indeed affect the performance of small and medium-sized enterprises. Innovation orientation and customer intensity have a positive relationship with the performance of small and medium-sized enterprises and are statistically significant, while risk-taking orientation has a negative relationship and is statistically significant. It is therefore recommended that business owners apply the concept of entrepreneurial marketing to leverage their limited resources to cope with competitors and maintain the sustainability of their business.

Keywords: entrepreneurial marketing; sustainability; performance; small and medium-sized enterprises

1 Introduction

The concept of marketing indicates how to recognize consumer wants and to match the marketing activities with other functional areas of a company in order to achieve an organizational objective through the fulfilment of customer needs. The concept of marketing is a widely studied field. However, since studies have been mainly focused on large industries, it is difficult to apply their findings directly to small and medium-sized enterprises. According to Chitrakar [1], the marketing activities employed by small and medium-sized enterprises are different from the conventional concept of marketing.

Researchers have suggested a new concept of marketing which best matches the characteristics of small and medium-sized enterprises. According to Stokes [2], entrepreneurial marketing is a holistic concept that best fits the marketing approaches of small and medium-sized enterprises. According to Becherer et al. [3], entrepreneurial marketing is a blend of entrepreneurship and marketing. Fink et al.[4], state that entrepreneurial marketing contemplates market orientation and entrepreneurial behaviours; it is an amalgamation of entrepreneurship and marketing disciplines.

Entrepreneurial marketing is a scientific concept that observes an entrepreneur ability in identifying business problems and exploiting current opportunities [5]. According to Morris et al. [6], entrepreneurial marketing is an opportunity driven and opportunity seeking way of thinking and acting. The author indicates that entrepreneurial marketing is a marketing discipline linked to entrepreneurial creative thinking and the utilization of an opportunity through innovative marketing activities. Entrepreneurial marketing is focused on creating a value proposition for current and potential customers and other stakeholders [7].

This study addresses the question of whether entrepreneurial marketing affects the performance of small and medium-sized enterprises operating in Ethiopia. Poor management, resource constraint, insufficient profit, low demand for products and services are some of the problems that small and medium-sized enterprises are facing today. Even though this study argues that entrepreneurial marketing has an impact on both large and small companies, the concept is more applicable to the problems of small and medium-sized enterprises with regards to

*Corresponding author: gemechu.mtu@gmail.com

https://doi.org/10.1051/shsconf/20219001018

determining their sustainability. Because entrepreneurial marketing focuses on new approaches to business solutions, resource leverage, risk management and value creation, it is very much dependent on the company's circumstances [8].

The subsequent parts of this contribution are structured as follows: overview of the concept of entrepreneurial marketing; extensive literature review of the variables used in this study; outline of the methodology and data used; results and discussion; conclusions and implications for further research.

2 Literature review

Entrepreneurship and marketing are traditionally considered as two different academic disciplines. Nowadays, entrepreneurial marketing is a separate discipline an entrepreneur can practice as a best business solution for small companies. Entrepreneurial marketing has been recognized and accepted not only by marketing and entrepreneurship disciplines, but also by companies looking for a competitive advantage (Schulte & Eggers, 2010). An entrepreneur can employ marketing activities in different forms, which is not the same as with conventional marketing theory [9].

According to Morrish et al. [10], entrepreneurial marketing is the process of regular assessment of the environment to find or establish new opportunities and exploiting the occasion effectively to renew competitive advantage. Organizations that implement entrepreneurial marketing are pre-emptive and creative in taking the lead or attracting customers rather than responding to customer needs [11]. Subsequently, entrepreneurial marketing enables companies to focus more attention on new market areas rather than providing services to current ones.

Some studies of entrepreneurial marketing claim that the lack of a conventional marketing strategy in small and medium-sized enterprises is considered an absence of marketing practice as a whole in such companies [12]. There is a basic difference between the traditional concept of marketing practiced by large companies and the new marketing structure of small and medium-sized enterprise [9]. As mentioned in the studies of Schindehutte et al. [11], the entrepreneurial marketing structure consists of seven elements. An overview of the seven elements is presented below.

Table 1. Entrepreneurial marketing and the tasks of entrepreneurship

Elements of entrepreneurial marketing	Entrepreneurial marketing as opportunity creation and discovery	Entrepreneurial marketing as opportunity assessment	Entrepreneurial marketing as opportunity exploitation
Customer intensity	Customer intensity helps create cognitive models that seek out opportunities that satisfy customers in unique ways	Customer intensity forces "customer value propositions" evaluations to be paramount in entrepreneurial opportunity assessment	Customer intensity enhances the ability of a company to exploit entrepreneurial opportunities by focusing on how value can be created for the customer that results in long-term profits for the firm
Value creation	An emphasis on the creation of a superior value proposition will impact the types of opportunities that it creates or discovers	An organizational focus on value creation will force opportunity assessment to consider the customer needs	As an organizational focus on value creation will impact how entrepreneurial opportunities are exploited
Resource leveraging	Dickson & Giglierano [13] notion of an entrepreneurial company attempting to minimize the commitment of its own organization's resources, while leveraging the use of external resources, specifically when resources devoted to R&D and other knowledge creation activities may diminish the company's ability to create or discover additional entrepreneurial opportunities	Resource leveraging may also diminish the company's distinctive capabilities to evaluate complex entrepreneurial opportunities	Resource leveraging may force a company to enter alliances with strategic partners to fully exploit the opportunity. Strategic partnering has many benefits including:(1) lowering project risks;(2) increasing intellectual capital; and (3) providing access to additional funding sources
Risk management	Dickson & Giglierano [13] "missing the boat vs sinking the boat" risk	Companies that are risk-accepting will evaluate opportunities using a risk-	Risk management and acceptance will result in a tendency to exploit innovation

https://doi.org/10.1051/shsconf/20219001018

	assessment suggests that opportunity discovery or creation is in fact attempting to manage risk	adjusted framework and will tend to accept speculative opportunities that offer the potential of building distinctive capabilities	
Innovation	The strategic leveraging of innovation should be the impetus for the discovery and creation of opportunities	Assessment should focus on how the opportunity leverages and builds the organization's innovation capabilities	Exploitation of opportunities should be based on leveraging product, process, administrative, or strategic innovation
Opportunity-driven	Opportunity creation and discovery must be driven by the assessment of the company's concept of strategy	Opportunity evaluation must be based on the strategic relevance of the opportunity	Opportunity exploitation can be based on either strategic on non-strategic motives
Proactive	Proactive companies attempt to create/discover new opportunities that enhance their distinctive capabilities	Proactive companies evaluate entrepreneurial opportunities on their ability to develop future competitive advantage	Proactive companies exploit opportunities in such a way as to enhance the opportunity to be a first product/market/ technology space

Source: Miles & Darroch [7].

2.1 Measurement of small and medium-sized enterprise performance

According to Radipere & Dhliwayo [14], business performance is described as how good an organization is accomplishing its objectives. Company performance can also be explained as the capacity of a company to satisfy the intent of the organization and main stakeholders [15]. The high performance of small and medium-sized enterprises in their business operation is a guarantee for sustainability and helps to exploit global market opportunities [16].

Performance measures have great significance in effective management of an organization and in the enhancement of the processes since only measurable things are manageable. Hence, the enhancement of the organizational performance requires some measurements to determine the impact of the level of organizational effectiveness upon business performance [17]. As they mentioned in their study, Return on Assets (ROA), Return on Equity (ROE), Return on Sales (ROS), Profit Margin (PM), Earnings per Share (EPS), Tobin-Q, Market Value Added (MVA) and Market-to-Book Value (MTBV), while others are not widely used. Therefore, the performance of small and medium-sized enterprises in this study is measured based on sales, customer satisfaction, and the profit of the company compared to other competitors.

3 Methodology

This study focuses on legally registered small and medium-sized enterprises operating in Ethiopia, specifically in the city of Jimma. A purposive sampling technique was employed to gather data from 100 business owners that have been on the market for more than 7 years. The primary data was gathered by self-administered questionnaire adopted from Becherer et al. [18]. To make it easier for respondents to answer the questions and enable an efficient use of statistics for data interpretation, a 5-point Likert scale was used: Strongly Disagree = 1, Disagree = 2, Neutral = 3, Agree = 4, Strongly Agree = 5.

Minitab and Statistical Software package for Social Science (SPSS) version 20 were employed to conduct multivariate statistical analysis (principal component analysis, factor analysis, cluster analysis, discriminant analysis), and other inferential statistics analysis, like multiple linear regression, to see the effects of the independent variables on the companies' performance (dependent variable). The general model for this specific study is as follows:

$$SMEP = \alpha + \beta 1(PA)it + \beta 2(OF)it + \beta 3(RTO)it + \beta 4\,(IO)it + \beta 5(CI)it + \beta 6(RL)it + \beta 7(VC)it + \boldsymbol{\varepsilon it}$$

Where:

SMEP = small and medium-sized enterprise performance (dependent variable); α is a constant term; $\beta 1 - \beta 7$ is coefficients for the respective explanatory variables; PA = Pro-activeness; OF = Opportunity Focused; RTO = Risk-Taking Orientation; IO = Innovation Oriented; CI = Customer Intensity; RL = Resource Leveraging; VC = Value Creation; ε = error term.

4 Results and discussion

In this section, the data gathered from the 100 respondents or company owners were analysed using the Minitab and SPSS statistical software packages. Inferential statistics analysis, like multiple linear regression, was applied to see the effect of the independent variables on the companies' performance (dependent variable), as well as multivariate statistical analysis (principal component analysis, factor analysis, cluster analysis, discriminant analysis).

Regression Equation

SMEP = 1.256- 0.0505 PA - 0.0361 OF - 0.1106 RTO + 0.8724 IO + 0.1239 CI + 0.0616 RL + 0.0073 VC

Table 2. Model summary[b]

Model	R	R Square	Adjusted R Square	Std. error of the estimate	Durbin-Watson
1	.914[a]	.836	.824	.15895	1.534

a. Predictors: (constant), VC, PA, IO, OF, RTO, CI, RL

b. Dependent variable: SMEP

Source: Authors - SPSS output.

As can be seen in Table 2 above, the regression model R square and adjusted R square statistics were 83.6% and 82.4% respectively. The result indicates that the changes in the independent variables explain 82.4% of the changes in the dependent variable. This means that the elements pro-activeness, opportunity focused, risk-taking orientation, innovation oriented, customer intensity, resource leveraging and value creation jointly explain 82.4% of the changes in the performance of small and medium-sized enterprises. The remaining 17.6% of the changes in their performance is therefore explained by other factors which were not included in this study. The result also indicates that all the variables included in this study are a good descriptive variable of the performance of small and medium-sized enterprises.

Table 3. ANOVA[a]

Model		Sum of Squares	df	Mean Square	F	Sig.
	Regression	11.866	7	1.695	67.089	.000[b]
1	Residual	2.324	92	.025		
	Total	14.190	99			

a. Dependent variable: SMEP

b. Predictors: (constant), VC, PA, IO, OF, RTO, CI, RL

Source: Authors - SPSS output.

From the statistical point of view, the ANOVA table shows the overall significance/acceptability of the model under use [19]. As shown in Table 3, the p-value is < 0.05, i.e. 0.000, which indicates that the variant explained by the model is not due to chance. This is also supported by the F-ratio=67.089, which confirms it is very unlikely to happen by chance.

Table 4. Coefficients

Term	Coef	SE Coef	T-Value	P-Value	VIF
Constant	1.256	0.403	3.12	0.002	
PA	-0.0505	0.0354	-1.42	0.158	1.03
OF	-0.0361	0.0746	-0.48	0.630	1.05
RTO	-0.1106	0.0333	-3.32	0.001	1.10
IO	0.8724	0.0478	18.26	0.000	1.17
CI	0.1239	0.0566	2.19	0.031	1.14
RL	0.0616	0.0409	1.51	0.135	1.19
VC	0.0073	0.0468	0.16	0.876	1.09

Source: Authors - Minitab output.

As can be seen in Table 4 above, innovation oriented (IO) has a positive relationship with the dependent variable and is statistically significant at the 1% significance level, i.e. (β=0.872; p-value=0.000). This indicates that as the innovation oriented variable changes, the performance of small and medium-sized enterprises changes in the same direction by 0.872. Risk-taking orientation (RTO) has a negative relationship with the dependent variable and is statistically significant at the 5% significance level, i.e. (β= -0.111; p-value=0.001). This indicates that as the risk-taking orientation variable changes, the performance of small and medium-sized enterprises changes in the opposite direction by 0.111. Customer intensity (CI) has a positive relationship with the dependent variable and is statistically significant at the 5% significance level, i.e. (β=0.124; p-value=0.031). This indicates that as the customer intensity variable changes, the performance of small and medium-sized enterprises changes in the same direction by 0.124.

Table 5. KMO and Bartlett's Test

Kaiser-Meyer-Olkin measure of sampling adequacy.		.515
Bartlett's test of sphericity	Approx. Chi-square	205.976
	df	28
	Sig.	.000

Source: Authors - SPSS output.

KMO measure: the overall measure of sampling adequacy (MSA) for the set of variables included in the analysis was 0.515, which exceeds the minimum requirement of 0.50 for overall MSA. It is therefore possible to conclude that the variables are correlated. The data also satisfy the criteria for appropriateness for factor analysis as the significance of Bartlett's test of sphericity is 0.000, which is lower than 0.05.

Table 6. Total variance explained

Component	Initial Eigenvalues			Extraction Sums of Squared Loadings			Rotation Sums of Squared Loadings		
	Total	% of Variance	Cumulative %	Total	% of Variance	Cumulative %	Total	% of Variance	Cumulative %
1	2.282	28.523	28.523	2.282	28.523	28.523	2.237	27.968	27.968
2	1.230	15.376	43.899	1.230	15.376	43.899	1.274	15.926	43.894
3	1.155	14.437	58.336	1.155	14.437	58.336	1.155	14.442	58.336
4	.994	12.423	70.759						
5	.876	10.953	81.712						
6	.816	10.200	91.912						
7	.557	6.961	98.873						
8	.090	1.127	100.000						

Extraction method: Principal Component Analysis

Source: Authors - SPSS output.

One of the most used techniques for dimension reduction or how many components (factors) to extract is known as Kaiser's criterion, or the eigenvalue rule. Using this rule, only factors with an eigenvalue of 1.0 or more are retained for further investigation. Accordingly, from Table 6 above, factors 1, 2 and 3 have an eigenvalue of 2.282, 1.230 and 1.155, respectively. Therefore, the three artificial components, i.e. Pro-activeness (PA), Opportunity Focused (OF) and Risk-Taking Orientation (RTO), explain 58.3% of the total variance. So, 58.3% of the information should be kept after the dimension reduction of the factors.

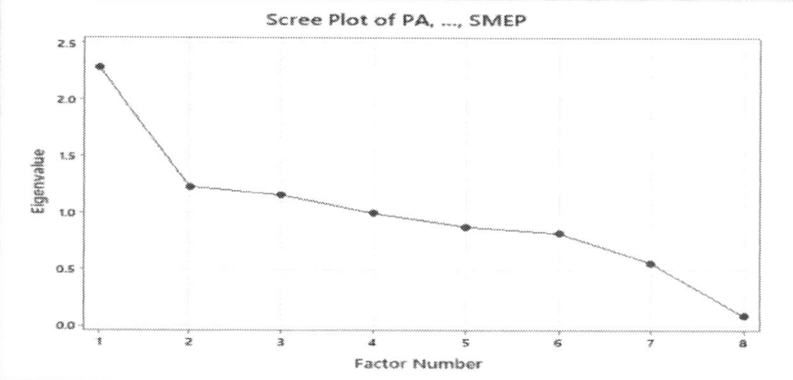

Figure 1. Scree plot of PA… SMEP

Source: Authors - Minitab output.

The scree plot in Figure 1 above graphically shows that factors 1, 2 and 3 have an eigenvalue greater than 1. This indicates that more than 50% of the total variance is explained by the three factors.

Table 7. Unrotated factor loadings and communalities

Variable	Factor1	Factor2	Factor3	Factor4	Factor5	Factor6	Factor7	Factor8	Communality
PA	0.003	0.549	0.012	0.718	0.133	0.383	-0.138	-0.016	1.000
OF	-0.037	0.518	0.439	-0.553	0.137	0.413	0.206	-0.003	1.000
RTO	-0.379	-0.239	-0.529	-0.261	0.528	0.322	-0.260	-0.035	1.000
IO	0.887	-0.129	-0.079	0.056	0.354	0.067	0.131	0.202	1.000
CI	0.368	0.658	-0.197	-0.272	0.024	-0.409	-0.388	0.019	1.000
RL	0.563	-0.277	0.119	-0.142	-0.484	0.435	-0.385	0.008	1.000
VC	-0.132	-0.265	0.789	0.081	0.391	-0.181	-0.312	0.001	1.000
SMEP	0.938	-0.084	-0.005	0.020	0.220	-0.044	0.119	-0.218	1.000
Variance	2.2819	1.2301	1.1550	0.9939	0.8762	0.8160	0.5569	0.0902	8.0000
% Var	0.285	0.154	0.144	0.124	0.110	0.102	0.070	0.011	1.000

Source: Authors - Minitab output.

Table 7 above shows matrix A: loadings and communalities for unrotated factors. Three factors are independent and significant as can be seen in the factors´ loadings (Factor 1, Factor 2 and Factor 3).

Table 8. Rotated factor loadings and communalities

Varimax Rotation

Variable	Factor1	Factor2	Factor3	Factor4	Factor5	Factor6	Factor7	Factor8	Communality
PA	0.001	0.050	-0.997	0.028	0.028	-0.019	-0.049	-0.002	1.000
OF	0.041	0.020	-0.019	0.076	-0.045	-0.995	0.005	-0.000	1.000
RTO	0.098	-0.986	0.052	-0.066	0.055	0.021	-0.087	-0.006	1.000
IO	-0.970	0.003	-0.027	0.053	0.021	0.028	0.123	-0.198	1.000
CI	-0.130	0.067	-0.029	0.980	0.103	-0.080	-0.018	0.004	1.000
RL	-0.216	0.091	0.053	-0.020	0.037	-0.006	0.970	0.003	1.000
VC	0.024	0.054	0.027	-0.099	-0.991	-0.046	-0.035	0.000	1.000
SMEP	-0.937	0.149	0.031	0.132	0.014	0.029	0.165	0.232	1.000
Variance	1.8939	1.0130	1.0024	1.0017	1.0011	1.0006	0.9945	0.0928	8.0000
% Var	0.237	0.127	0.125	0.125	0.125	0.125	0.124	0.012	1.000

Source: Authors - Minitab output.

As is evident from the rotated factor loadings and communalities (Varimax Rotation) in Table 7 above, the communalities are not changed, but there are six significant factors with variances greater than 1. The factor pattern structure has changed after rotation. Factor F1 has remained as is, a variable group, however, factor F2 has become a unique factor for the risk taking orientation (RTO) variable, factor F3 has become a unique factor for the pro-activeness (PA) variable, factor F4 has become a unique factor for the customer intensity (CI) variable, factor F5 has become a unique factor for the value creation (VC) variable, and factor F6 has become a unique factor for the opportunity focused (OP) variable. We can therefore conclude that the pro-activeness, risk taking orientation, customer intensity, value creation and opportunity focused variables are not correlated because the factors are supposed to be independent of one another.

5 Conclusions

The aim of this study was to determine the effect of entrepreneurial marketing on the performance of small and medium-sized enterprises in Ethiopia, specifically in the city of Jimma. Due to several constraints, small and medium-sized enterprises face marketing challenges [20] that force them to find alternative marketing strategies [21], which requires specific impact evaluation of variables [22]. According to the outcomes of the study presented in this contribution, certain elements of entrepreneurial marketing affect the performance of small and medium-sized enterprises, which is similar to the findings of Sherman et al. [23]. Innovation orientation and customer intensity have a positive relationship with the performance of small and medium-sized enterprises and are statistically significant, while risk-taking orientation has a negative relationship and is statistically significant. In general, non-traditional marketing strategies influence the performance of small and medium-sized enterprises.

Reference

1. S. Chitrakar, Market and Entrepreneurial Orientation of Nepalese Small and Medium Enterprises (SMEs). *Synopsis of Ph. D. Thesis Submitted to: Office of Dean Faculty of Management.* (2019)
2. D. Stokes, Putting Entrepreneurship into Marketing: The Processes of Entrepreneurial Marketing. *Journal of Research in Marketing and Entrepreneurship*. **2**(1), 1–16 (2000)
3. R. C. Becherer, P. Haynes, L. Fletcher, Paths to Profitability in Owner-Operated Firms: The Role of Entrepreneurial Marketing. *Journal of Business and Entrepreneurship*. **18**(1), 17 (2006)
4. M. Fink, M. Koller, J. Gartner, A. Floh, R. Harms, Effective entrepreneurial marketing on Facebook – A longitudinal study. *Journal of Business Research*. **113**, 149–157 (2018)
5. E. Hadiyati, Kajian Pendekatan Pemasaran Kewirausahaan dan Kinerja Penjualan Usaha Kecil [Study of Entrepreneurship Marketing Approach and Small Business Sales Performance]. *Jurnal Manajemen Dan Kewirausahaan*. **11**(2), 183–192 (2009)
6. M. H. Morris, M. Schindehutte, R. W. LaForge, Entrepreneurial Marketing: A Construct for Integrating Emerging Entrepreneurship and Marketing Perspectives. *Journal of Marketing Theory and Practice*. **10**(4), 1–19 (2002)
7. M. P. Miles, J. Darroch, Large firms, entrepreneurial marketing processes, and the cycle of competitive advantage. *European Journal of Marketing*. **40**(5–6), 485–501 (2006)
8. R. Schulte, F. Eggers, Entrepreneurial marketing and the role of information - Evidence from young service ventures. *International Journal of Entrepreneurship and Innovation Management*. **11**(1), 56–74 (2010)
9. E. J. Nijssen, *Entrepreneurial Marketing: an effectual approach*. 2nd ed. London and New York: Routledge Taylor & Francis Group (2014)
10. S. C. Morrish, M. P. Miles, J. H. Deacon, Entrepreneurial marketing: Acknowledging the entrepreneur and customer-centric interrelationship. *Journal of Strategic Marketing*. **18**(4), 303–316 (2010)
11. M. Schindehutte, M. H. Morris, A. Kocak, Understanding Market-Driving Behavior: The Role of Entrepreneurship. *Journal of Small Business Management*. **46**(1), 4–26 (2008)
12. S. Kraus, R. Harms, M. Fink, Entrepreneurial marketing: Moving beyond marketing in new ventures. *International Journal of Entrepreneurship and Innovation Management*. **11**(1), 19–34 (2010)
13. P. R. Dickson, J. J. Giglierano, Missing the Boat and Sinking the Boat : A Conceptual Model of Entrepreneurial Risk. *Journal of Marketing*. **50**(3), 58–71 (1986)

14. S. Radipere, S. Dhliwayo, The role of gender and education on small business performance in the South African small enterprise sector. *Mediterranean Journal of Social Sciences.* **5**(9), 104–110 (2014)

15. T. M. Smith, J. S. Reece, The relationship of strategy, fit, productivity, and business performance in a services setting. *Journal of Operations Management.* **17**(2), 145–161 (1999)

16. I. M. Aminu, M. N. M. Shariff, Influence of strategic orientation on SMEs access to finance in Nigeria. *Asian Social Science.* **11**(4), 298–309 (2015)

17. E. M. Al-Matari, A. K. Al-Swidi, F. H. B. Fadziah, The Measurements of Firm Performance's Dimensions. *Asian Journal of Finance & Accounting.* **6**(1), 24–49 (2014)

18. R. C. Becherer, M. M. Helms, J. P. McDonald, The effect of entrepreneurial marketing on outcome goals in SMEs. *New England Journal of Entrepreneurship.* **15**(1), 7–18 (2012)

19. C. Acton, R. L. Miller, J. Maltby, D. A. Fullerton, J. Campling, Analysis of Variance (ANOVA). *SPSS for Social Scientists.* **6**, 183–198 (2009)

20. S. C. Morrish, Entrepreneurial marketing: A strategy for the twenty-first century? *Journal of Research in Marketing and Entrepreneurship.* **13**(2), 110–119 (2011)

21. A. Maritz, H. Frederick, M. Valos, A discursive approach to entrepreneurial marketing: Integrating academic and practice theory. *Small Enterprise Research.* **17**(1), 74–86 (2010)

22. G. S. Mort, J. Weerawardena, P. Liesch, Advancing entrepreneurial marketing: Evidence from born global firms. *European Journal of Marketing.* **46**(3–4), 542–561 (2012)

23. H. Sherman, J. Shuart, L. Weinstein, New England Journal of Entrepreneurship. *New England Journal of Entrepreneurship.* **10**(2), 5 (2007)

Analytic Hierarchy Process in Multiple–Criteria Decision–Making: A Model Example

Anna Siekelova[1,*], *Ivana* Podhorska[1], *Jorma* J. Imppola[2]

[1] University of Zilina, Faculty of Operation and Economics of Transport and Communications, Department of Economics, Univerzitna 8215/1, 010 26 Zilina, Slovakia
[2] Seinäjoki University of Applied Sciences, Business and Culture, P. O. Box 412, FI-65101, Seinäjoki, Finland

Abstract. Managers have to make decisions several times a day. The decision-making process can be defined as an essential activity realized by managers every day. Decisions can be implemented intuitively, or by using relevant decision-making methods. This depends on the nature of the decision, as well as the intensity of its possible future effects. The theory of decision-making can be defined as a relatively young discipline. It can be stated that decision-making is no longer an intuitive process. Most decision-making situations are of a multiple criteria character. In this contribution, the authors focus on multiple-criteria decision-making, to which several methods can be applied. In the practical part, the authors use Saaty's method, also known as the Analytic Hierarchy Process. Saaty is considered to be the most important researcher dealing with the issue of multiple-criteria decision-making. The set multiple-criteria decision-making problem was to choose one business partner out of eight under consideration. The decision-making criteria included selected financial indicators and non-financial criteria. The aim of the contribution is to use the Analytic Hierarchy Process to assess potential business partners and to select an optimal candidate.
Keywords: multiple criteria; decision-making; Analytic Hierarchy Process; AHP

1 Introduction

Decision-making is described as one of the most important activities performed by business managers, and is also referred to as the core of management. The decision-making process can be defined as a set of activities that leads to the resolution of a decision-making problem that involves at least two meaningful alternatives, of which the selected one offers the best result with respect to the set goal, as well as the possibility of its implementation [1]. There are two sides to the decision-making process, namely [2]:

1. *The content side of decision-making* - this reflects the differences of individual decision-making processes.

2. *The formal-logical (procedural) side of decision-making* - this expresses that individual decision-making processes have certain common features, regardless of their content.

The formal-logical side of decision-making is a precondition for claiming that there is a framework procedure for resolving a decision-making situation with different content. It is the basis for decision-making theory, which aims to create appropriate procedures (methods) for solving decision-making situations with the highest degree of universality. One of the biggest issues that decision-making theory deals with is the issue of multiple-criteria decision-making. The beginnings of multiple-criteria decision-making (hereinafter "MCDM") can be seen in Benjamin Franklin's decision-making process. He used a simple paper system for the decision-making process for important issues. On one side of the sheet of paper, he wrote the arguments in favour of a decision; on the other side, he wrote the arguments against the decision. Gradually, he deleted the arguments on each side which were of relatively equal importance. When all the arguments on one side were deleted, the side which had remaining arguments was the side of the argument that he would support. In 1785, Marie-Jean-Antoine-Nicolas de Caritat published *"Essay on the Application of Analysis to the Probability in Majority Decisions"*. The most famous results of this study is the *Condorcet (voting) Paradox*. The development of studies dealing with multiple-criteria decision-making accelerated during the 1980s and early 1990s. Francis Edgeworth, a philosopher and political economist, significantly contributed to the development of statistical methods. He was the first who applied formal mathematical concepts to decision-making. The basis of modern MCDM theory is Pareto-optimality, also known as the concept of efficiency by Vilfredo Pareto. Frank Ramsey published an essay *"Truth and Probability"* dealing with the decision-making process with uncertain outcomes. It was the basis for later published study *"Foundation of Statistics"* by Leonard Savage. Studies by John von Neumann, Oskar Morgenstern, Gerard Debreu, as well as John Nash have become famous. They dealt with utility theory and influenced modern

* Corresponding author: anna.siekelova@fpedas.uniza.sk

economics. Ward Edwards is also known as the father of behavioural decision-making. His study focused on the process of how people make decisions or how they are able to improve the decision-making process. Another author, Herbert Simon, dealt with the rationality of the decision-making process. His study stated that people are not utility "maximisers", rather "satisfiers". They prefer decision-making based on a set of aspirational level criteria [3].

At present, the issue is being addressed by many researchers in various scientific fields. Liao and Wu [4] developed a comprehensive algorithm for multi-expert multiple-criteria decision-making problems, whereby quantitative and qualitative criteria in the form of benefits, costs or target types are considered. In another study, they focus on the integrated method for cognitive complex multi-expert multiple-criteria decision-making based on ELECTRE III with weighted Borda rule [5]. Fu, Chang and Yang [6] deal with multiple-criteria group decision-making based on group satisfaction. As mentioned above, there are a number of fields in which MCDM methods are used. Costache and Bui [7] used MCDM to identify areas prone to flash-flood phenomenon. Haddad, Sanders and Tewkesbury [8] evaluated four global market regions suitable for Boeing.

The definitions of MCDM vary and depend on the author. For the purposes of this contribution, the basic characteristics of multiple-criteria decision-making are considered to be as follows [9]:

- *Multiple criteria nature of the decision-making problem* - the decision maker assesses selected alternatives on the basis of several criteria.
- *Non-additivity of criteria* – the selected criteria are expressed in different units of measure.
- *Mixed set of criteria* – this means that some criteria are quantitative and some qualitative; the decision-maker wants to maximize some criteria (yield criteria) and minimize others (cost criteria).

As was mentioned above, the main characteristic of MCDM is the multiple criteria nature thereof. MCDM problems can be solved only on the basis of quantitative criteria of the yield type. However, although non-additivity and/or a mixed set of criteria are not a condition, these characteristics occur very often. They can be defined as complementary features of MCDM [10].

The general procedure for solving MCDM problems can be divided into three phases, namely:

1. Creation of a decision matrix for the MCDM problem. Formula 1 shows the general decision matrix for a MCDM problem.

$$\begin{array}{c} \\ a_1 \\ a_2 \\ \cdot \\ \cdot \\ a_p \end{array} \begin{array}{c} \begin{array}{ccccc} f_1 & f_2 & \cdot & \cdot & f_k \end{array} \\ \begin{bmatrix} y_{11} & y_{12} & \cdot & \cdot & y_{1k} \\ y_{21} & y_{22} & \cdot & \cdot & y_{2k} \\ \cdot & \cdot & \cdot & \cdot & \cdot \\ \cdot & \cdot & \cdot & \cdot & \cdot \\ y_{p1} & y_{p2} & \cdot & \cdot & y_{pk} \end{bmatrix} \end{array} \quad (1)$$

Where:

$f_1, f_2, \dots f_k$ criteria;

$a_1, a_2, \dots a_p$ alternative solutions;

$y_{11}, \dots y_{pk}$ the values of the alternatives achieved according to the individual criteria.

2. Determination of the weights of the individual criteria
3. Evaluation of alternative solutions and determination of preferential order

As previously stated, there are several methods used within MCDM. The different methods that can be used for determining criteria weights and for evaluating alternative solutions are presented in Figures 1 and 2. The classification was developed by the authors on the basis of published research [11,12].

 https://doi.org/10.1051/shsconf/20219001019

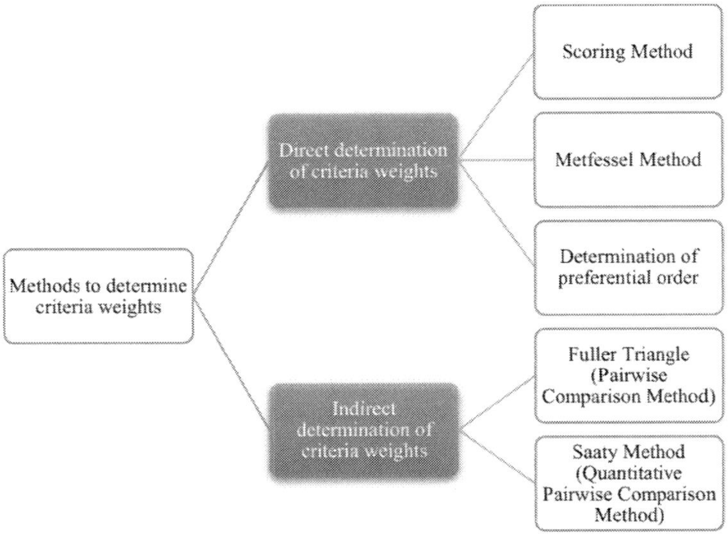

Figure 1. Methods used to determine criteria weights

Source: Authors.

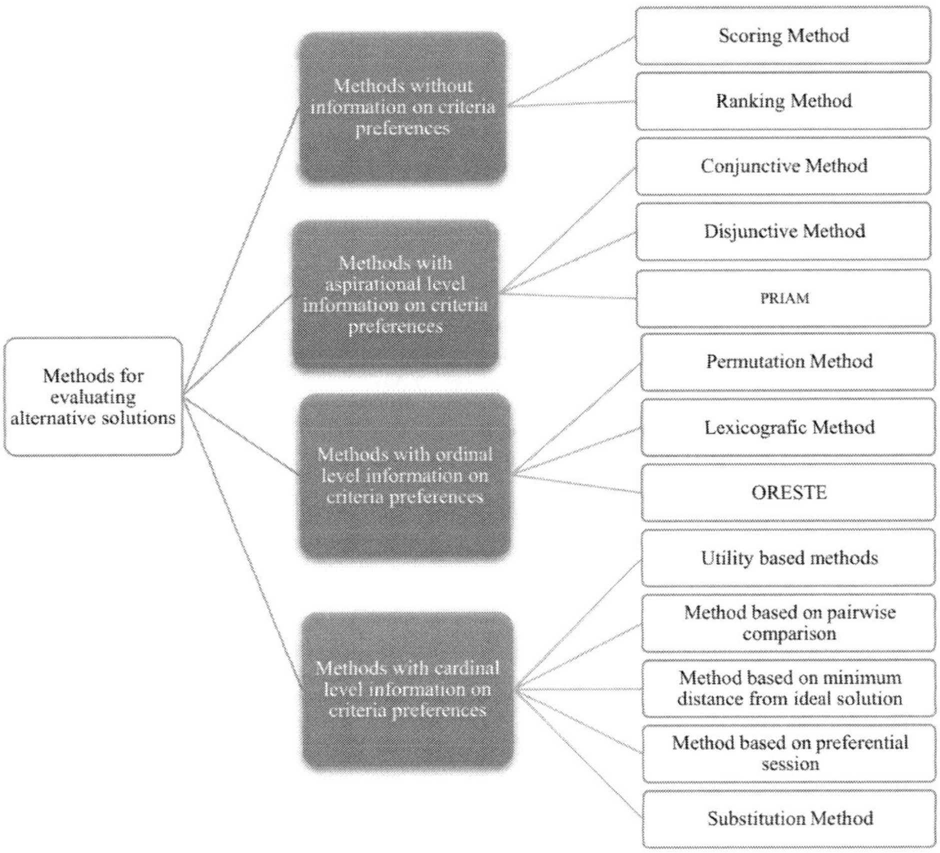

Figure 2. Methods used to evaluate alternative solutions

Source: Authors.

In this contribution, the authors focus on the application of the Analytic Hierarchy Process (hereinafter "AHP") on a model example of a MCDM problem. The AHP method is a pairwise comparison based methodology.

2 Methodology

AHP belongs to those method with a cardinal level of information on criteria preferences based on a pairwise comparison. In the 1970s, the principles of analytical thinking led to the development of a useful model for the quantitative solution of MCDM problems, also known as Analytical Hierarchy Process, by American professor Thomas Saaty. Saaty is considered to be the most important researcher dealing with the issue of MCDM. AHP consists of the decomposition of complex MCDM problems into partial components, which create the hierarchical structure of the problem. Subsequently, Saaty's quantitative pairwise comparison procedure is applied to each level of the hierarchical structure. Saaty and Kearns [13] distinguish three stages of the AHP method, namely:

Hierarchy

Priority

Consistency

The *hierarchy* of the AHP method usually consists of the goal, the alternatives for achieving the goal and the criteria that apply to the individual alternatives of the goal. It is a so-called three-level hierarchy, whereby the peak is the goal (the 1st level), followed by the criteria (the 2nd level), and lastly, the alternative solutions to the problem (the 3rd level). In specific cases, it is possible to divide the criteria into sub-criteria, i.e. a so-called four-level hierarchy.

In the AHP, priorities are used for the pairwise comparison of criteria, as well as for the pairwise comparison of the alternatives. Professor Saaty set the following scale for priority description.

"1" Both compared elements (criteria/alternatives) contribute equally to achieving the goal.

"3" Opinions and experiences gently prefer one element over another.

"5" Opinions and experiences strongly prefer one element over another.

"7" Opinions and experiences very strongly prefer one element over another.

"9" Opinions and experiences completely prefer one element over another.

It is also possible to use "2", "4", "6", or "8" to express an intermediate level of preference.

The consistency can be defined as follow. If the element k_i is s_{iq}- times more important than the element k_q, and the element k_q is s_{qj}- times more important than the element k_j, it should be able to state that k_i is $s_{ij} = s_{iq} * s_{qj}$-times more important than k_j. In practice, it is very difficult to meet this condition [14].

The AHP method can be divided into the following steps:

Define the problem to be solved

Choosing a business partner was defined as the MCDM problem. A selected company operating in the construction industry is looking for a new business partner. Eight business partners are under consideration. For the purposes of this contribution, the company, as well as potential business partners, are identified as follows: the company choosing the business partner is marked as *Company 1*, potential business partners are marked as PBP 1 – 8. In making its decision, Company 1 takes into account several financial indicators and non-financial criteria.

The selected and calculated financial indicators are presented in Table 1. The selected non-financial criteria were previous experience with a business partner, the length of time the business partner has been operating on the market, the costs of delivering materials, entries in the credit register or register of debtors.

Table 1. Selected financial indicators used to describe potential business partners

Financial indicator	Calculation
Quick ratio	$\frac{\text{Financial accounts + Short – term receivables}}{\text{Short – term foreign resources + Short – term accrued liabilities}}$
Short–term debt ratio	$\frac{\text{Short – term foreign resources + Short – term accrued liabilities}}{\text{Total assets}}$
Credit indebtedness	$\frac{\text{Credit + Loans}}{\text{Total assets}}$
Days sales outstanding	$\frac{\text{Trade receivables}}{\text{Revenues from sales of goods, products and services}}$
Days payable outstanding	$\frac{\text{Trade payable}}{\text{Total cost}}$
EBT in three years	Earnings before taxes
Return on assets	$\frac{\text{Earnings after taxes}}{\text{Total assets}}$
Return on sales	$\frac{\text{Earnings after taxes}}{\text{Sales}}$
Indicator characterizing the imminent decline of the company	$\frac{\text{Equity}}{\text{Liabilities}}$

Source: Authors.

1. *Develop hierarchical framework*

The three-level hierarchical framework was defined (see Figure 3).

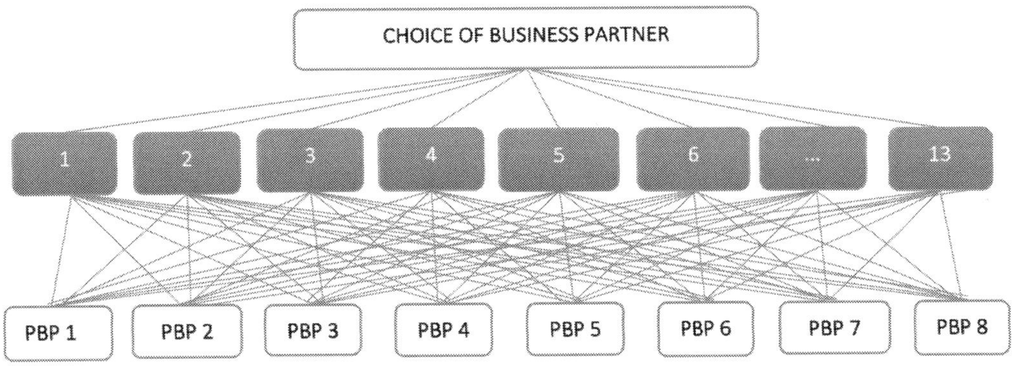

Figure 3. Hierarchical framework

Source: Authors.

Note: 1 – 4 describe the non-financial criteria in the following order: previous experience with a business partner (1); length of time the business partner has been operating on the market (2); costs of delivering materials (3); entries in the credit register or register of debtors (4); 5 – 13 describe the financial criteria in the same order as in Table 1.

2. *Construct pairwise comparison matrix*

Preferences between individual criteria are shown in the so-called pairwise comparison matrix (Saaty's matrix). Saaty's matrix $S = \{s_{ij}\}$ is a matrix that contains elements s_{ij}. S_{ij} expresses the intensity of the preference between the objects k_i and k_j. It holds that if the object k_i (the object in the row) is more important than k_j (the object in the column), then $s_{ij} \in \{1,2,3,4,5,6,7,8,9\}$, on the other hand $s_{ij} = \frac{1}{s_{ji}}$. The following formula is used to construct Saaty's matrix [15].

$$\begin{array}{cc} & \begin{array}{cc} k_i & k_j \end{array} \\ \begin{array}{c} s_i \\ s_j \end{array} & \begin{pmatrix} 1 & s_{ij} \\ 1/s_{ji} & 1 \end{pmatrix} \end{array} \quad (2)$$

Where:

k_i; k_j are criteria or alternatives;

s_{ij}; $1/s_{ji}$ expresses the intensity of the preference between the objects k_i and k_j.

Saaty's scale is used to express the intensity of the preference between the objects.

3. *Perform judgement for pairwise comparison*

The result of this step determines the weights of individual criteria (if criteria are compared) or the utility of the individual alternatives with respect to a given criterion (if alternatives are compared). Saaty defined a procedure which is based on the calculation of the eigenvector of a matrix according to the following formula.

$$S.v = \lambda_{max}.v \tag{3}$$

Where:

S Saaty's matrix;

v eigenvector of Saaty's matrix;

λ_{max} the largest eigenvalue of Saaty's matrix.

The values of the eigenvector must be standardized. This method of calculation is the most complex of all. Precisely because of its complexity, it is used only in specialized decision support programs. The geometric mean method is considered to be a simpler method that produces almost identical results, and for which specialized programs are not needed.

$$G_i = \sqrt[n]{\prod_{j=1}^{n} s_{ij}} \tag{4}$$

Where:

G_i geometric mean expresses the weights of individual criteria or determines the utility of individual alternatives;

s_{ij} elements of Saaty's matrix;

n number of criteria or alternatives.

4. *Check for consistency*

Consistency was described within the definition of the three stages of the AHP method. The consistency ratio expresses the level of consistency [16].

$$CR = \frac{CI}{RI} \tag{5}$$

$$CI = \frac{\lambda_{max} - k}{k - 1} \tag{6}$$

Where:

CR the consistency ratio;

CI the consistency index;

RI the random index;

λ_{max} the largest eigenvalue of Saaty's matrix;

k the number of objects - criteria/alternatives.

The random index was obtained by generating a large number of reciprocal random matrices (matrix elements came from a cardinal scale) and then averaging the eigenvalues of the matrices. A value of 1.56 was set for the number of 13 objects (the number of criteria). A value of 1.40 was set for the number of 8 objects (the number of alternatives – potential business partners) [16].

Steps 3 – 6 are performed for all levels of the hierarchy. By multiplying the utility of the individual alternatives with respect to the given criteria by the weights of these criteria, the total utility of the given alternative is determined. The final step is the selection of the optimal alternative based on the value of total utility.

3 Results and Discussion

A multiple-criteria matrix containing a description of the MCDM problem was created.

Table 2. Multiple-criteria matrix

Criteria/Alternatives	Non-financial criteria				Financial indicators								
	Previous experience	Length of operation	Costs of delivery	Entries in the credit/debtors registers	Quick ratio	Short–term debt ratio	Credit indebtedness	Days sales outstanding	Days payable outstanding	EBT	Return on assets	Return on sales	Decline of the company
PBP1	Yes	5	1.5	No	0.23	0.80	0.09	23	50	1	0,12	0,15	1.20
PBP2	Yes	8	1.3	No	0.40	0.55	0.10	15	48	1	0,09	0,10	0.95
PBP3	No	7	1.9	No	0.15	0.25	0.32	45	16	0	0,01	0,05	0.88
PBP4	No	3	1.1	No	0.25	0.40	0.12	33	89	0	0,02	0,06	2.1
PBP5	No	2	1.0	No	0.33	0.10	0.18	88	123	1	0,15	0,19	0.99
PBP6	Yes	4	1.4	No	0.59	0.85	0.26	62	23	1	0,18	0,25	1.85
PBP7	Yes	5	1.2	No	1.20	0.50	0.58	19	59	0	0,05	0,03	0.99
PBP8	No	9	1.9	No	0.90	0.75	0.42	37	28	0	0,09	0,02	1.01

Note: Previous experience is expressed verbally as yes or no; Length of operation is expressed in years; Costs of delivery are expressed in thousands of euros; Entries in the credit register or register of debtors are expressed verbally as yes or no; Financial indicators are expressed according to the calculation. If EBT was positive during the last 3 years, the value 1 is entered in the table, otherwise the value 0.

Source: Authors.

For the purposes of this contribution, the company, as well as potential business partners, are identified as follows: the company choosing the business partner is marked as *Company 1*, potential business partners are marked as PBP 1 – 8. In making its decision, Company 1 takes into account several financial indicators and non-financial criteria.

The selected and calculated financial indicators are presented in Table 1. The selected non-financial criteria were previous experience with a business partner, the length of time the business partner has been operating on the market, the costs of delivering materials, entries in the credit register or register of debtors.

The pairwise comparison matrix for the criteria was created. In the matrix, the criteria are numbered as in Figure 3.

Table 3. Pairwise comparison matrix for the criteria

Criteria	**1**	**2**	**3**	**4**	**5**	**6**	**7**	**8**	**9**	**10**	**11**	**12**	**13**
1	1	3	1/2	1/7	1/3	3	1/3	1	1	1/5	1	1	1/7
2	1/3	1	1/5	1/7	1/3	1/2	1/5	1/3	1/3	1/5	1/3	1/3	1/7
3	2	5	1	1/7	1/3	1	1/3	1	1	1/5	1	1	1/7
4	7	7	7	1	3	5	7	7	5	7	5	5	1
5	3	3	3	1/3	1	3	1	1/3	1/3	1/5	1	1	1/7
6	1/3	2	1	1/5	1/3	1	1/3	1/3	1/3	1/5	2	2	1/7
7	3	5	3	1/7	1	3	1	2	2	1/5	2	2	1/7
8	1	3	1	1/7	3	3	1/2	1	1	1/5	1/3	1/3	1/7
9	1	3	1	1/5	3	3	1/2	1	1	1/5	1	1	1/7
10	5	5	5	1/7	5	5	5	5	5	1	5	5	1/3
11	1	3	1	1/5	1	1/2	1/2	3	1	1/5	1	1	1/7
12	1	3	1	1/5	1	1/2	1/2	3	1	1/5	1	1	1/7
13	7	7	7	1	7	7	7	7	7	3	7	7	1

Source: Authors.

Saaty's matrix $S = \{s_{ij}\}$ is a matrix that contains elements s_{ij}. S_{ij} expresses the intensity of the preference between the objects – criteria. Formula 2 was used to construct Saaty's matrix. Saaty's scale was used to express the intensity of the preference between the objects.

According to Formula 4, the weights of the individual criteria were calculated.

Table 4. Standardized weights of the criteria

Criterion	Weights	Standardized weights	Criterion	Weights	Standardized weights
1	0.6210	0.0327	**8**	0.6757	0.0356
2	0.2920	0.0154	**9**	0.8211	0.0432
3	0.6603	0.0347	**10**	2.7287	0.1436
4	4.3834	0.2307	**11**	0.7154	0.0376
5	0.8278	0.0436	**12**	0.7154	0.0376
6	0.5169	0.0272	**13**	4.8616	0.2559
7	1.1823	0.0622			

Source: Authors.

According to Formula 3, the largest eigenvalue of Saaty's matrix λ_{max} was calculated, the consistency of which was subsequently checked according to Formulas 5 and 6.

$$S.v = \lambda_{max}.v \Longrightarrow \lambda_{max} = \frac{S.v}{v}$$

$$\lambda_{max} = 14.67$$

$$CI = \frac{14.67 - 13}{13 - 1} = 0.1392$$

$$CR = \frac{0.1392}{1.54} = 0.0903 < 0.1$$

The same procedure was chosen for the evaluation of the preferences between the individual alternatives in terms of criteria.

Table 5. Evaluation of the preferences between the individual alternatives in terms of previous experience

Alternative	PBP1	PBP2	PBP3	PBP4	PBP5	PBP6	PBP7	PBP8	Utility	Standardized utility	λ
PBP1	1	1	5	5	5	1	1	5	2.2361	0.2083	8.00
PBP2	1	1	5	5	5	1	1	5	2.2361	0.2083	8.00
PBP3	1/5	1/5	1	1	1	1/5	1/5	1	0.4472	0.0417	8.00
PBP4	1/5	1/5	1	1	1	1/5	1/5	1	0.4472	0.0417	8.00
PBP5	1/5	1/5	1	1	1	1/5	1/5	1	0.4472	0.0417	8.00
PBP6	1	1	5	5	5	1	1	5	2.2361	0.2083	8.00
PBP7	1	1	5	5	5	1	1	5	2.2361	0.2083	8.00
PBP8	1/5	1/5	1	1	1	1/5	1/5	1	0.4472	0.0417	8.00
Sum	X	X	X	X	X	X	X	X	10.7331	1	λ_{max}8

Source: Authors.

$$S.v = \lambda_{max}.v \qquad \lambda_{max} = \frac{S.v}{v}$$

$$\lambda_{max} = 8.00$$

$$CI = \frac{8 - 8}{8 - 1} = 0.00$$

$$CR = \frac{0.00}{1.40} = 0.00 < 0.1$$

By multiplying the utility of the individual alternatives with respect to the given criteria by the weights of these criteria, the total utility of the given alternative is determined. The final step is the selection of the optimal alternative based on the value of total utility. Based on the results of the AHP method, PBP1 was recommended to the company as the optimal business partner.

What follows is a summary of advantages and disadvantages of the AHP method:

• The strength of the AHP method is that no special knowledge is required to apply the method, unlike some other methods of multiple-criteria decision-making. Quantitative pairwise comparisons made by the decision-maker are relatively simple and acceptable, as only two elements are always used, where the strength of the preference is relatively easy to determine. The intensity of importance is determined verbally by the decision-maker, who then assigns the appropriate numerical degree to the verbal preference according to Saaty's scale. A great benefit of the method, in the first phase of the decision-making process, is the creation of the hierarchical structure. This makes the problem clearer, making the individual parts of the problem more understandable, which contributes to a more comprehensive understanding of the problem as a whole. By creating and analysing the hierarchy, it is possible to see if all the important aspects of the problem have been included in the model. The AHP method can work with different types of criteria. Therefore, the method can combine criteria defined both verbally or numerically without any additional modifications. The method offers verification of the consistency of the decision-maker's judgments. The positive thing about consistency is that a certain degree of inconsistency is also accepted because it is not always possible to achieve full consistency. In the case of a large inconsistency of the matrix, it is possible to rethink the individual comparisons. This leads to better concentration, as well as the decision-maker being able to find new information in the reassessment, which may lead to them determining a different intensity of importance. The advantage of the AHP method is also the fact that it has a broad spectrum of applications. For example, the method can also be used for problems where there is a risk.

• The disadvantage of the analytical hierarchical process may be the number of pairwise comparisons. This is because it is necessary to compare the individual criteria and then compare the alternatives with respect to a specific criterion. If the decision-making problem contains a large number of criteria as well as alternatives, the decision-making process can be time consuming, and over time, the inconsistency of the matrices may begin to increase due to the loss of attention and lack of concentration of the subject. A minor drawback is also the basic scale (nine-point scale). When comparing the criteria, respectively alternatives, the decision-maker must determine the strength of the preference between the two criteria, respectively alternatives. With a large number, whether criteria or alternatives, a given scale may not be enough if it is necessary to sufficiently differentiate the individual elements. This shortcoming is followed by another weakness in the form of consistency, which is a relatively common problem when using the AHP method. If the nine-point scale is not enough, the full consistency of the matrix cannot be achieved either. On the other hand, the method also respects a certain degree of inconsistency, which can be considered an advantage. The analytical hierarchical process has a major drawback in that if the decision-maker uses the method to produce the results, but decides to add a new alternative to the model or remove the old alternative, the preferential order of other alternatives can change without changing the values of pairwise comparisons with regards to individual criteria or their preferences. Long analysis has shown that the order of the alternatives may change due to the lack of consistency of the intensities. Although it may sound nonsense to introduce or remove alternatives from the evaluated model, in practice such a need may arise.

4 Conclusion

The aim of the contribution was to use the Analytic Hierarchy Process to assess potential business partners and select the optimal alternative. MCDM is the most common type of decision-making process in business practice. AHP is an MCDM method originally developed by Prof. Thomas Saaty. The method derives ratio scales from paired comparisons. The ratio scales are derived from the principal eigenvectors and the consistency index is derived from the principal eigenvalue. The AHP method can be defined as one of the most commonly used MCDM methods in practice. The problem to be solved, as well as the steps of AHP method, are described in the methodology part of contribution. The problem to be solved was defined as follows: a selected company operating in the construction industry is looking for a new business partner. Eight business partners are under consideration. Using the AHP method, PBP1 was recommended to the company as the optimal business partner.

There are a variety of software applications available for the AHP method, of which Expert Choice was the first to be developed. Expert Choice was created in the 1980s by Thomas Saaty, the creator of the AHP method, in order to help organizations around the world make better, more transparent, and faster decisions. Expert Choice combines team collaboration tools and proven math techniques to enable the best decision to be made to achieve a goal.

Despite the shortcomings of the AHP method, it is one of the most popular and one of the most objective methods of multiple-criteria decision-making.

References

1. F. C. Zola, J. C. Colmenero, F. V. Aragao, T. Rodrigues, A. Braghini, Multicriterial model for selecting a charcoal kiln. *Energy*. **190**(1), 1–14 (2020)
2. E. Mastrocinque, F. J. Ramirez, A. Honrubia-Escribano, D. T. Pham, An AHP-based multi-criteria model for sustainable supply chain development in the renewable energy sector. *Expert Systems with Applications*. **150**, 59–76 (2020)
3. M. Koksalan, J. Wallenius, S. Zionts, *Multiple Criteria Decision Making: From Early History to the 21st Century*. Singapur: World Scientific. 1–198 (2011)
4. H. C. Liao, X. L. Wu, DNMA: A double normalization-based multiple aggregation method for multi-expert multi-criteria decision making. *Omega-International Journal of Management Science*. **94**, 23–38. (2020)
5. H. C. Liao, X. L. Wu, X. M. Mi, F. Herrera, An integrated method for cognitive complex multiple experts multiple criteria decision making based on ELECTRE III with weighted Borda rule. *Omega-International Journal of Management Science*. **93**, 44–52 (2020)
6. C. Fu, W. J. Chang, S. L. Yang, Multiple criteria group decision making based on group satisfaction. *Information Sciences*. **518**, 309–329 (2020)
7. R. Costache, C. T. Bui, Identification of areas prone to flash-flood phenomena using multiple-criteria decision-making, bivariate statistics, machine learning and their ensembles. *Science of the Total Environment*. **712**, 269-281 (2020)
8. M. Haddad, D. Sanders, G. Tewkesbury, Selecting a discrete multiple criteria decision making method for Boeing to rank four global market regions. *Transportation Research Part A: Policy and Practice*. **134**, 1–15 (2020)
9. D. Lamba, D. K. Yadav, A. Barve, G. Panda, Prioritizing barriers in reverse logistics of E-commerce supply chain using fuzzy-analytic hierarchy process. *Electronic Commerce Research*. **20**(2), 381–403 (2020)
10. B. U. Elvarsson, S. Agnarsson, S. Gudmundsdottir, J. Vidarsson, Using multi-criteria analysis the assess impacts of change in ecosystem-based fisheries management: The case of the Icelandic cod. *Marine Policy*. **116**, 120–132 (2020)
11. J. Benitez, S. Carpitella, A. Certa, J. Izquierdo, Constrained consistency enforcement in AHP. *Applied Mathematics and Computation*. **380**, 85–101 (2020)
12. T. Si, C. B. Wang, R. Q. Liu, Y. S. Guo, S. Yue, Y. J. Ren, Multi-criteria comprehensive energy efficiency assessment based on fuzzy-AHP method: A case study of post-treatment technologies for coal-fired units. *Energy*. 200, 159–172 (2020)
13. T. L. Saaty, K. P. Kearns, *Analytical Planning. The Organization of Systems*. Oxford: Pergamon Press. (1985)
14. T. L. Saaty, *Multicriteria Decision Making: The Analytic Hierarchy Process. Planning, Priority Setting, Resource Allocation*. New York: McGraw-Hill. (1980)
15. T. L. Saaty, Decision making with the analytic hierarchy process. *International Journal of Services Sciences*, **1**(1), 83–98 (2008)
16. T. L. Saaty, *Fundamentals of Decision Making and Priority Theory with the Analytic Hierarchy Process*. Pittsburgh: RWS Publications. (2000)

Legal Issues of Teleworking

Pavel Sládek[1,*], *Tomáš* Sigmund[2]

[1] Institute of Technology and Business, Faculty of Corporate Strategy, Okružní 517/10, 37001 České Budějovice, Czech Republic
[2] Prague University of Economy and Management, Department of Systems Analysis, W. Churchill Sq 1938-4, Prague 13067 3, Czech Republic

Abstract. Teleworking is a specific type of work where the employee works outside their usual place of work. A characteristic of this activity is the reduction of the need to commute to their employer's workplace and the performance of work during fixed working hours. The following prerequisites are important for real telework: a computer. The worker is in a different place than the employer or the customer. Communication takes place at a distance via telecommunication networks. Instead of a fixed time regime, teleworking allows the employee to individually optimize the performance of their work according to their own needs in terms of time and place (the workplace is most often home). We use the method of identification, analysis, comparison and synthesis of various legal regulations for teleworking all around the world. In teleworking, the powers of the employer and its employees are not equal. That is why it is necessary to create national laws that will better protect the rights of teleworkers.
Keywords: teleworking; home office; homeworking; European Framework Agreement on Telework; telecommunication networks

1 Introduction

Teleworking (nowadays, the term home office is more commonly used) is a specific style of work that has recently become more and more relevant. The main feature of this activity is the reduction of the need to commute to the employer's workplace and the performance of work during fixed working hours. The advantage of working remotely is the ability of the employee to optimize the performance of their work according to their own needs in terms of time and place. Teleworking has also close links to other modern telecommunications technologies that help its successful implementation. The current dynamic development of telecommunications creates conditions for increasingly simpler and better possibilities of communication with the world from home through various devices, from telephone connections to video conferencing. Managers distinguish between home office and homeworking. Home office is much more common and affects a large number of employees. The employee stays at home, usually due to a child's illness. This way of working is often limited in time. Another option is homeworking - the employee works at home and occasionally commutes to work for better communication and improved social relationships.

2 Methodology

We use the method of identification, analysis, comparison and synthesis of various legal regulations for teleworking all around the world. Due to the rapid development and implementation of teleworking, its use among employees is increasing, which creates the necessity for legal regulation. We compared regulations and drew conclusions accordingly.

3 Results and Discussion

3.1 Teleworking in the Czech Republic

With regards to European legislation on teleworking, the European Framework Agreement on Telework was adopted in 2002. Based on this agreement, individual European countries have developed their own legislation. However, these laws have been interpreted differently in each country [1]. This difference was due to different political interests in each country. The Czech Labour Code 262/2006 [2] regulates the specifics of distance work, but its full implementation is often problematic. The practice of teleworking on the Czech labour market is regulated within the framework of the

* Corresponding author: sladek@mail.vstecb.cz

Labour Code, specifically Section 317. The use of this paragraph is possible only if the employee does not work at the employer's workplace and sets the schedule themselves.

3.2 Teleworking in Slovakia and Poland

The legislation in neighbouring countries, such as Slovakia and Poland, is more complex. According to Horňák [3], Slovak legislators have succeeded much better in applying European framework agreements on telework to national law. The Slovak provision is marked as § 52 of Act No. 311/2001 Coll. [4], and in many ways reflects our Act 262/2006 Coll. In it, it is explicitly written that if the employer provides the employee with technical equipment, it must take care of that equipment. Paragraphs 3 and 4 also contain a very important provision.

(3) The employer shall take measures which prevent the isolation of an employee working at home (homeworking or teleworking) away from other employees and give them the opportunity to meet other employees.

(4) The employee's working conditions for homeworking and teleworking must not disadvantage the employee in comparison with a comparable employee at the employer's workplace. (Translated by the author) [4]

These provisions are missing in Czech legislation, although according to Horňák [3] they can be deduced from the general provisions of the Labour Code or the Discrimination Act. From the point of view of corporate social responsibility, these issues are very important.

In Poland, Articles 67^5 to 67^{17} of the Law of 26 June 1974, Dz.U. 1974 No. 24 item 141, Labour Code [5] regulate teleworking. The mentioned parts of the Code come from the amendment of 24th August 2007 (Dz.U. No. 181 pos. 1288). As Polish trade unions hold a stronger position than Czech trade unions, they also have greater competences in the field of teleworking. According to the Polish provisions, the worker also has better opportunities to negotiate with the employer compared to the situation in the Czech Republic. This is because teleworkers in Poland enjoy greater protection.

A common problem is that employees are forced into teleworking. In Poland, an employee has the option to return to the standard way of working if this remote style of work does not suit them.

Art. 67^9 requires the consent of the employee to change the conditions of work, in the case specified in Art. 67^7 § 3, as well as states that the cessation of work in the form of telework, based on the principles set out in Art. 67^8, may not constitute a reason for justifying the termination of the employment contract by the employer. (Translated by the author) [5]

The employer also has the right to inspect the place where the employee works, even if it concerns the employee's home. However, the employer must have the written consent of the employee to do so and must not infringe on the privacy of the teleworker and their family.

Art. 67^{14}

§ 1. The employer has the right to control the performance of work by the teleworker in their place of work.

§ 2. If the work is performed in the teleworker's home, the employer has the right to check:

1) the performance of work;

2) the inventory, and check if the employee maintains, services and/or repairs the entrusted equipment, as well as its installation;,

3) occupational safety and health issues - with the prior consent of the teleworker expressed in writing, or by means of electronic communication or similar means of personal remote communication. (Translated by the author) [5]

Like Slovak law, Paragraphs 3 and 4 of the Polish law protects teleworkers against discrimination. The worker must not be discriminated against, for example, in terms of starting and leaving employment, working conditions, promotion and access to training in order to increase professional qualifications compared to other employees employed to carry out the same or similar work.

3.3 Teleworking in the world

Despite the rapid development in the field of information and communication technologies, France is not as developed in the field of teleworking as other similarly developed European countries. Therefore, a law supporting teleworking was adopted in 2012 [6]. Despite this law, France faces many problems with this type of work. However, teleworkers are still not as well regarded as traditional workers. French labour law prohibits teleworkers from having employees themselves, with the exception of the help of their spouse, dependent children or support staff.

The employee also has the right to "refuse to take the place of a distance worker and that is not a reason for terminating the employment contract" [7].

However, an exception applies here in the case of exceptional circumstances. It is explicitly stated that telework may be ordered without the employee's consent, e.g. in the event of a threat of an epidemic. Within the context of the current quarantine (October 2020), this provision is very topical. Another problem addressed in French law is the obligation to employ the disabled. This is perceived negatively because it leads to greater isolation of the disabled. The same division exists in Quebec law. On the one hand, remote work can be perceived as an adaptation measure, whereas on the other hand, it is important to minimize the risks of isolation.

Another element that hinders the development of teleworking in France is the concept of French management [7], which is considered to be hierarchical and does not adapt to the free organization of the work of the teleworkers towards management, who control this exclusively through well-formulated goals. "The absence of goal-based management, managers' fear of losing power, and the fear of abuse are major obstacles to organization and management." [7] The development of telework is also burdening the legal environment, especially in small and medium-sized enterprises, where there is a risk of litigation because telework is not so often used in the French environment.

In Spain, the status of teleworkers is described in the 2012 law in Article 13 concerning the characteristics of teleworking. This law also addresses the health aspects of employment as well as distance employment. With regards to the ethical aspects of teleworking, in 2017, research was conducted into the gender aspects of telework. The conclusion was that it "perpetuates the asymmetry between the sexes: men in decision-making posts and women as subordinates." [8] These problems also exist in the area of traditional employment.

Another European country that has a law on teleworking is Romania. Act 81/2018 applies to specific professions, such as brokers, sales representatives, analysts, programmers, accountants, financial and tax advisers, translators, etc. [9]. This regulation on teleworking is in line with European legislation, which meets the needs and requirements of the labour market. However, there are certain problems with regards to the issue of occupational safety and health. Other problems are the recording and checking of working hours [10]. According to the current 2020 article on reducing the impact of the crisis in Romania caused by Covid-19, teleworking is popular. This way of working has also become an important motivator for Romanian employees [11]. In order for employees to take full advantage of teleworking, the main precondition is the existence of the law. However, teleworkers feel disadvantaged in the labour market in Romania. Even though lawyers agree that the law is alright [12], the fact is that these workers often receive lower pay, which later disadvantages them with regards to social benefits.

Legislation is being amended around the world. However, in Brazil [13], for example, the benefits of this law are more in its pure existence because it represents progress in recognizing this way of working as a viable way. It is contained in Act 13.467 of 2017, but addresses this issue very little. In Colombia, teleworking has been regulated since 2008 by Act 1221, which, according to the authors, supports job creation and obliges employers to educate employees. [14] Another interesting dimension of this issue is the transnational issue of teleworking. Telework is very often seen as cyberspace or virtual work. This situation enables the application of more than one law in a contract [15]. This option should be avoided in particular by the employee (as the weaker link) because they stand very little chance of winning a dispute. This type of teleworking is characterized by the requirement to have knowledge of foreign languages. [16]

4 Conclusion

In teleworking, the powers of the employer and the employee are not equal, which is why some legal regulation is necessary. Teleworking further exacerbates employment inequalities, particularly if the employee is a woman - mother [17]. In Europe, teleworking is regulated quite uniformly because of EU regulations, even though differences between countries exist. In other countries, however, detailed regulation is often lacking, as a result of which workers are often left stranded.

With globalization and the increasing power of international companies, conflicts between various legal systems may arise.

References

1. A. Belzunegui-Eraso, A. Erro-Garces, Teleworking in the Context of the Covid-19 Crisis. *Sustainability*. **12**(9), 3662 (2020)
2. Czech Republic, *Zákoník práce [Labour Code] Zákon [Law] 262/2006 Sb.* (2006)
3. L. Hornak, Legal regulation of teleworking in the selected neighboor countries as an inspiration for czech legislation. *International Multidisciplinary Conference on the Foreign Impacts on National Law*. 50-58 (2014)
4. Czech Republic, *Zákoník práce [Labour Code] Zákon [Law]* 311/2001 (2001)
5. Poland, *Ustawa z dnia 26 czerwca 1974 r. Kodeks pracy [The Act of June 26, 1974, Labor Code]*. [online]. Available at: http://isap.sejm.gov.pl/isap.nsf/DocDetails.xsp?id=WDU19740240141

6. I. Desbarats, Between the protection of workers and the performance of the company: the gestation period of a French right to telework. *Cahiers de Droit*. **54**(2–3), 337–358 (2013)

7. République Francaise, *Télétravail* [*Teleworking*] (Articles L1222–9 à L1222–11) [online]. Available at: https://www.legifrance.gouv.fr/

8. A. Galvez, F.Tirado, J. M. Alcaraz, "Oh! Teleworking!" Regimes of engagement and the lived experience of female Spanish teleworkers. *Business Ethics-a European Review*. **29**(1), 180–192 (2019)

9. A. A. Davidescu, A. M. Adriana, S. A. Apostu, Work Flexibility, Job Satisfaction, and Job Performance among Romanian Employees-Implications for Sustainable Human Resource Management. *Sustainability*. **12**(15) (2020)

10. V. Teofila, A. Torres, J. Ignacio, V. Torres, Telework and occupational health: medical-legal and labor aspects. *Revista Ces Derecho*. **9**(2), 287–297 (2018)

11. R. Anderco, Teleworking. *8th International Conference on Perspectives of Business Law in the Third Millennium*. 196-202 (2018)

12. M. E. Marica, Hoeworking/teleworking. Atypical forms of employment: between usefulness and precariousness. *Juridical Tribune-Tribuna Juridica*. **8** (1), 73–85 (2018)

13. B. R. Hazan, P. Morato, A. Duarte, Teleworking in Brazilian Law: Grounds, Perspectives and Changes Promoted by Labor Reform. *Lex Humana*. **10**(1), 1-23 (2018)

14. M.Gutierrez, P. Claudia, R. A.Capone, Socio-economic and legal impact of teleworking in Barranquilla 2014-2015. *Justicia*. **22**(32), 197–213 (2017)

15. D. P. Fincato, M. de Bitencourt, Cyber as Workplace: the Problem (or Solution?) of Transnational Teleworking. *Quaestio Iuris*. **8**(4), 2209–2235 (2015)

16. Z. Caha, The Importance of Corporate Language Education in a Globalised Economy. *32nd Conference of the International-Business-Information-Management-Association (IBIMA)*. **32** (2018)

17. M. Hilbrecht, S. M. Shaw, L. C. Johnson, J. Andrey, 'I'm home for the kids': Contradictory implications for work-life balance of teleworking mothers. *Gender Work and Organization*. **15**(5), 454–476 (2008)

Smart Innovations as Accelerators for SMEs in Rural Areas

Dominika Šulyová[1,*], *Marie* Gabryšová[2], *Josef* Vodák[1]

[1]University of Žilina, Faculty of Management and Informatics, Univerzitná 8215/1 Žilina, Slovak Republic
[2]Katowice Business University, ul. Harcerzy-Wrzesnia 3, Katowice, Poland

Abstract. Innovative approaches are driving the stabilization, development, and competitiveness of small and medium-sized enterprises (SMEs). The introduction of smart concepts has a favourable impact on their activities. At present, urban areas and urban projects are more preferred and supported. The importance and benefits of rural areas and SMEs' engagement therein is underestimated in terms of country development at the European level. This contribution aims to point out innovative approaches that contribute to the development of SMEs and smart cities or regions using best practice examples from abroad with a focus on specific rural areas in the Czech Republic—where this issue is gaining momentum. The analysis of best practice examples and expert opinions in this field was used to make comparisons and to summarise the knowledge gained. This contribution also serves as a basis for mapping the factors that can support regional development and growth, as well as provides recommendations for future follow-ups and new concepts for the coming years in Czech rural areas, or areas that are culturally similar, such as Slovak rural areas.
Keywords: Smart cities; SMEs; accelerators; rural areas; Czech Republic

1 Introduction

Nick Skillicorn argues that innovation 'represents a transformation of the original idea to the final solution that needs to be seen from a customer perspective' [1]. In most cases, companies require creativity from their employees, but they do not provide sufficient financial and time support at the management level. Although an idea arises, it is not implemented, or it is implemented only internally without considering customer preferences [1,2,3]. Innovation in the sense of 'applied unpublished ideas hitherto, which must meet the implementation conditions, otherwise it is just a form of an idea' is a definition according to the expert David Burkus. In his opinion, SMEs make a mistake when dealing with innovation management because they see innovation as only being linked with a final product, ignoring the fact that innovation is more than just a smart product or service. The change will affect the entire company and all stakeholders internally or externally [1].

Stephen Shapiro perceives the concept of innovation very simply in terms of relevance. If SMEs want to remain stable and successful, they should be flexible and adapt to new market requirements. Pete Foley defines an intuitive idea as one that is aligned with a new concept and communicated effectively as innovation. Within constraints, they are in favour of ensuring a balanced share of small and large ideas that elevate average businesses above the standard best practices in the field [1]. Gijs van Wulfen defined a relevant offer as one that works on an efficient business model as an innovation for customer segments. McFarthing, Brands, Hobcraft, and Shipulsky see perceptions of innovation as 'new products and services that add value to SME management, reflect customer and stakeholder requirements, meet a strategic goal, and improve market position, including profit' [1]. Paul Sloane combines innovation with an element of creativity, which is the basis for creating new concepts; however, its implementation conditions the emergence of innovation. At the same time, it helps generate value in the form of higher sales, lower costs, or a combination of the two [1]. Lindegaard refuses to define this concept and sees it as being overrated. Boyd favours the idea that an actual change in technology or management is innovation only if it is surprising and sudden. Greber recognizes the organic value of the term in terms of creative thinking and social aspects vis-a-vis stakeholders when building trust and relationships. Bebra summarizes opinions by saying that innovation gives the world a future [1]. The expert definitions and opinions above point to the fact that—as with the Smart City issue—no comprehensive and general view of terminology and understanding of innovation exists. The common views of those 15 experts are [1, 4]:

- ideas (60%);
- idea implementation in practice (60%);
- generation of value-added, creation of final solution (40%);
- a different perspective, thinking (27%);

* Corresponding author: dominika.sulyova@fri.uniza.sk

- advancement (13%);
- the unimportance of the definition (13%);
- focus on a new market, product or service (7%).

In summarizing these elements, the term innovation can be seen as 'a value-generating challenge for all stakeholders' [1,5,6,7]. In this contribution, the term will be seen as the adoption of new intelligent changes by companies at the local or national level when developing concepts of smart cities or villages.

2 Theoretical background

2.1 Rural area smart innovation implementation

The development of small and medium-sized enterprises in rural areas depends on the local conditions and features of each region. According to Steiner, two areas have been identified that need to be addressed in the management of smart villages, i.e. cooperation and education [8]. Cooperation is related to stakeholder engagement in building smart development concepts. The economy of the selected rural region is accelerated by the knowledge and know-how acquired in data form that is distributed through the network [8].

Smart region knowledge databases are a key to succeeding with the innovation management and the activities associated therewith. Collective learning, predictions, social factors, culture, and public opinion, as well as natural benefits, resources, and human capital are indicators that accelerate the growth of rural areas [8]. Local identity and adaptation to change are implemented through development programmes, such as the European Union's LEADER or Local Action Group. Their role is to [8, 9, 10]:

- accelerate economic and demographic development and growth;
- support the innovation activities in the regions, cities, or villages selected;
- mobilize capacities, knowledge, human potential, partnerships;
- support and give preference to a multidimensional approach;
- implement a development strategy, assess its progress through the monitoring of several criteria, i.e. implementation of a smart product or service, integration of creative and innovative management forms and decision-making processes, selection of new methods, and activities for a specific area;
- apply for grants and select suitable projects for rural areas.

The implementation of the Europe 2020 project, which forms part of European Regional Policy, is an adapted version of the 1990 plan for the 21st century, which reconciles the key role of innovation in regional development better. The EU´s priority is to implement smart specialization concepts for economic, employment, and GDP growth. Insufficient interconnection between R&D, innovation resources, and processes with the current rural areas economic structure restricts this progress [11]. Sustainable development and support for villages is underestimated, although they form an important part of a holistic approach to a country's development. The city of Kentucky in the USA is an example of best practice in harmonizing and adopting smart concepts in regions. Local SMEs have used the principle of diversification. The region has a natural advantage as a result of its tobacco and maize agriculture. In the 18th century, Kentucky was among the world's top wine producers, but 1920 laws restricted its production and sought to promote tobacco production instead. In a drive to look for new alternative ways to gain a competitive market advantage, local SMEs expanded their activities to growing vines. The small wineries provide customers with a specific value-added, quality logistics, niche specialization, and a wide range of products. The agricultural strategy is based on setting up small family-run organic vineyards, with access for tourists and free tastings. The local businesses are not only intended for domestic visitors but also potential clients from abroad. Innovation is supported by communities, and the new elements have an impact on technologies, cultural factors, socio-behavioural factors, and management acceptance [11]. SMEs operate on a similar basis in the Piedmont Alps. The innovative approach is linked to the ecological utilization of the mountain landscapes natural resources, resulting in a high quality of life and a healthy lifestyle. The smart development strategy is based on the following success factors [11]:

- resource revitalization;
- environmental protection;
- rejection of the consumerist way of life;
- taking advantage of opportunities connected with introducing new technologies (mountain Wi-Fi internet connection and sensors);
- migration of people from cities who prefer living in rural areas for their attractiveness and benefits.

The rural Langhe region of Italy suffered massive depopulation after World War II. Over the last 20 years, strategic management has sought to combine local traditions with the agricultural sector and support the diversification of the area through the Integrated Territorial Program 2007. At the same time, product innovation has created a competitive advantage for the rural area and led to the emergence of a smart region. These benefits have sparked dialogue on qualitative environmental development, socio-cultural communities, and agricultural associations. It follows that rural areas can be perceived as a source of innovative specialization [11]. North America has built its concept of innovation on a negative trend—the increase in childhood obesity. Rural agricultural businesses began to receive support to provide quality natural diets. The aim was to build a system that would meet the criteria of sustainability, humanity, and economic efficiency. Cities, villages, regions, and SMEs were invited to join this process [11]. According to Mitra, smart concepts bring forth the following benefits for SMEs [12]:

- opportunities;
- better infrastructure, supply chain;
- virtual jobs;
- knowledge database support vis sensor data.

According to Da Pires et al., the innovation implementation framework that acts as an accelerator for SMEs should contain an analysis of available resources in the selected city, village, or region and identify stakeholders, as well as the expectations and benefits thereof. In addition, the selected area needs to be connected and aligned with the economic, social, knowledge, and natural conditions of the rural area, i.e. classification of specific and high-income areas [11, 13].

2.2 Smart approaches and impacts on SMEs

The implementation of smart cities, villages and regions is strongly dependent on the private sector. Kummitha and Krishna's study confirms that IoT technology collects data that business managers use to support management and decision-making processes. Businesses create initiatives in the field of technological and social change within the framework of maintaining or increasing competitiveness in the market. Smart clusters therefore influence SMEs' business models. For the effective creation of smart approaches, it is necessary to have four stakeholders involved in the implementation process [14, 15, 16]:

- Government - resource allocation, market management, legislation;
- SMEs - support for social interventions, sources of innovation and technical solutions;
- Public - participation, trust, support, cooperation in finding solutions and generating ideas in people-oriented centrist models.

Small cities and regions, or rural areas, act both as a regulator and a supporter of public relations and cooperation. The tax regime creates a suitable background for businesses that use the data collected to modify their business models and search for opportunities to increase their competitive advantage and value-added for customers (Figure 1).

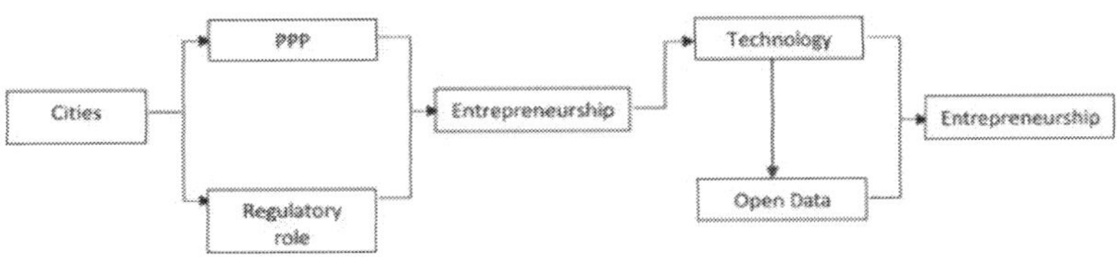

Figure 1. Smart grouping as an accelerator of business opportunities

Source: [14].

Smart City Morocco researchers conducted a study in 2018 that worked with the following two hypotheses [17]:

H1: Implementation of smart solutions will support not only the competitiveness of companies, but also of the region where they operate.

H2: Smart approaches in cities or villages provide opportunities for SMEs.

The results were graphically represented by a model that examined positive aspects of smart approaches on SMEs (Figure 2).

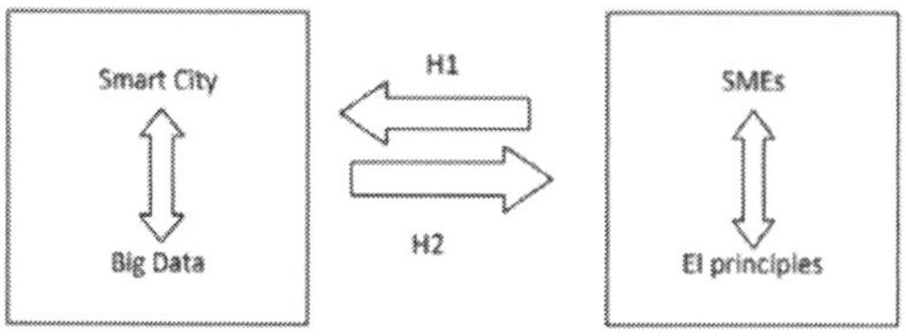

Figure 2. A model examining the positive impact of smart approaches on SMEs

Source: [14].

The survey of 210 SMEs confirmed both hypotheses. After the implementation of the smart concepts, the quality of the products of the participating companies increased by 67%, with managers also indicating an increase in competitiveness by around 91%. Implementation constraints came in the form of cultural and organizational differences and management styles, i.e. the human factor. The final results confirmed both hypotheses.

3 Methodology

A secondary analysis of world best practice in the field of rural area smart innovation implementation was used for this purpose. According to the survey, South Korea has the longest experience (10 years) in this area. Its models are surprisingly close to the local conditions in the Czech Republic, which is this contributions area of focus. This contribution examines the development of innovation implementation among SMEs in the Czech Republic in the years 2007–2013, 2016, and 2020. In addition to the secondary analysis, best practices from South Korea regarding smart innovation implementation in rural areas is compared with those in the Czech Republic, including a comparison of different expert opinions on the definition of innovation. A summary of the lessons learned follows in the Discussion section, which contains a summary of common and differing elements, including recommendations and implementation constraints for developing and stabilizing SMEs in rural areas through innovation accelerators.

4 Results

4.1 South Korea rural area innovations

Rural development is progressing differently in the countries of the EU and North America. Europeans are more inclined towards incorporating social cohesion and cultural differentiation into their models and approaches, while Americans favour the market as a mechanism for creating equality through competition. In its approach, South Korea leans towards the European models. Rural development investment plans for the agricultural sector were initiated in 1991, with the latest dating back to 2017. The development over time and the budgets thereof are shown in the following table [18].

Table 1. Investment plan for the innovative development of rural areas in South Korea

Phase	Titles	Period	Main objectives	Budget
1.	Implementation plan for rural agriculture	1991-1996	increasing the competitiveness of the sector	USD 47 billion
2.	Regulatory plan for rural agricultural development	1997-2003	increasing the competitiveness of the sector + rural area	USD 37 billion
3.	Detailed plan for agricultural development with the support of rural communities	2004-2017	sector development, quality of life and business	USD 104 billion

Source: [18].

The primary reasons for the emergence of these projects were the high level of poverty, the ageing of the rural population, and expensive services in the 1950s and 1960s. The productivity of the agricultural sector was low, with the vast majority of people living below the poverty line [19].

 https://doi.org/10.1051/shsconf/20219001021

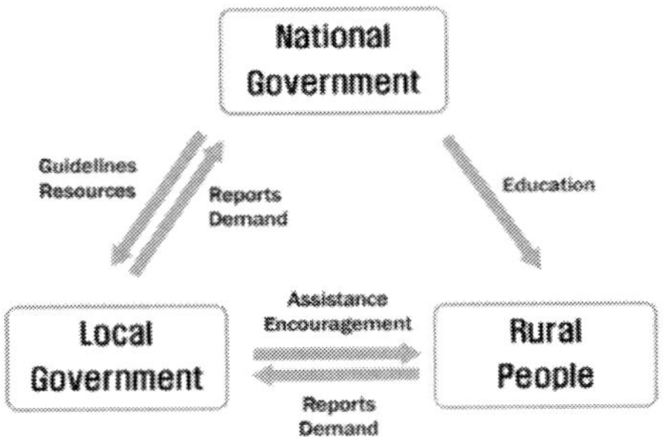

Figure 3. Interactive structure between rural development stakeholders in South Korea

Source: [20].

Representatives of strategic (national) management, i.e. government, regional authorities, and the people living in rural areas play a key role in these processes (Figure 3). The role of the region's representatives is to initiate discussions with stakeholders, monitor the project's success, and to provide services and support to the public. People contribute to the functioning of initiatives through their pro-active approach and participation in events [20]. The structure of the interaction is graphically represented in Figure 3. Strategic management allocates resources to the region, controls reports, and provides the public with educational opportunities in rural areas [20]. A rural area in South Korea is a village with a population of about 100,000 people. Very small rural areas range between 50,000 and 200,000 people. According to research studies carried out in 2017, rural areas in Mexico, Slovakia, Belgium and the USA have the highest potential (positive part in Figure 4). Values of around zero were achieved by, for example, the Czech Republic, Great Britain and Chile [21].

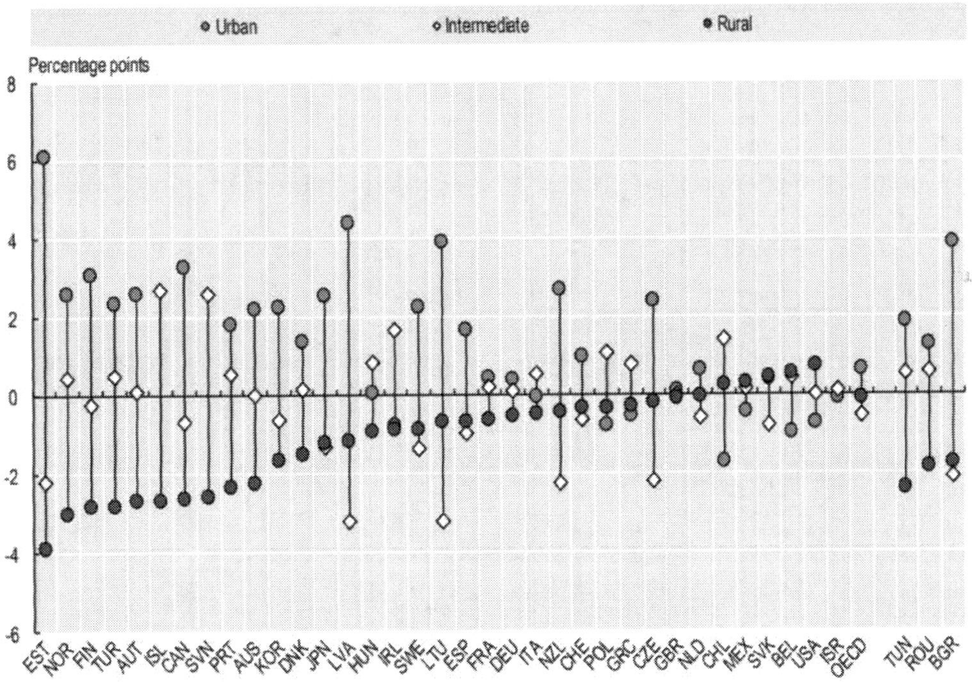

Figure 4. Change in population numbers vs. urban rural areas from 2010 to 2017

Source: [21].

The highest urban-to-rural population was identified in Estonia between the years 2000 and 2017 at a value of +6% (see Figure 4).

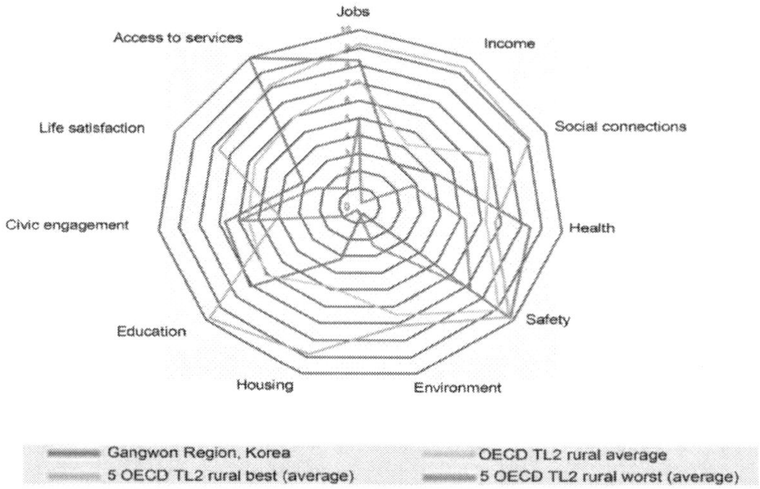

Figure 5. Comparison of rural areas in South Korea and OECD countries based on best practice

Source: [21].

Based on best practice, the rural Gangwon region of South Korean excels in terms service access, security, health and public involvement. With regards to income, work, housing, education, and the environment, the region achieved higher values from the best practice benchmarking process (see Figure 5) than the OECD [21].

A key element is vision sharing and ensuring effective internal and external communication. A study of rural areas in South Korea found that [21]:

- development has been carried out for the last 60 years, creating a long-term tradition and experience-based knowledge management;
- diversification of economic, financial and social development has increased GDP, national incomes and reduced regional disparities;
- solving rural problems supports the development of the country as a whole;
- changes in the demographic curve predict an ageing population, which implies the need to educate the younger generations and implement growth strategies following the new trends;
- for centralized management of rural areas it is adequate to strengthen cooperation between ministries and departments;
- compared to urban areas, lower quality of life still prevails in rural areas.

4.2 Czech Republic Rural Area Innovations

The countries of the European Union, including the Czech Republic, are facing problems in the form of population migration to cities, changes in the demographic curve, or digital transformation issues [22].

The building of new technology-based smart rural areas and social innovation is a rising trend in the Czech Republic, in particular where it concerns the energy, agricultural, and forestry sectors (drones for the removal of bark beetle, etc.). The key elements of their success lie in the people involved, their creativity and attitude towards the region [22]. The countryside therefore represents a strong element of Czech development strategy. In 2016, the majority of areas contributed to improving the agricultural sector and villages generated competitive advantages mainly through their natural character and charm, i.e. *genius loci* [23].

The existing projects aim to motivate people to seek new solutions and implement innovations on the basis of the principle of getting back to the original rural roots [24].

The rural area development plan for the period 2014-2020 focused primarily on supporting the sustainability of the limited resource management, i.e. to ensure efficient farm production so as to improve soil biodiversity by 25% and save water resources by 12%. The project involved 3,500 agrarian and 1,450 forestry programs. Small rural farmers make up 89% of the Czech agricultural sector, producing only 30% of the total volume. The demand for regional products is also falling as a result of the growing number of people (up to 80%) commuting every day to work in cities [25].

Smart principles in rural development consist of six priority initiatives. The first is the transfer of knowledge from the agriculture and forestry industries, and support for rural innovation. The second is to generate start-ups for 1,250 young farmers, who will drive the growth of the agricultural and food sectors, and in forestry, to reconstruct 830 km of forest roads for efficient logging and to invest in IoT technologies [25].

In addition to the cultivation of organic crops, the third priority is to produce quality rural livestock and products. Relevant activities should satisfy the following three priorities [25, 26]:

- healthy ecosystem and preservation of biodiversity;
- mitigation of adverse impacts on the environment;
- creation of new jobs—approx. 1,100 jobs through Local Action Group programmes.

The European Commission has set aside €3.5 billion to meet these priorities, of which €1.2 billion will come from the Czech budget and the rest will come from European Union resources [27]. The stakeholders include 180 Czech companies that are incorporated in the LAG database, which was updated in 2020 [28]. An overview of smart rural development projects in the Czech Republic is presented in the following tables.

Table 2. SME rural development support projects (Part 1)

Project	Region/area	Year(s)	Main findings
Diversification of the rural economy	Varnsdorf	2007 – 2013	Creative thinking and use of brownfield sites can significantly support rural development.
Diversification of the rural economy	Jetřichovec	2007 – 2013	Pig suspensions should serve as heating fuel; silage of high quality is suitable for feeding animals, and low-quality ones for biogas; the best alternative is corn silage.
Restructuring, developing physical potential and promoting innovation	Krucemburk, Kuřim	2007 – 2013	New meat production marketing strategies need to be introduced.
Implementing local development strategies	Czech Republic	2007 – 2013	Growing importance of and need to work with communities to build knowledge and experience management–best practice for others.
Support for young farmers	Vysočina, Hradec Králové, Plzeň and Olomouc	2018	Practical understanding of projects through personal visits to farms; building a knowledge base and support.
Enhancing farm viability and competitiveness	Černov	2014 – 2020	Monitoring non-economic factors, such as animal welfare on farms and organizational practices is important.
Climate change adaptation	Šardice	2014 – 2020	Management of limited resources, such as water, land and energy, including the implementation and support of educational activities in this area, is important.
Agri-food chain integration & quality	Písek, Podivín	2014 – 2020	Cooperation with science and research teams is the key to successfully meeting the specific objectives of rural development.
Water management	Bahna I	2014 – 2020	Igniting public interest in the protection of water resources and to see sustainability as a lifestyle trend is essential.

Source: [29].

Table 3. SME rural development support projects (Part 2)

Project	Region/area	Year(s)	Main Findings
Diversification and job creation	Pardubice	2014 – 2020	Cooperation with the LAG programme led to the creation of new jobs and regional growth.
Farm performance, restructuring & modernisation	Záhoří	2014 – 2020	Water resources can be protected by motivating private owners to change their work practices and personal attitudes.
Restoration, preservation & enhancement of biodiversity	Kašperské Hory, incl. relevant areas in Czech Republic	2014 – 2020	Organic farms were criticized for using grants without bringing relevant results to the public. Farm management needs to be changed for rural development.
	Doudleby nad Orlicí		Improving animal welfare on pig farms generates a competitive advantage.

Source: [29].

To ensure the successful implementation of projects, the objectives and achievements need to be monitored, which is carried out by the Monitoring Committee of the European Union for the Rural Development Programme [30].

Support for rural development in the Czech Republic is an interesting trend. People are beginning to prefer a healthy lifestyle, which increases mobility to villages, which has a positive effect on tourism. For SMEs, rural development represents an opportunity to supply quality products and provide smart services with the support of state aid and EU grants, thereby ensuring the sustainability of smart principles. Project scheduling and implementation need to incorporate the monitoring of results, which brings the following benefits [30]:

- the data collected and analysed are used to monitor progress and evaluate results;
- allocation of powers and responsibilities for tasks is streamlined;
- effective public and stakeholder communication is ensured;
- models, processes and projects are continuously improved;
- current data are a necessary basis for drawing up subsequent rural development policies and supporting SMEs in the region–across the board or locally.

5 Discussion

Experts define the concept of innovation from different perspectives. On the basis of a secondary analysis of their opinions it is possible to deduce the common and differing elements of their understanding of innovation, which includes the authors´ own opinion on the issue. The authors agree with the views of Skillicorn and Burkus, who say that 'the idea must be put into practice; otherwise it is not an innovation'. The situation is similar when perceiving strategy. Ways to achieve the objectives, i.e. strategies, do not exist unless they are implemented in practice. If they act only on a theoretical level, they cannot generate benefits, such as profits, market position, or competitive advantage. Skillicorn, McFarthing, Brands, Hobcraft, and Shipulsky make up the largest pool of authors who share the view that 'innovation affects both the internal and external environment'. On the contrary, Shapiro and Foley argue that 'innovation has the most impact on the market, i.e. on the external environment'. The majority of authors align themselves with the first opinion, since innovation affects management, employees, business models, services and products, which in turn affects customer satisfaction, sales, profit and the market. Skillicorn and Gijs van Wulfen combine the concept with customer orientation. The authors agree with Graber, McFarthing, Brands, Hobcraft, and Shipulsky that 'innovation affects all stakeholders, not just customers'. This common understanding is shared by authors such as Babra, Bakrus and Shapiro. Opinions vary when discussing Lindegaard's opinion, who does not want to define innovation, or Boyd, who sees innovation merely as something surprising and sudden. The authors understand the process of defining concepts as an important element of the creation of awareness and knowledge management. Innovation can also come in the form of an evolutionary process, with many managers recognizing and preferring the gradual introduction of innovative changes for better participation and change in the thinking of their employees. If innovation is not rapid and comes in the form of re-engineering, it does not mean that it is not a new creative idea. A summary of the common and differing views is presented in the following table. The views are ranked according to priority, whereby overlapping views have a higher priority, differing ones a lower priority (ranked towards the bottom of Table 4).

Table 4. Innovation features

Key findings	Author(s)
Innovation is not just about products.	Burkus; Šulyová, Vodák
The idea must be implemented in practice.	Skillicorn; Burkus; Šulyová, Vodák
Innovation has an impact on both the internal and external environment.	Skillicorn; McFarthing, Brands, Hobcraft & Shipulski, Šulyová, Vodák
Innovation affects all stakeholders, not just customers.	Graber; McFarthing, Brands, Hobcraft & Shipulski, Šulyová, Vodák
Innovation is linked to creativity.	Sloan; Graber; Šulyová, Vodák
Innovation is linked to flexibility.	Shapiro; Šulyová, Vodák
Innovation is the future.	Babra, Šulyová, Vodák
Innovation is synonymous with relevance.	Shapiro
Innovation must be sudden and surprising.	Boyd
Innovation is linked mainly to the market (external environment).	Shapiro; Foley
Customer orientation.	Skillicorn; Gijs van Wulfen
Innovation does not need to be defined.	Lindegaard

Source: modified on the basis of Nick Skillicorn, 2016.

The secondary analysis conducted showed the positive impact on SMEs of building smart city concepts, which is supported by Miter´s claims. The benefits can also be seen in rural areas, creating a competitive advantage for the entire region. The common elements are, inter alia, efforts to protect the environment; utilize natural advantages in the form of mountains, meadows, or pastures; and to generate an effective type of biosphere and plant or animal diversity. Supporting these smart concepts will increase rural traffic, employment, lower urban mobility, and generate higher levels of gross domestic product. The key elements of the structure are the definition of the vision, its communication to all stakeholders, and the alignment of global factors with the specific benefits for the region. By comparing the implementation of rural area innovations in South Korea and the Czech Republic, the following supporting factors have been identified and prioritized (Table 5), whereby, once again, overlapping factors have a higher priority, differing ones a lower priority.

Table 5. South Korea vs. Czech Republic – factors supporting rural smart approaches

Supporting factor	South Korea	Czech Republic
Diversification of development areas is important.	Yes	Yes
Development structure depends on monitoring.	Yes	Yes
Focus on agricultural and forestry industries.	Yes	Yes
Trend correlates with changes in the demographic curve.	Yes	Yes
Development based on long traditions.	Yes	No
Public mobility trend from cities to rural areas is increasing.	No	Yes
Growing poverty of rural areas.	Yes	No
Desire for a healthy lifestyle.	No	Yes
Centralized rural area management	Yes	No
Management requires subsidies from the Czech government/EU.	No	Yes
Strong cooperation between the state, the regions and the general public.	Yes	No

Source: Authors.

The common factors supporting rural development in South Korea and the Czech Republic are the diversification of development areas, the correlation of trends with changes in demographic curves, and preference for investment in the agricultural and forestry industries. The traditional approach to rural poverty-based development prevails in South Korea. In the Czech Republic, the main theme is increasing the mobility rate to rural areas and the promotion of a healthy lifestyle. Management in South Korea is more centralized but also strongly connected with public communities. The Czech Republic uses subsidies from the Czech government or the EU, which may be a negative factor rather than a positive one (Table 6).

Table 6. Recommendations and constraints on SME innovation development, stabilization, and smart approach implementation

Recommendations	Constraints
It is adequate to see rural development as a tradition, and not just as a trend.	Alignment of concepts to local conditions.
The diversification of smart projects in the regions is a key factor to success.	Lack of funding for rural areas.
Traditionally supported industries, such as agriculture or forestry, should embrace new sectors, such as breweries.	The need to draw on limited European Structural and Cohesion Funds that arrive late. To obtain funding, it is necessary to meet pre-determined administrative requirements.
Cooperation with SMEs will bring investment for the implementation of development projects in rural areas.	Low awareness or little interest from the public in the implementation of rural development projects.

Source: Authors.

The secondary analysis and the results obtained confirm the suitability of implementing innovations through SME development projects in rural areas.

6 Conclusions

Building smart cities, villages and regions is an issue that is gaining traction in European countries. The key to success is innovation, which must be perceived within the context of changes in both the internal and external environments. The main findings from the secondary analysis of the associated benefits for stakeholders (SMEs, management, public) of participating in such concepts are:

- higher economic growth, employment, innovative ideas;
- monitoring, building knowledge bases, best practices for culturally and regionally similar countries;
- predictions, use of natural conditions, and the promotion of diversity.

The secondary analysis of best practice confirmed the importance of rural growth and development in South Korea's regions. Although the given territories differ in terms of culture, when applying smart concepts to rural areas, common factors prevail, such as diversification, focus on the agricultural and forestry industries, or the impact of global trends—changes in demographics and the increasing migration of people towards rural areas in search of a healthy lifestyle.

It is imperative to learn and direct the innovative activities of companies, villages or cities towards the benefit of all stakeholders and to eliminate the narrow perception of innovation as purely being technological in nature. By developing and supporting existing projects in the Czech Republic, their expansion to new potential areas, and the establishment of a traditional approach to this issue, this trend can be stabilized and used by SMEs as a basis for gaining a competitive advantage.

References

1. N. Skillicorn, *What is Innovation? 15 experts shares their innovation definition* [online]. Available at: https://www.ideatovalue.com/inno/nickskillicorn/2016/03/innovation-15-experts-share-innovation-definition/ (2020)
2. M. Kubina, V. Lendel, Successful Application of Social CRM in The Company. *Procedia Economics and Finance.* **23**, 1190–1194 (2015)
3. F. Pollák, P. Dorčák, P. Markovič, Reputation Management. In: Umut Ayman, Anil Kemal Kaya, *Promotion and Marketing Communications*, [online]. Available at: https://www.intechopen.com/books/promotion-and-marketing-communications/reputation-management (2019)
4. P. Dorčák, P. Markovič, F. Pollák, Multifactor analysis of online reputation as a tool for enhancing competitiveness of subjects from automotive industry. *Journal Economics.* **65**(2), 173–186 (2017)
5. V. Lendel, M. Varmus, Identification of the main problems of implementing the innovation strategy in Slovak businesses. *Acta Universitatis Agriculturae et Silviculturae Mendelianae Brunensis.* **60**(4), 221–234 (2012)
6. M. Holubcik, Theoretical knowledge in terms of forming cooperation. *Selected papers of 5th world conference on business, economics and management: New trends and issues proceedings on humanities and social sciences (BEM-2016).* **4** (2017)

7. D. Zrakova, M. Kubina, G. Koman, 2. Influence of information-communication system to reputation management of a company. *Procedia Engineering.* **192**, 1000–1005 (2017)

8. M. Pechrová, A. Kolářová, Assessing of the Projects Promoting Innovations in Rural Areas in the Czech Republic. *Agrarian Perspectives: European Association of Agricultural Economist.* **22**(18–19) (2012)

9. C. Cunha, E. Kastenholz, M. J. Carneiro, Entrepreneurs in rural tourism: Do lifestyle motivations contribute to management practices that enhance sustainable entrepreneurial ecosystems? *Journal of hospitality and tourism management.* **44,** 215–226 (2020)

10. W. Dhewanto, S, Ratnaningtyas, A. Permatasari, G. Anggadwita, E. A. Prasetio, Rural Entrepreneurship: Towards collaborative participative models for economic sustainability. *Entrepreneurship and Sustainability issue.* **8**(1), 705–724 (2020)

11. R. Da Pires, M. Pertoldi, J. Edwards, F. B. Hegyi, *Smart Specialisation and Innovation in Rural Areas.* Luxembourg: Publications Office of the European Union (2014)

12. J. Mitra, *Smart Cities: Ecology, Technology, Entrepreneurship and Citizenship. University of Essex* [online]. Available at: https://www.climateneutral.eu/wp-content/uploads/2017/10/Jay-Mitra_Smart-Cities-for-Evolving-Entrepreneurship.pdf (2017)

13. J. Holly, A. Kona, Will the Internationally Recognizable Country Identity of Slovakia "Work" or will the Current Attempts at Unifying the Presentation of Slovakia Abroad and the Search for an Attractive Narrative End in Failure. *Mediterran Journal of Social Sciences.* **6**(6), 368–371 (2015)

14. R. Krishna, R. Kummitha, Smart Cities and entrepreneurship: An agenda for future research. *Technological Forecasting and Social Change.* **149** (2019)

15. J. Straková, P. Pártlová, J. Váchal, M. Vochozka, Excellent top manager system (ETMS) - Quality management tool tertiary education. *Proceedings of the 29th International Business Information Management Association Conference - Education Excellence and Innovation Management through Vision 2020 – Regional Development Sustainability to Global Economic Growth.* **29** (2017).

16. R. Vachalova, S. Matejkova, J. Vachal, P. Partlova, M. Dumbrovsky, L. Jurik, Evaluation methods of agricultural potential in rural areas including environmental function. *Annals of Warsaw University of Life Sciences – SGGW Land Reclamation.* **43**(1), 87–97 (2011)

17. H. Rochdane, S. Hamdani, Role of Small and Medium Enterprises in the Development of Smart Cities: Case of Casablanca Smart-City Project – Morocco. *International Journal of e-Education, e-Business, e-Management and e-Learning.* **8**(3), (2019)

18. J. Hwang, J. Park, S. Lee, The Impact of the Comprehensive Rural Village Development Program on Rural Sustainability in Korea. *Sustainability.* **10**(7), (2018)

19. J. Craigie, *The self-help rural communities of South Korea* [online]. Available at: https://www.ruralnetwork.scot/self-help-rural-communities-south-korea (2019)

20. S. H. Park, *Rural Development of Korea: Historical Evolution and Performance* [online]. Available at: https://krei.re.kr/eng/selectBbsNttView.do?key=368&bbsNo=192&nttNo=37509&searchCtgry=&searchCnd=all&searchKrwd=&pageIndex=5&integrDeptCode= (2018)

21. E. Garcilazo, L. S. Springare, M. Sasaki, D. O. Higuera, *Rural study of Korea. OECD Regional Development Working Papers No. 2019/05.* Paris: OECD Publishing (2019)

22. O. Plevák, *Smart village is a remedy for outflow of people from rural areas* [online]. Available at: https://www.euractiv.com/section/agriculture-food/news/smart-village-is-a-remedy-for-outflow-of-people-from-rural-areas/ (2020)

23. Vaishar, J. Zapletalová, E. Nováková, Between Urban and Rural: Sustainability of Small Towns in the Czech Republic. *European Countryside.* **8**(4), (2016)

24. AEIDL, *Smart villages, a remedy for rural areas in the Czech Republic* [online]. Available at: https://www.aeidl.eu/en/news/latest-news/5293-smart-villages-a-remedy-for-rural-areas-in-the-czech-republic.html (2020)

25. European Commission, *Factsheet on 2014-2020 Rural Development Programme for the Czech Republic* [online]. Available at: https://ec.europa.eu/info/sites/info/files/food-farming-fisheries/key_policies/documents/rdp-factsheet-czech-republic_en.pdf (2020)

26. D. Popjakova, M. Blazek, Verification of counterurbanisation processes: Example of the Ceske Budejovice region. *Bulletin of geography-socio-economic series.* **27**(27), 153–169 (2020)

27. eAGRI Venkov, *Program rozvoje venkova 2014-2020* [*Rural Development Program 2014-2020*]. [online]. Available at: http://eagri.cz/public/web/mze/venkov/program-rozvoje-venkova/prv-2014-2020/ (2020)

28. European Network for Rural Development, *LAG Database* [online]. Available at: https://enrd.ec.europa.eu/leader-clld/lag-database/_en?f%5B0%5D=im_field_enrd_lag_country%3A19213 (2020)

29. European Network for Rural Development, *Project in Czech Republic* [online]. Available at: https://enrd.ec.europa.eu/projects-practice/_en?f%5B0%5D=sm_enrd_eu_countries%3ACzech%20Republic (2020)

30. eAGRI Dotace, *Hodnocení a monitoring* [Evaluation and monitoring]. [online]. Available at: http://eagri.cz/public/web/mze/dotace/program-rozvoje-venkova-na-obdobi-2014/hodnoceni-a-monitoring/ (2020)

Events as an Effective Marketing Communication Tool

Martin Vítek[1,*]

[1]Institute of Technology and Business, Faculty of Corporate Strategy, Department of Tourism and Marketing, Nemanická 436/7, České Budějovice, Czech Republic

Abstract. A brewery wanted to know how a new beer they were launching would be received by consumers. It was therefore decided to organize an event during which the author of this contribution conducted research. The aim of the research was to find out whether the new product, a low alcohol beer, will attract consumers with its taste, whether its price was set correctly, and whether the targeting and marketing communication tools used were appropriate. The results of the research confirmed that the taste of the new product as well as its price corresponded to the imagination of consumers and that the communication mix was appropriate, but required adjustment to make it even more precise.

Keywords: event; event marketing; marketing communication; new product; communication mix

1 Introduction and literature review

The marketing communication mix of companies includes in varying degrees, according to the specifics of a particular organization, advertising, direct marketing, sales support, public relations, event marketing and sponsorship, personal sales and online communication. However, as the pace of life accelerates, so the forms and content of individual communication channels need to be adapted. At the same time, it is necessary to keep in mind that the target group is always exposed to the marketing message in a certain communication context, which may strengthen or weaken it [1].

Event marketing is a very specific communication tool. It is based on the direct emotional involvement of the target group, thereby positively affecting the relationship to a particular product or brand. What is important are "experience worlds" that appeal to one or more senses. This includes the use of certain colours, shapes, melodies and smells during events in order to "enhance" the product and give it tactile benefits. This is an active workload in which the target group meets at a specific place and at the same time and perceives a multisensory experience [2].

The success of event marketing is based on the experience of its participants on the basis of secondary - learned motivation. This is formed during a person´s development on the basis of conditioned emotional reactions, which is reflected in tendencies to certain appetites [3].

One of the most attractive events are congresses. Participants are enriched through their travel experiences [4]. Congress tourism is a collection of activities, services, economic and social sectors that provide and ensure travel experiences. Since the1950s, the tourism sector has become a rapidly developing and expanding sector of the world economy. In addition to its foreign exchange-earnings effect, tourism is an alternative source of growth for national economies, contributing significantly to national income, employment and the balance of payments [5].

Likewise, exhibitions and trade fairs are also of interest. These are characterized by the confrontation of supply and demand in a specific field of human activity, their periodic nature, their predetermined goal, and rich accompanying programme.

For example, some cities with an interesting history provide grounds for various events to celebrate their founding and therefore become one large exhibition with many exhibits. Hosting special events can be a unique opportunity for boosting tourism development, resulting in loyalty and attachment towards the destination and its assets [6].

Within this context, the event image is transferred to city image. This city image plays a pivotal role in mediating the effect of the event image, on participants satisfaction, and their intention to revisit the host city [7].

The organization of various festivals also falls under event marketing. The word festival usually refers to an organized set of special social events, such as music or film performances. Festival began to be used as a noun in the 16th century, when it acquired the meaning of a feast or celebration. The term festival as we know it today means a regular recurring programme consisting of cultural performances, exhibitions and competitions. The Latin word festívus, means cheerful,

* Corresponding author: vitek@mail.vstecb.cz

 https://doi.org/10.1051/shsconf/20219001022

joyful, holiday day. The English word festive is derived from the word festívus and means festive, holiday. The use of festivals as promoters of tourist destinations is huge [8].

There are many types of festivals in the world, with most countries celebrating important events or traditions with traditional cultural events and activities. Most festivals offer a variety of speciality foods and drinks to enhance the cultural experience. Satisfaction with the festival experience also significantly influences visitors' attachment to a place and their intentions to recommend a destination [9].

Worldwide, the number and variety of cultural festivals have grown dramatically. Many areas see festivals as an important way to attract tourists, and their spending, to a region, resulting in a positive economic impact [10].

Film, theatre and music festivals can be included among the very popular types of festivals. Each type of festival has its specifics. Hosting film festivals has become a prevailing practice to promote culture or festival tourism [11].

Music festivals can be self-serving, for example, to promote a specific music genre, but it is also very often used as a comprehensive tool for marketing communication. They are used by corporate entities to promote a product or brand, but also used, for example, by cities for their celebrations. In such cases, it is usually multi-genre, intended for a larger number of visitors and therefore includes a larger number of target groups. In the field of marketing, the informational function of music festivals cannot be overlooked [12]. Participants of music festivals who experience changes in their self-construct related to self-growth and self-development from their participation in a music festival, use music to facilitate self-awareness and social relatedness [13]. For the organizing entity, however, the most important thing is always the intensive communication with each visitor, not only through music, but a comprehensive experience that can address all senses.

Rock festivals occupy an important place in contemporary consumer culture [14]. In Germany, on the other hand, music and folklore play an important role. A typical example is Oktoberfest, which is one of the most famous beer festivals in the world [8]. The deliberate combination of consuming good beer and a pleasant musical experience deepens the positive experience and perception of a particular brand of drink. Such events offer lovers of craft beer a comfortable but regulated social and physical environment for personal and collective experiences [15].

The town of České Budějovice is home to the brewery Budvar, which many years ago decided that, in addition to its premium brand, it would produce beers that were simpler, low in alcohol and more affordable. On the occasion of the launch of a new beer, the brewery, in cooperation with the author of this contribution, decided to implement an advertising campaign, at the end of which a music festival was organized. This type of event is seen as an important communication tool for strengthening and enriching the mutual business relationships with suppliers and customers [16].

2 Methodology

The joint intention of the beer producer and their marketing agency was to organize an event that would support a new product launch. The event took place in summer 2019, during which the author of this contribution and their team carried out research.

It should be noted that the brewery had already done its own research before launching the new beer, to which the author of this contribution did not have access. The research presented here is therefore truly independent, with no team member knowing even basic information about who the target group for the new product was. The only thing the researchers knew was that the taste of the beer had been gradually tuned according to the tastes of specially selected beer consumers, whose sensory properties and evaluation abilities had previously been very carefully verified.

The aim of the research was to find out whether the new product would attract festival visitors with its sensory properties, whether its price was pitched correctly, and whether the selected marketing communication channels were appropriate and effective.

Three research questions were formulated:

1. *What was the composition of the visitors of the music festival?*
2. *How did the visitors perceive the different variables of the communication mix which were used by the brewery to promote the event and launch the new beer?*
3. *What qualities and characteristics does the beer have in terms of taste and is it reasonably priced?*

The brewery set aside suitable premises at the festival venue and organized the cooperation of other services. The author of this contribution and their team collected the data and then processed it. After mutual assessment, the brewery approved the use of personal questioning as the research method. A structured questionnaire was drawn up consisting of 20 questions split into two parts, each part to be filled in at a different time.

The first part of the questionnaire was filled in when entering the venue and the second upon leaving. The questionnaire contained open questions, closed-ended questions with the option "other", and closed questions for qualitative and quantitative data. The lower age limit of the respondents was set at 18 years, no upper limit was applied.

https://doi.org/10.1051/shsconf/20219001022

At all access points, the agency placed stands with various attractions and simple competitions, which were used to attract people to participate in the personal questioning. Participants were motivated to participate by giving them a voucher for two free beers. When handing over the vouchers, the assistants asked if anyone would be interested in participating in the questionnaire survey and if so, the assistants began to act as interviewers and recorded the respondents' answers on the record sheet. All participants could also win interesting prizes. These were nice gift items with the brewery's logo.

The selection of respondents was not random. They were chosen from those who showed active participation in the competitions, under the assumption that after the music part of the festival they would be willing to complete the questionnaire. Each active participant was assigned an identification number that corresponded to the questionnaire number. After the end of the music, all the numbers were drawn, and the winners were asked final questions during the awards ceremony. A total of 214 respondents aged 18 to 57 participated in the research.

3 Results and Discussion

3.1 Characteristics of respondents

What was the composition of the festival visitors? The respondents consisted of 117 men and 97 women. Although a balanced number was not achieved, this difference is not important for the targeting of this research.

In order to improve the communication mix for the future, the socio-economic part of the questionnaire sought to establish where the visitors came from.

According to the postcode, a table was compiled showing the number of visitors according to the number of inhabitants of the municipalities and cities from which they came from (Table 1).

Table 1. Respondents by size of town/village they came from

Number of residents	Respondents/visitors	
	n	%
more than 80,000 residents	160	74.8
10,000 – 79,999 residents	17	7.9
4,000 – 9,999 residents	16	7.5
smaller villages	21	9.8
Total	**214**	**100.0**

Source: Author.

Additional data on the selected group of visitors enables the brewery's marketers to compare the results with their target group(s). In accordance with the requirements of the brewery, the level of education and job classification of the visitors was established. This helps to identify and better define particular target groups (Tables 2 and 3).

Table 2. Educational attainment of respondents'

Education	Respondents/visitors	
	n	%
University graduate	57	26.6
High school graduate	148	69.2
Elementary school leaver	9	4.2
Total	**214**	**100.0**

Source: Author.

Table 3. Job classification of respondents

Job classification	Respondents/visitors	
	n	%
Manual worker/craftsman	64	29.9
Administration	118	55.1
Manager/businessman	32	15.0
Total	**214**	**100.0**

Source: Author.

3.2 Efficacy of marketing communication mix

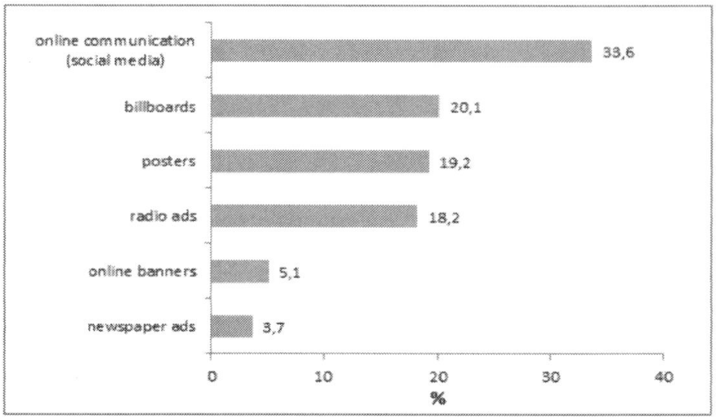

Figure 1. Efficacy of marketing communication mix

Source: Author.

Figure 1 provides the answers to the question, through which marketing communication tools did the respondents learn about the analysed event. In addition to these general findings, the results of the questionnaire survey revealed that for respondents from large cities with a population of more than 10,000 residents, billboards and online advertising were the most attractive, while respondents from cities and municipalities with smaller populations were most attracted by posters and radio advertising. This implies that part of the funds from other marketing communication channels can be redirected towards a billboard campaign in large cities and towards the use of posters in villages. For a better overview, the perceived efficacy of the advertising undertaken from the perspective of respondents is presented in the table below.

The research results show that the respondents were very attracted by radio advertising. Almost everyone agreed that this was due to the funny pre-recorded spots by well-known members of the music groups that performed at the festival. Most of the respondents who said the radio campaign appealed to them, stated that they were travelling to work by car at the time of hearing it. For future radio advertising campaigns, the brewery should increase the frequency of commercials when people are either going to or from work.

3.3 Evaluation of beer taste

The preference for beer taste is different in men and women, so the sensory impressions were divided. The evaluation of sensory values showed that the degree of bitterness of the beer was appreciated by 87% of the respondents. The beer's density was perceived by 20% of men to be not quite sufficient. However, they agreed that it was a beer with a low alcohol content and therefore fewer substances were released into it at the time of fermentation than stronger beers, for which the fermentation time is significantly longer. The overall taste of the beer was evaluated as medium bitterness, slightly harsh in character, medium-bodied, and with medium-strong sharpness.

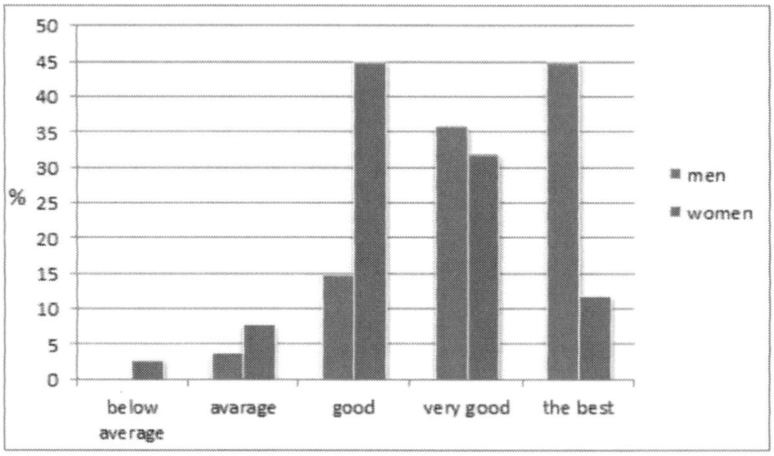

Figure 2. Evaluation of beer taste (in %)

Source: Author.

The price of the beer was perceived by most of the women interviewed as reasonable. In contrast, the opinions of the men differed. Men from cities felt that the price could be higher, whereas those from the countryside felt that it corresponded with their ideas or could be lower. In total, however, it was confirmed that the price of the beer set by the brewery corresponded to the expectations of the majority of respondents.

4 Conclusion

The main purpose of this contribution was to demonstrate how it is possible to carry out research into a select element of marketing communication - an event, which was organized as part of the presentation of a new product. The goal of the research was fulfilled, not only in its basic part, but also in the related context.

The results of the research will help the brewery verify whether they are communicating properly with their customers. The brewery will also be able to compare their previous findings regarding their target groups. With regards to the taste of the beer, it was established that it was excellent. With regards to price, it was established that this corresponds with consumers´ expectations.

In addition to fulfilling the basic goals of the research, it was found that the music festival, as a marketing tool, was appropriate to the target group(s), as was reflected in the positive evaluation of 97% of the respondents.

References

1. M. Karlíček, *Marketingová komunikace [Marketing comunication]*. Praha: Grada Publishing (2016)
2. J. Vysekalová, *Emoce v marketingu. [Emotions in marketing]*. Praha: Grada Publishing (2014)
3. M. Nakonečný, *Motivace chování [Behavior motivation]*. Praha: Triton (2014)
4. Ch. Goeldner, R. Brent, *Cestovní ruch: principy, příklady, trendy [Tourism: principles, examples, trends]*. Brno: BizBooks (2014)
5. H. O. Saridogan, The Impacts of Tourism Revenues on Economic Growth: Unit Root and Cointegration Tests with Structural Break for Turkey. *Eskisehir Osmangazi Universitesi IIBF Dergisi- Eskisehir Osmangazi Universitesi journal of economics and administrative sciences*. **15**(3), 829–848 (2020)
6. G. Mainolfi, V. Marino, Destination beliefs, event satisfaction and post-visit product receptivity in event marketing. Results from a tourism experience. *Journal of business research*. **116**, 669–710 (2020)
7. H. Li, CH. Lien, SW. Wang, T. Wang, WW. Dong, Event and city image: the effect on revisit intention. *Tourism review*. In print. (2020)
8. C. B. Kraus, T. F. Fiuza, F. D. Zucco, C. M. B. de Quadros, Satisfaction and engagement of festival visitors: Empirical analysis of an Oktorberfest. *Podium-sport leisure and tourism review*. **9**(2), 308–336 (2020)
9. O. Culha, The effect of food festival quality on place attachment and destination recommendation intention through festival experience and festival satisfaction: The case of the Didim International Olive Festival. *Journal of convention & event tourism*. (2020)
10. J. D. Snowball, G. G. Antrobus, Festival value in multicultural contexts: City festivals in South Africa. *Tourism economics*. (2020)
11. G, Kendall, J. H. T. Chan, M. C. H. Yeung, K.K. Law, Do film festivals attract tourists? *Current issues in tourism*. (2020)
12. J. D. Montoro-Pons, M. Cuadrado-Garcia, Music festivals as mediators and their influence on consumer awareness. *Poetics*. **80** (2020)
13. R. Lawendowski, T. Besta, Is participation in music festivals a self-expansion opportunity? Identity, self-perception, and the importance of music's functions. *Musicae Scientiae*. **24**, 206–226 (2020)
14. D. Chaney, Rock festivals as marketplace icons. *Consumption markets culture*. **23**(3), 215–222 (2020)
15. J. Ikaheimo, Exclusive craft beer events: liminoid spaces of performative craft consumption. *Food culture & society*. **23**(3), 296–314 (2020)
16. H. Lošťáková, *Nástroje na posilování vztahů se zákazníky na B2B trhu. [Tools for strengthening customer relationships in the B2B market]*. Praha: Grada Publishing (2017)

The Push and Pull Factors Affecting the Migration of Mongolians to the Republic of South Korea

Altanchimeg Zanabazar[1,*], *Nam Son* Kho[2], *Sarantuya* Jigjiddorj[1]

[1]National University of Mongolia, Business School, Mongolia
[2]Mokpo National University, Business School, South Korea

Abstract. Migration, which is a rational labour issue within society, has become a hot topic globally. In recent years there has been an increase in labour migration due to globalization, disparities in the development levels of countries, culture and environment, social instability, political regimes and war.
The objective of the presented study was to research the push & pull factors affecting the migration of Mongolians to the Republic of South Korea. The survey was conducted among 498 Mongolians living and working in the Jeollanamedu and Jeollabugdu districts of the metropolitan cities of Incheon and Ulsan, as well as in the capital city Seoul.
Numerous statistical techniques were applied to test the reliability and validity of the data, as well as factor analysis to confirm that the concepts of each variable were correctly measured, correlation analysis to assess the strength of the relationship between the independent and dependent variables, and multiple regression analysis to predict the value of variables used in the study.
The results of the study showed that the major push factor was an economic one, for which the main motives for migration were low or instable income, the economic downturn and poverty. Likewise, the pull factors were also economic, whereby the motivation to migrate to South Korea was driven by high wages, wellbeing, the opportunity to save and/ or social factors, such as access to quality education, the cultural experience and joining family members who had already settled in the country.
Keywords: Migration; pull factor; push factor; employee turnover

1 Introduction

Globalization has changed people's perception of the world. An increase in access to information through social media and other means of mass communication has contributed dramatically to changing people's perceptions about lifestyles and wellbeing, thereby increasing their awareness of the disparities and disadvantages of living in poor countries, which in turn has led to a change in their aspirations, hopes and dreams.

These changes often make people consider emigration in pursuit of security and wellbeing. Younger generations in particular are keener on the migration option because of the opportunities it presents to prevent the daily struggle at home. Wage discrepancies between countries is a key reason for international labour migration, as is the desire among students with a view to enrolling in tertiary education institutions in developed countries for a shorter or longer period of time [1].

Economic conditions serve as a push factor for developing countries and as a pull factor for industrialized countries.

Despite the social and economic stability achieved in Mongolia in recent years, the income of its citizens still remains low. Limited job opportunities has exacerbated the situation and led to an increase in migration from rural areas to urban ones, as have the country's continuing socio-economic problems, which has resulted in an increase in official and unofficial migration to foreign countries [2]. Migration has had an important economic impact on Mongolia in terms of human resource turnover [3].

The objective of the study: The current study aims to determine the push and pull factors affecting the migration of Mongolians to the Republic of South Korea.

* Corresponding author: altanchimegz@num.edu.mn

2 Methodology

Migration is the movement of people from one geographical location to another, or from one administrative territory to another, or leaving the homeland permanently or temporarily for an overseas country or territory.

The chance of estimating the level of migration using a universal method are slight. As a result, researchers have developed estimation criteria that reflect the history, traditions and geographical location of the country.

Crossing territorial boundaries from one place to another is referred to as out-migration. The movement within a country, i.e. moving home to another area, is referred to as in-migration. In contrast, the movement from one country to another to access job opportunities or for socio-economic reasons is referred as emigration.

Despite economic factors, migration as a phenomenon is closely associated with politics, culture, environment, health, education and transportation.

Osaki and Safrida state that migration in industrialized societies is created due to labour demand and the people involved do so voluntarily or as a result of certain circumstances [4, 5]. According to the latest concepts, the requirements of the labour market is considered the main factor for migration, whereby the push factors of the home land and the pull factors of the destination [6] country play a crucial role. The extensive use of pull and push factors has been evidenced frequently through research, whereby various theories are used to explain migration.

2.1 Theories on push and pull factors of migration

Push and pull factor theory is widely used in migration research and was introduced by Ernst Ravenstein. In his work "Laws of Migration", he explained the theory of step migration [6], whereby migration occurs gradually and step by step geographically. Subsequent researchers have used the theory in their works to explain patterns of geographical migration, be it within the homeland or to a host country.

Researchers have identified the factors of migration that push people to move away from their homeland, and likewise the pull factors, including the positive aspects of different geographical areas which may encourage people to be attracted to other locations and seek opportunities elsewhere.

Both push and pull factors must be present for migration to occur [7]. According to the concept of push and pull factors, push factors are those life situations in one's present homeland that generate dissatisfaction, such as poverty, unemployment, rapid population growth, political repression, low social status. In contrast, the pull factors that make migration appealing are well-being, job opportunities, political freedom, education and the welfare system of destination country [8]. Understanding the migration factors helps identify patterns of migration.

The 'push-pull' theory considers population movement to be the result of two forces acting in different directions. One is the power to promote population movements, whereas the other is the power of impeding population flows. On the whole, 'push' power tends to be more dominant compared to 'pull' power [9].

Lee concludes that migration is a clever decision for those who move to other destination countries to benefit from the opportunities that are not sufficient in their homeland, but that these opportunities are closely related to demographics and economic factors [10].

According to Lee's theory, unemployment, poverty, low wages and human rights issues in the homeland are push factors, while appealing conditions in the host country like high life expectancy, economic stability, low birth rate and the availability of 3D jobs are pull factors.

The stimuli for migration can be categorized as follows:

1. Factors relating to the homeland;
2. Factors relating to the host country;
3. External challenges; and
4. Personal characteristics.

Abella adds that significant changes in the demographics in developing and developed countries leads to migration [11]. Developed countries have ageing populations are decreasing birth rates. Developed countries therefore seek to attract people from developing countries to fill job vacancies and maintain productivity. Within this context, the growing need for caregivers and nurses in a rapidly ageing society is a strong pull factor for female migrant workers to go and work overseas.

Like other researchers, Richard concludes that migration is driven both by push and pull factors [12], for which the reasons are economic, safety concerns, the environment, and social and political issues.

Push factors represent a particular challenge for a country. In general, negative conditions cause migration, such as low productivity, unemployment, unfavourable economic circumstances, natural disasters, pollution and resource depletion [13].

Another study on the push and pull factors affecting female migrant workers in Indramayu, Indonesia, concluded that economic reasons and the unemployment rate often make female employees consider the option of working abroad. Interestingly, according to the study, the successful choice of destination country does not only lead to economic benefits, but also an upgrade of the social status of the migrant workers' household [14].

A different study on the pull factors of migration revealed that higher salaries for the same workload in destination countries also contributed to the consideration of migration [15].

Social reasons for considering migration, as the case study in Myanmar shows, are improvements in living standards for people in rural areas and access to better public services [16].

Despite the voluntary nature of migration, persecution, war, crop failure and famine, or religious conflict can be a push factor. The roots of other push factors may include racial and ethnic discrimination and violence.

Pull factors are those in the destination country that appeal to the people to leave their homeland. These factors are economic and job opportunities, higher wages, favourable conditions for work and life [13]. Other pull factors can be the attractiveness of geographical areas such as peace, warm climate, less pollution and occurrence of natural disasters, comfortable life, for example for retired people with sufficient financial resources.

According to Xiang et.al, labour migration has become increasingly popular in many parts of Asia, however the risks associated with social protection are an issue [17].

According to [18], the reasons for migration to another country fall into two categories, namely economic and non-economic, which influences the type of migration: demand-pull; supply-push; and networks (see Table 1).

Table 1. Factors encourage migration

Type of migration	**Demand - pull**	**Supply - push**	**Network / other**
Economic	Labour recruitment	Unemployment, underemployment or low wages	Job and wage information flows
Non-economic	Family unification (Family members join spouse, or children with parents)	Escaping from war, conflicts	Communications, transportation, aid organizations and desire for new experiences/adventure

Source: [18].

In their research, Muhammad and Yuherina gave consideration to the economic, social and cultural push factors for migration [14].

On the other hand, researchers like Van Hear, et.al reviewed existing factors affecting migration, approaching the issue from a different angle, namely migration policy within the political system based on the case of Afghan and Somali movements [19].

Parkins in her article "Push and Pull factors of Migration" identified four factors that influence migration: crime, discrimination, occupation and skills, of economic and social opportunities. In the study, in order to ensure comparability, the researcher classified the respondents into four categories: those who had not yet migrated; people considering migration; people in the process of migrating; and people who have already migrated [20].

Researchers frequently consider economic, demographic, socio-cultural, political and miscellaneous factors as the key factors affecting migration. What follows is a summary of their findings.

Economic factors: The majority of researchers see this as key. In contrast to the developed world, developing countries face a number of challenges - unemployment, low wages, limited job opportunity, poverty – which often serve as push factors. The majority of people moving from developing countries to industrialized ones are therefore migrants seeking economic benefits.

Demographic factors: These include personal characteristics, including age, gender, social origins, education level, race, family and ethnicity [21].

Socio-cultural factors: This concerns, for example, the younger generations desire to obtain diverse cultural experiences through studying abroad and living for a shorter or longer period in overseas countries.

Political factors: Political crises, military coups, discernment, lack of political freedom, social origins, political persecution often makes people consider migration as a means to survive.

Miscellaneous factors: These tend to be personal motives like institutional stability, the degree to which it is possible to disagree with the authorities of state, employment, promotion criteria, links to family and friends, and skills match [21].

Study model: Based on the literature review, the following research design was developed, including the respective hypotheses.

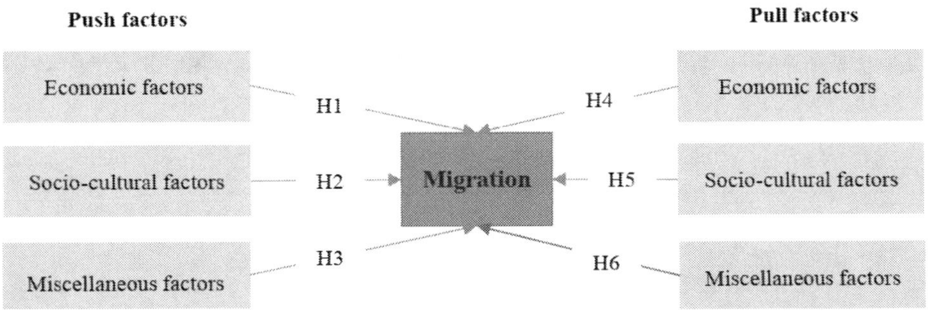

Figure 1. Research Design

Source: Authors.

Hypotheses for the study:

The following hypothesis were formulated within the framework of the study:

H1: Economic factors serve as push factors for Mongolia.

H2: Socio-cultural factors serve as push factors for Mongolia.

H3: Miscellaneous factors serve as push factors for Mongolia.

H4: Economic factors serve as pull factors for the Republic of South Korea.

H5: Socio-cultural factors serve as pull factors for the Republic of South Korea.

H6: Miscellaneous factors serve as pull factors for the Republic of South Korea.

3 Results and Discussion

A questionnaire survey was conducted among Mongolians living and working in the Jeollanamedu and Jeollabugdu districts of the metropolitan cities of Incheon and Ulsan, as well as in the capital city of South Korea, Seoul. Numerous analyses were carried out to verify the proposed hypotheses of the study, including factor analysis, reliability test, correlation and regression analysis using Smart PLS 3.0.

The survey was conducted from 15 June to 25 August 2019. Six hundred fifty questionnaires were distributed to potential respondents. In total, 498 (76.6%) filled in questionnaires were returned. The questionnaire consisted of 27 questions divided into 6 sections, with a 5 point Likert scale for each question. The demographic characteristics of the respondents was collated, including age, gender, duration of stay and residence status (see Table 2).

Table 2. Background information of respondents

Regions the participants live		Age		Gender		Duration of stay		Residence status	
Region	**%**	**Age**	**%**	**Gender**	**%**	**Years**	**%**	**Status**	**%**
Ulsan	16.3	Up to 20	7.1	Male	59.7	3 months	9.7	Student	27.5
Seoul	35.4	21-25	15.0			3 -6 months	18.0	Contractual worker	20.0
						6 months up to -1 year	15.9	Tourist	12.7
Jeollabugdu	15.0	26-35	49.8			2-3 years	27.7	Permanent resident	13.7
Jeollanamedu	25.1	36-45	22.7	Female	40.3	4-6 years	15.0	Medical treatment	0.6
						7-10 years	8.2	Illegal	25.5
Incheon	8.2	Over 45	5.4			11 and more years	5.6		

Source: Authors.

The background information of the respondents reveals that 59.7% were male and 40.3% female. By age, 71.9% of the respondents were 18-35 years old, which means that the majority of the migrants living and working in the Republic of South Korea are younger people. By duration of stay, 43.5% stayed up to 1 year, 27.7% 2-3 years, 13.8% 4 and more years, and 15.0% up to 6 years.

By residence status, 27.5% were student's, 20.0% worked under contract, 13.7% were permanent residents, and 25.5% were illegal immigrants.

3.1 Reliability analysis of the variables

A reliability test was conducted to ensure the trustworthiness of the tool. Cronbach´s α indicated 0.7 points [22], which confirms the reliability of the questionnaire for further analysis (see Table 3).

Table 3. Reliability analysis of push and pull factor variables

	Factors	**Number of questions**	**Cronbach's α**
Push factors	Economic factors	5	.861
	Socio-cultural factors	3	.844
	Miscellaneous factors	2	.909
Pull factors	Economic factors	4	.892
	Socio-cultural factors	4	.701
	Miscellaneous factors	2	.926
	Migration	7	.873

Source: Authors.

Factor analysis was subsequently carried out to describe the push and pull variables in terms of other factors. To measure the internal consistency in scale items, a CR (Composite Reliability) test was performed. The results of each item presented a coefficient of 0.7 [23], which means the reliability of the construct is high. For the measurement of the amount of variance that is captured by a construct, AVE (Average Variance Extracted) was applied [23,24].

Table 3.1 Reliability analysis of push factor variables

Factors		Economic factors	Socio-cultural factors	Miscellaneous factors	Cronbach's alpha	CR	AVE
Economic factors	EF1-1	.918			.861	.818	.698
	EF1-2	.814					
	EF1-3	.918					
	EF1-4	.814					
	EF1-5	.477					
Socio-cultural factors	SCF1-1		.918		.844	.905	.731
	SCF1-2		.811				
	SCF1-3		.918				
Miscellaneous factors	MF1-1			.974	.909	.819	.776

Source: Authors.

Table 3.2 Reliability analysis of pull factor variables

Factors		Economic factors	Socio-cultural factors	Miscellaneous factors	Cronbach's alpha	CR	AVE
Economic factors	EF2-1	.866			.892	.821	.701
	EF2-2	.843					
	EF2-3	.866					
	EF2-4	.843					
Socio-cultural factors	SCF2-1		.481		.701	.895	.711
	SCF2-2		.430				
	SCF2-3		.863				
	SCF2-4		.737				
Miscellaneous factors	MF2-1			.895	.926	.756	.685

Source: Authors.

https://doi.org/10.1051/shsconf/20219001023

3.2 Factor analysis

Factors analysis was carried out to identify the correlation between the push and pull factor variables. The results are summarized in Tables 4 and 5.

Table 4. Correlation analysis for push factors

		EF1-1	SCF1	MF1	MIG
EF1	Pearson Correlation	1			
SCF1	Pearson Correlation	961**	1		
	Sig. (2-tailed)	.000			
MF1	Pearson Correlation	.108**	.119**	1	
	Sig. (2-tailed)	.000	.000		
MIG	Pearson Correlation	.908**	.904**	.106**	1
	Sig. (2-tailed)	.000	.000	.022	

**Correlation is significant at the 0.01 level (2-tailed).

Source: Authors.

Regarding the push factors, for the Mongolians living and working in the Republic of South Korea the economic factors are statistically significance [$r = .908$, $N = 466$, $p < 0.01$], as are the socio-cultural factors [$r = .904$, $N = 466$, $p < 0.01$], both of which played a key role in their migration to the country. However, other factors such as social instability and air pollution were less significance [$r = .106$, $N = 466$, $p < 0.22$] for their decision to migrate.

Table 5. Correlation analysis for pull factors

		EF2	SCF2	MF2	MIG
EF2	Pearson Correlation	1			
SCF2	Pearson Correlation	.717**	1		
	Sig. (2-tailed)	.000			
MF2	Pearson Correlation	.136**	.085**	1	
	Sig. (2-tailed)	.003	.006		
MIG	Pearson Correlation	.912**	.760**	.136**	1
	Sig. (2-tailed)	.000	.000	.003	

** Correlation is significant at the 0.01 level (2-tailed).

Source: Authors.

A statistically significant pull factor for Mongolians to migrate to the Republic of South Korea were the economic factors [$r = .912$, $N = 466$, $p < 0.01$], and slightly less significant were socio-cultural factors [$r = .760$, $N = 466$, $p < 0.01$]. Environmental safety, favourable climate and quality of life were less significant to their migration compared to the aforementioned factors [$r = .136$, $N = 466$, $p < 0.05$].

3.3 Regression analysis

To prove the proposed hypotheses, multifactor regression analysis was carried out for each of the factors (Tables 6 and 7).

Table 6. Multifactor regression analysis for push factors

Model Summary

Model	R	R Square	Adjusted R Square	Std. error of the estimate
1	.915[a]	.837	.836	1.5126

a. Predictors: (Constant), EF1, SCF1,MF1
Source: Authors.

The results of the analysis shows that the correlation between the push factors for Mongolia and the migration of people was 83.7% (R Square=.837).

ANOVA[a]

	Model	Sum of squares	df	Mean square	F	Sig.
	Regression	5424.259	3	1808.086	790.273	.000[b]
1	Residual	1057.022	462	2.288		
	Total	6481.281	465			

a. Dependent Variable: MIG
b. Predictors: (Constant), EF1, SCF1,MF1
Source: Authors.

Coefficients[a]

	Model	Unstandardized coefficients		Standardized coefficients	t	Sig.
		B	Std. error	Beta		
1	(Constant)	4.009	.693		5.784	.000
	EF1	.671	.089	.512	7.536	.000
	SCF1	.848	.140	.412	6.056	.000
	ESF1	.005	.054	.002	.086	.931

a. Dependent Variable: MIG

Source: Authors.

The results reveal that the main push factors for Mongolians to migrate to the Republic of South Korea are economic ($\beta = 0.512$ $p<0.001$) and socio-cultural ($\beta = 0.412$ $p<0.001$). The positive beta coefficient of the related variables and the statistical significance $p < 0.01$ means hypotheses 1 and 2 are confirmed. Poor retention rates and increased staff turnover in companies in Mongolia is partly related to the economic factors of the migration.

On the other hand, the environment and safety concerns ($\beta = 0.002$ $p>0.005$) were not significant for migration. Although the beta coefficient for the variables was positive, statistically it was less significant ($p>0.005$), which means that hypothesis 3 is refuted, i.e. these factors are not a push factor for Mongolians to migrate.

Table 7. Multifactor regression analysis for pull factors

Model Summary

Model	R	R Square	Adjusted R square	Std. error of the estimate
1	.925[a]	.856	.855	1.4231

a. Predictors: (Constant), EF2, SCF2, MF2

Source: Authors

The results of the analysis shows that the correlation between the pull factors for Mongolia and the migration of people was 85.6% (R Square=.856).

ANOVA[a]

	Model	Sum of squares	df	Mean square	F	Sig.
	Regression	5545.607	3	1848.536	912.736	.000[b]
1	Residual	935.675	462	2.027		
	Total	6481.281	465			

a. Dependent Variable: MIG
b. Predictors: (Constant), EF2, SCF2, MF2

Source: Authors

Coefficients [a]

	Model	Unstandardized coefficients		Standardized coefficients	t	Sig.
		B	Std. error	Beta		
1	(Constant)	3.985	.638		6.244	.000
	EF2	.819	.038	.755	29.591	.000
	SCF2	.360	.042	.317	8.552	.000
	MF2	.041	.048	.015	.847	.397

a. Dependent Variable: MIG

Source: Authors

The results reveal that the main pull factors for Mongolians to migrate to the Republic of South Korea are economic ($\beta = 0.755$ $p<0.001$) and socio-cultural ($\beta = 0.317$ $p<0.001$).

The respondents noted some positive sides to living and working in the Republic of South Korea such as remittance, saving money, learning about a foreign culture and obtaining or improving professional skills.

The dependence of the related variables all presented positive (beta coefficients) and proved to be statistically significant ($p <0.01$), which means the proposed hypotheses 5 and 6 are confirmed. Conversely, the environment and safety concerns ($\beta = 0.015$ $p>0.005$) had no impact on migration. For hypothesis 6, although the beta coefficients presented positive values for the factors, they were statistically insignificant ($p>0.005$), which means it was not supported or the factors do not serve as a pull factor with regards to migration to the Republic of the South Korea.

4 Conclusion

Within the social context of migration, human resources and labour issues are at stake. In the modern world, the migration of labour is an important issue for policy makers, researchers and other stakeholders. The potential growth and development of the labour force and the optimal use thereof are essential to a country´s development. The migration of labour plays a particularly important role in supporting the growth of an economy, increasing the skills level of the labour force and meeting the labour market demands of a country. On the other hand, migration causes a number of problems for an economy and society, such as a brain drain or the loss of highly skilled technical labour [25].

Despite recent economic and social improvements in Mongolia, little has changed in terms of incomes and job opportunities. As a result, in-migration from rural to urban areas and migration to foreign countries remains high. In facing these various challenges, the younger and middle-age generations have chosen to seek higher paid jobs in foreign labour markets. The current situation is having an impact on the retention of skilled labour at the local business level.

For Mongolians, the main reason to migrate to the Republic of South Korea is economic. Unemployment, unstable low incomes, the economic crisis, poverty, poor job opportunities and living standards all serve as push factors for migration.

In addition, the research results indicate that socio-cultural factors, such as a lack of accessibility to healthcare, separation from family, and family conflicts also serve as push factors. However, factors like social stability and air pollution have no effect.

According to the survey results, the pull factors for migration to the Republic of South Korea are the comparatively high wages, a comfortable livelihood, accessibility to quality healthcare, the possibility to gain a good education and the cultural experience.

References

1. G. J. Borjas, Immigration in High-Skill Labor Markets: The Impact of Foreign Students on the Earnings of Doctorates. *National Bureau of Economic Research Working Paper 12085* (2006)
2. Z. Altanchimeg, D. Battuya, J. Tungalag, The Current Circumstances and Challenges of Migrant Labor Force of Mongolia in North Eastern Asia. *Jurnal ilmiah peuradeun, The International Journal of Social Sciences.* **4**(1), 27–38 (2016)
3. International Organization for Migration, Mongolia: internal migration study. [online]. Available at: https://publications.iom.int/system/files/pdf/mongolia_internal_migration_study.pdf (2018)
4. K. Osaki, Migrant remittances in Thailand: Economic necessity orocial norm? *Journal of Population Research.* **20**(2), 203–222 (2003)
5. E. Safrida, Dampak Kebijakan Migrasi Internal Terhadap Perilaku Pasar Kerja Di Indonesia [The Impact of Internal Migration Policies on Labor Market Behavior in Indonesia]. *Forum Pascasarjana.* **9**(2), 150–158 (2008)
6. E. G. Ravenstein, The laws of migration. *Journal of Royal Statistical Society of London.* **48**(2), 167–235 (1885)
7. S. D. Kline, Push and pull factors in international nurse migration. *Journal of nursing scholarship.* **35**(2), 107-111 (2003)
8. G. Dorigo, W. Tobler, Push-Pull Migration Laws. *Annals of the Association of American Geographers.* **73**(1), 1–17. (1983)
9. D. J. Bague, *Principles of Demography.* New York: John Wiley and Sons Inc. (1956)
10. E. S. Lee, A Theory of Migration. *Demography.* **3**(1), 47–57 (1966)
11. M. Abella, Social Issues in the Management of Labour Migration in Asia and Pacific. *Asia-Pacific population journal.* **20**(3), 61–86 (2005)

12. R. C. Jones, Harbingers of Migration Regression: Global Trends and a Mexican Case Study. *Social Science Quarterly*. **97**(2), 293–310 (2016)

13. W. Adams, *"Introduction", in W. Adams (ed.), The Brain Drain.* New York: Macmillan (1968)

14. M. Iqbal, Y. Gusman, Pull and Push Factors of Indonesian Women Migrant Workers from Indramayu (West Java) to Work Abroad. *Mediterranean Journal of Social Sciences*. **6**(5), 167–173 (2015)

15. F. Djafar, Dynamics of Push and Pull Factors of Migrant Workers in Developing Countries: The Case of Indonesian Workers in Malaysia. *Journal of Economics and Behavioral Studies*. **4**(12), 703–711 (2012)

16. K. K. Thet, *Pull and Push Factors of Migration: A Case Study in the Urban Area of Monywa Township, Myanmar* [online]. Available at: https://www.worldofstatistics.org/files/2014/03/Pull-and-Push-Factors-of-Migration-Thet.pdf (2012)

17. B. Xiang, J. Lindquist, Migration Infrastructure. *International Migration Review*. **48**, 122–148 (2014)

18. P. L. Martin, G. Zurcher, Managing Migration: The Global Challenge [online]. *Population Bulletin*. **63**(1), 1–19. Available at: https://www.prb.org/wp-content/uploads/2008/03/63.1migration.pdf (2008)

19. N. Van Hear, O. Bakewell, K. Long, Push-pull plus: reconsidering the drivers of migration. *Journal of Ethnic and Migration Studies*. **44**(6): Special Issue: Aspiration, Desire and the Drivers of Migration, 927–944 (2017)

20. N. C. Parkins, Push and Pull Factors of Migration. *American Review of Political Economy*. **8**(2), 6–24 (2010)

21. K. G. Singh, Push and Pull factors of Migration: A case of Brick Kiln Industry of Punjab State. *Asia Pacific Journal of Social Sciences*. **1**(1), 82–116 (2009)

22. J. C. Nunnally, *Psychometric theory. 2nd ed.* New York: McGraw-Hill (1978)

23. J. F. Hair, *Multivariate Data Analysis, 7th Edition.* London: Pearson (2010)

24. C. Fornell, D. F. Larcker, Evaluating structural equation models with unobservable variables and measurement error. *Journal of Marketing Research*. **18**(1), 39–50 (1981)

25. C. Jandos, Improving Mongolia's foreign migration. *Demography*. 44–49 (2014)

Motivational Factors Used to Stimulate the Creativity of Employees in Companies Operating in the Czech Republic

Miluše Balková[1,*], *Lenka* Ližbětinová[1]

[1]Institute of Technology and Business, Faculty of Corporate Strategy, Okružní 517/10, 370 01 Českе Budějovice, Czech Republic

Abstract. Current circumstances are forcing companies to react quickly to the changes that are taking place. Maintaining competitiveness and survival depends on adaptability and the search for innovations. The goal of modern human resource management is to stimulate employees to be creative and utilise their potential. The aim of this study is to find out what motivational factors Czech companies use to stimulate and encourage the creativity of their employees and whether the chosen factors for achieving this are related to company size.
Keywords: creativity; management; employees; motivation; HRM

1 Introduction

In today's globalized world, businesses are increasingly exposed to turbulent change, which places high demands on the ability to adapt quickly and innovate. This process can either generate opportunities for business success, or pose a huge threat to a business´ very existence [1]. The ability to respond correctly to market changes depends on, among other things, a company's know-how and the creativity of its human resources. The originality of the ideas that people in a company are able to generate differentiates a company´s products or services from those of its competitors. Supporting and using creativity in a company is therefore becoming a great challenge for managers at all levels.

The key drivers of competitive advantage are inimitable intangible resources [2], such as a company's ability to foster innovation and creativity [3,4,5]. The prerequisite is an emphasis on supporting the creativity of employees and setting up a management system that utilizes the internal motivation of employees. The art of offering innovative products needs creative employees who are stimulated to develop their creative potential and generate creative ideas. Ideas are subsequently a source of innovation and competitive advantage [6]. Within industry, there are sectors which, by their very nature, require a high level of creativity from their employees (creative industries), while in other industries, employee creativity is not much required ("non-creative" sectors) and the fulfilment of mere tasks is expected. Due to the different work requirements, the management systems in the two groups differ greatly [7]. Industry 4.0 is changing approaches to human resource management, whereby creativity is becoming a key element [8,9].

The aim of this study is to determine what motivational factors Czech companies apply to stimulate the creativity of their employees and whether this is related to company size.

2 Concepts and Basic Theories

Employee creativity is a key factor in a company's success. In their analytical study, Purdenko et al. [10] state that under modern market conditions, companies can increase their level of competitiveness through the implementation of policies that support their employees' creative activity, and that in doing so it will increase the effectiveness of the creative management system of the company as a whole.

In studies, a creative personality is associated with such traits as independence, entrepreneurship, radicality, openness to experience [11]. According to Fedor [12], there is a relationship between creativity, values, personality traits and conflicting personality behaviours. A positive correlation is observed with the values of independence and success and with the values of modesty and conformism. Creativity is also positively associated with strategies for managing decision-making, taking responsibility and distancing; influence - seek social support, remove obligations and reassessment [12, 13, 14].

The creative behaviour of employees significantly influences the approach of a superior employee. The results of a questionnaire survey conducted by Nazir et al. [15] show that authoritarian leadership has an effective impact on the

* Corresponding author: balkovam@mail.vstecb.cz

creative behaviour of workers. To stimulate innovation [16], managers should provide employees with a safe workplace environment in which they can safely take risks and show creativity, which ultimately contributes to increasing employee and organizational performance [15].

Al-Tit et al. [17] conducted a questionnaire survey of industrial companies and, based on the analysis of responses, concluded that employee pro-activity is a function of three categories of HR practices. First, the employee should be supported to improve their skills. Second, the employee should be motivated and given the opportunity to share knowledge. Third, employees should be encouraged to participate in problem-solving in the presence of the lead manager, in order to improve the quality of relationships and to recognize team members, their support and effective communication [17]. According to Henri [18], there are two ways in which creative employees can be stimulated in a company. This includes the creation of working conditions [19] for creativity and the opportunity to become a leader in the field. The author mostly talks about specialists who can draw on their deep knowledge and further develop it. This statement is deduced from the trend of today's innovative economy, which forces individuals to produce something new all the time.

Madjar et al. [20], on the basis of questionnaires and interviews, examined the relationship between creative performance and the extent to which employees received support for creativity from both work (supervisors/co-workers) and non-work (family/friendly) sources. The results showed that the support of both work and non-work sources significantly and independently contribute to creative performance. These relationships create a positive mood, whereby employees with less creative personalities respond the most positively to support from non-work sources [20].

A lot of research has been conducted into identifying the factors that lead to a creative outcome, but little that leads employees to start the creative process. This issue was addressed by Unsworth et al. [21], who used semi-structured interviews to explore the factors and processes that are involved in this phenomenon. They found that general work motivation, creativity requirements, cultural support for creativity, time resources and independence are among the stimuli that determine whether an event creator pays off. Dul et al. [22] also confirm that a creative work environment encourages creative performance. Gong et al. [23] state that employees' creativity is positively linked to further training opportunities as well as the appraisal of their performance [24].

This study uses the responses to a questionnaire survey to determine the relationship between company size and the selected motivational factors for stimulating and supporting employee creativity. The research questions were formulated as follows: What motivational factors are most often applied to stimulate the creativity of employees? Are the selected motivational factors for stimulating employee creativity related to company size?

3 Methodology

To fulfil the aim of the study, a questionnaire survey was conducted among 372 Czech companies to determine what motivational factors they apply to stimulate and support the creativity of employees. The questionnaire was addressed to HR professionals. A stratified random selection according to company size and regional location was used to create the basic sample (return rate of questionnaires was 54%). The company size was determined through the questionnaire, and the individual motivational factors evaluated on the basis of a 5-point scale, whereby 1 = we do not apply it and 5 = we apply it in full (see Table 1).

Table 1. Evaluation scale

Evaluation scale					
	☹ 1	2	3	4	5 ☺
Application in the company	We do not apply it	We do not apply it yet, but are considering doing so	I cannot assess the actual situation	We partially apply it	We apply it in full

Source: Authors.

The assessed motivational factors that are considered important for the stimulation of creativity in a company were: A1 - meaningfulness of work (employees know the meaning of their work); A2 - involvement (giving space to assert oneself); A3 – enthusiasm (shared interest and enjoyment of work); A4 – cooperation; A5 - recognition (public, personal); A6 - open communication; A7 – support; A8 – autonomy; A9 - emphasis on health; A10 – trust; A11 - responsibility (inner sense of responsibility). The responses were analysed using SPSS Statistics 26. First, the companies were divided according to size, then the average assigned values for the individual motivational factors were calculated according to the size of the company. Finally, an ANOVA test was performed at a significance level of 5%.

4 Results and Discussion

The companies that took part in the research were of varying sizes, of which 31.2% were large companies, 26% were medium-sized companies, 25% were small companies, and 17.7% were micro-companies (see Table 2).

Table 2. Structure of the research sample

Company size	Frequency	%	Valid %	Cumulative %
Micro-	66	17.7	17.7	17.7
Small	93	25.0	25.0	42.7
Medium–sized	97	26.1	26.1	68.8
Large	116	31.2	31.2	100.0
Total	**372**	**100.0**	**100.0**	

Source: Authors.

Based on the average values assigned by the HR professionals according to the evaluation scale, the most used motivational factor is meaningfulness of work. However, this is, on average, only partially applied (value 4.21). The second most frequently partially applied motivational factor is emphasis on health (value 4.14). Motivation through cooperation (value 3.98) and responsibility (value 3.97) were also allocated significant values. Unfortunately, most respondents could not assess the current situation with regards to the application of the identified motivational factors in their company (see Table 3). According to the results, it can be deduced that in order to stimulate creativity, Czech companies believe they must set common goals, the achievement of which is a shared responsibility, and place emphasis on health and a safe workplace environment. However, research shows that creativity is significantly motivated by independence, recognition of performance and enjoyment of work [11], which is an approach that is less widely applied in the Czech Republic. The reason for this is historical, in the sense that people in the Czech Republic were educated for almost fifty years to live in groups (especially in school and work). This finding could be another subject of research.

Table 3. Arithmetic averages for the identified motivational factors by company size

Motivational factor/ Company size	A1	A2	A3	A4	A5	A6	A7	A8	A9	A10	A11
Micro-	4.21	3.70	3.35	4.02	3.61	3.65	3.74	3.58	4.26	4.03	4.02
Small	4.25	3.67	3.45	4.06	3.67	3.85	3.78	3.53	4.34	3.92	3.85
Medium–sized	4.31	3.57	3.43	3.94	3.51	3.76	3.78	3.46	4.06	3.85	4.12
Large	4.09	3.69	3.38	3.93	3.35	3.66	3.63	3.46	3.97	3.74	3.91
Total	**4.21**	**3.65**	**3.41**	**3.98**	**3.52**	**3.73**	**3.73**	**3.50**	**4.14**	**3.87**	**3.97**

Source: Authors.

The data (see Table 4) from the ANOVA test (at a significance level of 5%) proved that company size has a statistically significant effect on the emphasis on health (p-value 0.012). The relationship between company size and the other motivational factors proved to be insignificant (see Table 4). From Table 3 follows, that as company size grows the application of motivational factors becomes statistically less significant. The greater frequency with which the selected motivational factors are applied in micro- and small companies corresponds more to the individual work performance required of employees and the achievement of common goals than the average. This creates more space for the utilization of the creativity of individual employees. In contrast, HR professionals in large companies are losing the ability to assess whether these motivations factors are actually applied.

Table 4. ANOVA test results for motivational factors by company size

ANOVA test						
Dependent variable		Sum of Squares	Df	Mean Square	F	P-value
A1 Meaningfulness of work	Between Groups	2.624	3	0.875	1.171	0.321
	Within Groups	275.021	368	0.747		
	Total	277.645	371			
A2 Involvement	Between Groups	1.018	3	0.339	0.432	0.730
	Within Groups	289.248	368	0.786		
	Total	290.266	371			
A3 Enthusiasm	Between Groups	0.565	3	0.188	0.183	0.908
	Within Groups	379.142	368	1.030		
	Total	379.707	371			
A4 Cooperation	Between Groups	1.193	3	0.398	0.439	0.725
	Within Groups	333.675	368	0.907		
	Total	334.868	371			
A5 Recognition	Between Groups	5.723	3	1.908	1.795	0.148
	Within Groups	391.180	368	1.063		
	Total	396.903	371			
A6 Open communication	Between Groups	2.488	3	0.829	0.762	0.516
	Within Groups	400.631	368	1.089		
	Total	403.118	371			
A7 Support	Between Groups	1.744	3	0.581	0.637	0.592
	Within Groups	335.834	368	0.913		

	Total	337.578	371			
A8 Autonomy	Between Groups	0.785	3	0.262	0.270	0.847
	Within Groups	356.212	368	0.968		
	Total	356.997	371			
A9 Emphasis on health	Between Groups	8.569	3	2.856	2.751	**0.043**
	Within Groups	382.162	368	1.038		
	Total	390.731	371			
A10 Trust	Between Groups	3.945	3	1.315	1.385	0.247
	Within Groups	349.334	368	0.949		
	Total	353.280	371			
A11 Responsibility	Between Groups	4.144	3	1.381	1.467	0.223
	Within Groups	346.531	368	0.942		
	Total	350.675	371			

*Note: **bold** highlights statistically significant results at a significance level of 5%*

Source: Authors.

5 Conclusion

In Czech companies, the motivational factors meaningfulness of work and emphasis on health are the most frequently used to stimulate the creativity of employees, as are cooperation and responsibility - an inner sense of responsibility. According to the results, it can be deduced that in order to stimulate creativity, Czech companies believe employees must work in a team with set common goals, the achievement of which is a shared responsibility, and place emphasis on health and a safe workplace environment. Unfortunately, although research shows that creativity is significantly motivated by independence, recognition of performance and enjoyment of work, none of the respondents felt they could evaluate the situation in their company, which is reflected in the fact that this approach is less widely applied in the Czech Republic.

Company size has a statistically significant effect on the emphasis on health. As company size grows, the application of these motivational factors decreases. Micro- and small companies seemingly make more use of individual performance incentives within the context of achieving common goals than the average. The greater ability to apply these motivational factors suggests a better ability to use the creative potential of individual employees compared to large companies where HR professionals seem to lose track of the motivational factors they actually apply.

The preference for collective motivation over individual motivation in Czech companies could be the subject of further research.

References

1. S. Cardinaleschi, M. Damiani, F. Pompei, Knowledge-intensive sectors and the role of collective performance-related pay. *Industry and Innovation*. **27**(5), 480–512 (2020)
2. J. Hyršlová, J. Chocholac, D. Sommerauerová, P. Štěpán, *Sustainability and transport construction companies*. International Days of Statistics and Economics 2019 (2019)
3. R. K. Scott, Creative Employees: A Challenge to Managers. *Journal of Creative Behavior*. **29**(1), 64–71 (1995)
4. H.-J. Cho, V. Pucik, Relationship between innovativeness, quality, growth, profitability and market value. *Strategic Management Journal*. **26**(6), 555–575 (2005)
5. J. Poór, A. D. Engle, Á. Kovács, A. Albrychiewicz-Slocinska, Z. Caha, V. Kumpikaite-Valiuniene, Z. Horbulák, Initial Findings for Labour Markets in the Czech Republic, Hungary, Poland and Slovakia. *Central European Journal of Labour Law Personnel Management*. **3**(1), 47-60 (2020)
6. S. Jovčić, P. Pruša, M. Dobrodolac, L. Švadlenka, A proposal for a decision-making tool in third-party logistics (3PL) provider selection based on multi-criteria analysis and the fuzzy approach. *Sustainability*. **11**(15) (2019)
7. I. Grabner, *Managing Organizational Creativity Motivational aspects of management control systems for creative employees*. Wien: Wirtschaftsuniversität Wien. (2007)
8. A. Kucharčíková, M. Mičiak, M. Hitka, Evaluating the effectiveness of investment in human capital in e-business enterprise in the context of sustainability. *Sustainability*. **10**(9) (2018)
9. D. Jašková, Assessment of Social Development in Slovakia in the Context of Human Resources. *Central European Journal of Labour Law Personnel Management*. **2**(2), 21–32 (2019)
10. E. Purdenko, O. Matusova, V. Andryeyeva, Assessment of Employees' Creative Activity in Modern Conditions of Intellectualization of the Economy. *Baltic Journal of Economic Studies*. **5**(1), 182 (2019)

11. B. Barbot, M. Besancon, T. Lubart, Creative Potential in educational settings: Its natuer measure, and Nurture. *Education 3-13.* **43**(4), 1-11 (2015)

12. A. Fedorova, The Connection Between Worker's Creativity, Values and Conflict Behavior. *Psychology. Journal of Higher School of Economics.* **16**(1), 191–203 (2019)

13. E. Nedeliakova, M. Panak, J. Ponicki, R. Sousek, Progressive Management Tools for Quality Improvement. In: *International Conference on Engineering Science and Management,* 195–198 (2016)

14. M. Uhlerová, The Role of Trade Unions and Social Dialogue During the Crisis: the Case of Slovakia. *Central European Journal of Labour Law Personnel Management.* **3**(1), 61–76 (2020)

15. S. Nazir, A. Shafi, M. A. Asadullah, W. Qun, S. Khadim, Linking paternalistic leadership to follower's innovative work behavior: the influence of leader–member exchange and employee voice. *European Journal of Innovation Management* (2020)

16. S. Lorincová, J. Schmidtová, J. Javorcíková, The impact of the working position on the level of employee motivation in Slovak furniture companies. *Acta Faculatis Xylologiae.* **60**(2), 211–223 (2018)

17. A. A. Al-Tit, The impact of AMO-HR systems on proactive employee behavior: The mediating contribution of leader-member and team-member exchange. *International Journal of Engineering Business Management.* **12**, 1–13 (2020)

18. J.-F. Henri, Organizational culture and performance measurement systems. *Accounting Organizations and Society.* **31**(1), 77–103 (2006)

19. S. Lorincová, K. Stachová, Z. Stacho, Z. Joniaková, Defining the differences in corporate culture in wood-processing and forest enterprises. *Bioresources.* **15**(2), 3320–3343 (2020)

20. N. Madjar, G. R. Oldham, M. G. Pratt, There's No Place like Home? The Contributions of Work and Nonwork Creativity Support to Employees' Creative Performance. *The Academy of Management Journal.* **45**(4), 757–767 (2002)

21. K. L. Unsworth, C. W. Clegg, Why do employees undertake creative action? *Journal of Occupational and Organizational Psychology.* **83**(1), 77–99 (2010)

22. J. Dul, C. Ceylan, Work environments for employee creativity." *Ergonomics.* **54**(1), 12–20 (2011)

23. Y. Gong, Employee learning orientation, transformational leadership and employee creativity: The mediating role of employee creative self-efficacy. *Development and Learning in Organizations.* **24**(2), 366–372 (2010)

24. L. Mura, B. Gontkovicova, E. D. Spisakova, Z. Hajduova, Position of employee benefits in remuneration structure. *Transformations in Business and Economics.* **18**(2), 156–173 (2019)

Transformation of Managerial Competencies within the Context of Global Challenges

Viktoriya Bondarenko[1,*], *Olesia* Diugowanets[1], *Oksana* Kurei[1]

[1]Uzhhorod National University, Faculty of History and International Relations, Narodna Square 3, 88000 Uzhhorod, Ukraine

Abstract. This contribution presents an analysis of the transformational changes in the global environment on the content of modern models of managerial competencies. Theoretical approaches to the definition of managerial competencies according to international standards are studied, and on this basis a hierarchy of competencies, within the context of their acquisition through professional training, is determined. The predictions and changes in the basic competencies of managers in the near future (2025) are analysed. Sets of general managerial (organizational) competencies (hard skills) and non-specialized supra-professional competencies (soft skills) are defined and compared. Digital competences and intercultural competences are defined as conceptually new skills required by a modern effective manager.
Keywords: management; competence; model; hard skills; soft skills

1 Introduction

The process of the high-quality professional training of future managers within the context of the latest global challenges, including COVID-19, requires a thorough rethink of the competency model of a successful manager. The transformation of the global environment has affected various areas of business and requires urgent critical analysis and the search for additional supra-professional skills for modern leaders, providing them with self-development, cognitive thinking, the ability to adapt to market turbulence and digitalization.

A systematic analysis of scientific papers reveals the defocus and one-sidedness of the research into the topic of this study. This includes analyses of the theoretical models of a successful manager [1, 2], predictions about future managerial skills [3-5], consideration of the impact on businesses and management structures of the post COVID-19 era, the requirements being placed on managers in the era of intellectual development under Industry 4.0 [6-8], and the determination and attempted definition of the essence of digital literacy and digitalization of the global economy [9-11].

In our opinion, digital and intercultural competencies are conceptually new managerial skills that are required to overcome the above challenges due to global economic developments. The digital competencies of managers should be considered from at least three vantage points: the ability to perceive new technologies; the ability to utilise them at work; and the ability to transfer these technologies when training staff and stakeholders. Acquiring the competence of cultural self-awareness is a fundamental skill that helps prevent cultural conflicts and helps managers to understand the impact of their own cultural beliefs (personal self-awareness) and values on stakeholders, taking into account management practices in different cultures.

However, despite the large number of scientific papers, and given the recent challenges of globalization, the system of competencies of a modern successful manager has yet to be generalized; there are no definitions of the new supra - professional skills managers should acquire and possess and which should be incorporated into managerial training programmes. The purpose of this study is therefore to clarify and define the competencies of a successful manager, thereby setting clear priorities for the training of a competitive manager in the face of recent global challenges.

2 Methodology

For the purpose of the research, the main stages of the study were set as follows: 1) systematic and structural analysis of the theoretical approaches for determining the competencies of managers according to international standards, and determination of the hierarchy of managerial competencies required within the context of their acquisition during training; 2) analysis of predictions and changes in basic managerial competencies in the near future (2025) to determine the priorities for the professional training of competitive managers; 3) analysis of "post COVID-19" analytical reviews by

* Corresponding author: viktoriya.bondarenko@gmail.com

top international consultancy firms and other organizations to identify new managerial competencies and determine the content thereof.

For each of these stages, the following research methods were used: system analysis method (competency models were considered as a complex system of interrelated skills that together ensure the success of a leader in practice); abstract-and-logical analysis (the formation of hypotheses, conclusions based on evidence, confirmation of the author's opinions by other authors, analytical and empirical research results, which reduce the probable level of subjectivity of the research results); and the abstraction method (i.e. distraction from additional related information in order to determine the top key competencies of a manager in the "post COVID-19" era). In addition to this systematic approach, general scientific approaches were also used: chronological (while studying the changes in the skills of a successful manager over time, from the initial class models to predictive models); and terminological (while clarifying the essence of digital competencies and cultural identity).

The information basis of the study consisted of scientific papers, theoretical models of managerial competencies, empirical research, analytical reviews and predictions by international organizations and companies. To obtain objective and reliable results from the scientific research and to confirm the hypotheses, the following criteria were used to select the source base: representativeness, time, and the level of generalization of information based on empirical data.

3 Results and Discussion

Ukraine's integration into the global economy has led to a wide range of changes that have complicated management problems, intensified competition, and paved the way for an innovative breakthrough in Ukraine's economy. Ensuring the requirements for its successful transition requires the high-quality training of (future) managers who are able to think critically, strategically, and are able to implement a modern management philosophy. This includes:

- providing the conditions for the self-realization of an individual`s creative potential;
- developing the link between fundamental knowledge and highly professional practical skills and abilities;
- the adaptation of staff to the environment with constantly changing responsibilities;
- the reorientation of employees' values and strengthening the role of work ethics;
- effective problem solving related to global competition and the informatization of innovations [12];
- the ability to properly perform duties under the constant pressure of political, social, economic and psychological disincentives (stress, pressure, uncertainty in society and in specific groups);
- increasing the role of the manager and deepening the communication with staff, as well as competence in the implementation of management tasks (contributes to a significant improvement in the mental health of all employees) [13];
- providing behavioural, affective and cognitive dimensions to interpersonal communications [14];
- increasing interest in self-education, a high level of self-discipline;
- the constant strengthening of the business reputation of managers, which requires managerial success, extensive experience and deep knowledge of their field of activity, high moral qualities, knowledge of foreign languages, etc.

The competencies approach as a personnel management tool gives a clear definition of the professional and behavioural requirements for an employee depending on their managerial level, profession, position and tasks. In both Ukrainian and foreign approaches, there is a distinction between the concepts of "competency" and "competence". The main difference between them is as follows: competency is a concept that defines the field of professional activity in which a person is competent, it is a description of work related tasks or expected results; competence is a concept that highlights aspects of human behaviour associated with the performance of work and defines the main characteristics of the individual who has achieved or is able to achieve good results within the activity [15].

Competency models are a key tool in human resource management systems and practices. Over the past 30 years, public administration systems around the world have undergone major reforms, and competency management has already been implemented in the public sector in some OECD countries [16].

The manager's competency model is a complete set of competencies and indicators of the behaviour necessary for the successful performance of functional responsibilities by an employee [17].

Recent research on managerial competency models suggests the use of generalized blocks of managerial behaviour, which together form a corporate model of their competencies (corporate competencies of the manager). In this regard, three options are used for building models of managerial competencies:

1) universal competency models of the most successful companies;

2) universal models of competencies adjusted in accordance with organizational goals and corporate cultural values;

3) building a competency model from scratch based on the development strategy, mission and corporate culture of an organization with the help of external consultants or independently [18].

Universal (standardized, basic) models of competencies have been published in scientific literature [5, 19-24]. Furthermore, a characteristic trend for models of organizational competencies is both the use by companies of only general competencies and the development of managerial competencies for managers of different levels or special competencies for employees of key departments.

To identify essential skills, existing models of managerial competencies were analysed, thereby applying the method of systematization for the formation of a comprehensive modern model of competencies within the context of global challenges (see Table 1).

Table 1. System analysis of competency models
(according to international standards; basic skills without consideration for the challenges of the global environment)

Competency model	**Highlighted competencies**
SHL's "Eight Competencies" model (international management standards - specialist, manager, team leader)	- entrepreneurial qualities (business, creativity, determination, strategy) - qualities of the executive (leadership, planning and organization, quality orientation, persuasiveness) - professional qualities (special knowledge, analysis and problem solving, oral and written communication) - personal qualities (interpersonal understanding, flexibility, stability, personal motivation)
	*SHL consultants use the method of direct attributes based on cards with universal competencies models for certain categories of staff: **UCF model** (Universal Competency Framework), **IMC model** (Inventory of Management Competencies), **CC model** (Customer Contact), **WSC model** (Work Styles Competency), **PMC model** (Perspectives on Management Competencies), **DDA model** (Director's Development Audit)*
"20 Facets" model (areas of the manager`s activity through competencies)	management skills; motivation; decision-making skills; individual traits; interpersonal skills
Whiddett & Hoolyforde model (a competency model development based on company resources)	- business development (personal development, ideas generation and justification) - achievement of results (planning, management clarity, goal setting) - analysis - work with information (analysis and decision making) - people (teamwork, influence, relationship management)
SHRM model (Society for Human Resource Management)	business knowledge; personal capacity; strategic contribution; practical experience; possession of management technologies
CFT model (Corporate Financial Technologies)	system thinking; result orientation; analytical skills; creativity; flexibility; quick and adequate response; ability to learn; organizational skills; delegation of powers; ability to time-plan; ability to listen to others, maintain feedback; presentation and negotiation skills; ability to retransmit knowledge and skills

Source: Authors.

An analysis of the standardized models in Table 1 shows that most approaches consist of 4-12 competencies, which are often focused on ensuring the financial success of a company and team building, which are implemented through leadership, organizational skills and communication. Therefore, a general model of managerial competencies should include: (1) hard skills; and (2) soft skills [25].

Hard skills are necessary for the successful achievement of business goals. They are mandatory for management and, until recently, occupied key positions in managerial competency models. Management competencies include, for example, strategic vision, generation of effective decisions, task formulation and the planning of their implementation stages, organization of implementation processes, management of subordinates, professional external and internal communication skills, and achieving results.

Soft skills are important both in professional activities and in everyday life. Soft skills allow managers to be successful regardless of the specifics of their activities and the direction of their work. Traditionally, the following types of non-specialized supra-professional competencies are distinguished:

- communicative competencies (ability to listen, persuade, argue, conduct presentations and negotiations, networking, self-presentation, public speaking, business correspondence);

- self-management competencies (management of emotions, time, self-development, stress, planning and goal setting, feedback use, focus on services);

https://doi.org/10.1051/shsconf/20219002002

- cognitive competencies (creative and critical thinking, cognitive flexibility, structural and logical thinking, decision making, tactical and strategic thinking, information retrieval and analysis, project thinking).

The higher the professional level of the manager, the greater the role soft skills play. As a result, general professional competencies take a back seat.

Hard skills apply to a specific group of positions. As a rule, they do not play a key role for administrative management because the higher the professional level of the manager, the more subordinates and employees can advise him on various professional issues. It should be noted that managerial competencies do not become irrelevant [5]. On the contrary, their role is still important. However, having non-specialized supra-professional competencies, especially of the cognitive type, will allow administrative and crisis management to apply their hard skills more effectively.

The modern world of business is changing rapidly, so it is important to pay attention not only to the competence of employees, but also to predict the competencies that will guarantee success in the future [4]. According to a World Economic Forum study, the competencies needed to perform most work will change significantly by 2022, especially for managers at all levels of government. It is expected that the average global "skills stability" - the share of core competencies that will remain unchanged - will be about 58%. Competencies such as technological design, which emphasizes the growing demand for various forms of technological competence, will be in demand. However, along with knowledge of the latest technologies, personal competencies (creativity, originality and initiative, critical thinking, persuasion and negotiation) will play a significant role [26].

According to the previously referenced global prediction of managerial competency models for 2015-2020, 35% of key required competencies have already changed [27]. In 2020, instead of managerial competencies, non-specialized supra-professional competencies of the cognitive type have come to the fore. Changes in the list of priorities are the result of understanding current trends in the business environment. It has become clear that in 2020, competitive advantage and leadership will come to those companies with cognitive competencies, namely thinking skills and creativity, because it is on these abilities that an organization depends to effectively adapt to change. Consequently, these competencies have become a priority in development programmes for modern management (Tab. 2).

Table 2. Changes in key competencies in 2020 and predictions for 2025

Preliminary prediction for 2020	**2020**	**Prediction for 2025**
1. Complex problem-solving	1. Analytical (critical) thinking and analysis	1. Analytical thinking and innovation (NEW)
2. Critical thinking	2. Complex solution of complex problems	2. Active learning and learning strategies
3. Creativity	3. Innovative technologies (NEW)	3. Complex problem-solving
4. People management	4. Active learning and the latest learning strategies (NEW)	4. Critical thinking and analysis
5. Coordinating with other people	5. Creativity, originality and initiative	5. Creativity, originality and initiative
6. Emotional intelligence	6. Attention to detail, reliability (NEW)	6. Leadership and social influence
7. Expert assessment and decision-making	7. Emotional intelligence	7. Technology use, monitoring and control (NEW)
8. Focus on services	8. Thinking, solving problems and ideas (NEW)	8. Technology design and programming (NEW)
9. Negotiating skills	9. Leadership and social impact (NEW)	9. Resilience, stress tolerance and flexibility (NEW)
10. Cognitive flexibility	10. Cognitive flexibility	10. Reasoning, problem-solving and ideation (NEW)

Source: Authors.

The World Economic Forum 2020 data shows changes in the most popular competencies of future employees over the next 5 years [26]. The pandemic-related global recession of 2020 and the shift towards the expected introduction of new technologies, automation and further innovations by 2025 will form the core competencies or even skill groups of future managers. These include analytical thinking and innovation, integrated problem-solving, prudence, as well as self-management skills, such as active learning, resilience, flexibility, creativity, originality and initiative (see Table 2).

In the midst of the pandemic recession, and given the shift to distance working, online learning and training with digital tools will increase, thereby presenting a window of opportunity for the (re)training of future managers, with new educational technology tools and the use of both formal and informal methods of acquiring professional skills.

As is evident in Table 2, critical thinking, as predicted, tops the rankings for the most important competency in both 2020 and 2025. Since the 1990s, critical thinking and the appropriate training skills have become extremely relevant. This was due to the urgent need for managers to make informed decisions on various aspects of company management, as well as the ability to perceive and analyse the information received. Complicated business processes require non-standard solutions. As a result, creativity, originality and initiative are now becoming key competencies of management and companies. These competencies not only enable adaptation to change, but also actively shape new industries and markets.

Importantly, emotional intelligence and cognitive flexibility have become new to core competencies in 2020. Cognitive flexibility is the mental ability to switch from one thought to another, as well as to think about several things at once [28]. From the standpoint of administrative management, cognitive flexibility can be a person's ability to adapt thinking in the transition from an old situation to a new one, as well as the ability to overcome habitual reactions and thoughts in new conditions. Moreover, cognitive flexibility involves the simultaneous consideration of many aspects of the phenomenon, such as two parameters of an object or many aspects of a complex situation. In an open, large, and highly versatile world, this ability becomes extremely important. With its help, managers are able to apply their creativity and solve complex problems.

It is the shared opinion of the authors that it is advisable to emphasize and develop the emotional competencies of a manager in connection with the growing role of emotional intelligence, i.e. the ability to use one´s own behaviour and mood to influence the work of subordinates. Under the emotional competencies of a manager is understood: the ability to adequately tolerate a stressful situation without getting into a state of stress; the ability to make independent decisions regardless of external instructions or orders; the ability to defend their own point of view without violating the rights and interests of subordinates; the ability to solve problems of a professional and personal nature, taking into account the current situation in the company; the ability to consciously empathize with the experience of the interlocutor, given their state of mind; the ability to control impulsivity, i.e. actions without sufficient conscious control, being influenced by circumstances or emotions; and the ability to soberly assess reality regardless of the circumstances [3].

These theoretical and consistent predictions of a successful manager´s skills have been significantly influenced by the COVID 19 pandemic, which on a global scale can be described in one term as a "reset" of the business and global economic system as a whole.

A report from the Accenture consulting company [8] on the impact of the coronavirus pandemic shows that viable business strategies should provide flexible methods of managing employees, creating digital jobs; differentiation and stable supply chains for customers; stable IT infrastructure and systems; digital channels and e-commerce platforms. In other words, industrial enterprises need to scale their digital innovations across production and all their ecosystems [8].

The McKinsey global consultancy brand, in its analytics within the context of COVID-19, and in particular in its article "Digital strategy in a time of crisis", notes that every company should: review the strategies of digital transformation; form anti-crisis headquarters (or "nerve centres"), which should quickly review the portfolio of value proposals and offer new solutions; and build new business models where the most vulnerable or responsible elements look different. In practical terms, this refers to supply chains, data security, remote control modes, manpower, and of course, larger-scale automation.

Within this context, McKinsey discusses behavioural change in terms of the courage it takes to propose change and quickly revisit old skills and paradigms, as well as in terms of the need to learn quickly, experiment and test hypotheses and new solutions, simplify and focus at the same time and constantly [6].

The case of the British company KHC emphasizes that today we are not considering the transition to digital technology, but already thinking about how to do it faster and scale it up throughout a company. Prior to doing so, KHC invested heavily in digital platforms, which provided the necessary flexibility and efficiency in response to COVID-19 [7]. The company's management emphasizes: "Today you need zero tolerance for previously established systems or approaches, and you must constantly ask yourself new, important questions about the future. Be ruthless. Choose a technology partner who is willing to invest in technology just like you." [7].

In summarizing the analytical reports, it becomes clear that enterprises of all activities are becoming more digital and are being reorganized into new flexible structures. Digital transformation or Industry 4.0 in the post-Covid-19 era significantly change the essence of business activities. Company managements are being forced into taking this into account, and managers must possess digital competencies. Since, in fact, such competencies are new components in the model of a successful leader, there are differences in their understanding. Some [11] argue that these competencies are closely linked to leadership skills that drive change. Digital leadership is revealed in two ways: digital competencies / literacy and strong leadership skills that drive a company's digital transformation. The study hypothesizes that digital leadership is the foundation of a manager's psychological well-being [11].

According to a survey conducted by the British Computer Association, only 52% of employers believe that their employees have digital literacy skills, 42% of students at one Italian university do not know about the dangers of free Wi-Fi, and 50% of students download files without checking their content and source [10].

Summing up, it can be concluded that the digital competencies of a manager include: a technological dimension: through which managers are able to explore and solve technical problems in a new way; a cognitive dimension: the ability to read, analyse, select and interpret information on the basis of relevance, reliability and suitability; a social dimension: the ability to teach and upskill employees in digital literacy, constructively and responsibly, and the ability to be a leader-motivator and generator of change and innovation based on digital technologies.

Analytical reviews on the impact of the coronavirus pandemic also touch on issues of corporate, team and community culture as key foundations for the adaptation to change. In addition, a critical analysis of different research papers has identified the new competencies of successful managers of global companies, such as intercultural competencies, cultural identity, intercultural communication, global thinking, and more.

International business experts agree that in order to succeed in global business, managers need flexibility and an appropriate attitude to practices and values that can be radically different from those to which they are accustomed. Being a modern manager with global intercultural competencies is not only where they manage people, but also how they do it. [9].

As globalization intensifies, managers are confronted with an environment that is becoming more complex, dynamic, uncertain, and highly competitive than ever before. The starting point for overcoming these challenges is to acquire the competence of cultural self-awareness - an understanding of the impact of one's own cultural conditioning, assessment of one's own cultural beliefs, values, and a deep understanding of how companies and management practices differ in different cultures [29].

In addition, multinational corporations have become one of the most important driving forces of constant globalization and development of the world economy. By doing business around the world and across borders in diversified environments, they have spawned and emphasized concepts such as "global managers" and intercultural management [30].

4 Conclusion

The study presents a systematic vision of the competency model of a successful manager, with clear priorities for their training within the context of recent global challenges. As a result of the research, the following results were obtained.

Global challenges, such as the COVID-19 pandemic, the digital transformation, demographic change, and environmental challenges, have had a significant impact on societies and economic systems at all levels. They have created both difficulties and new business opportunities and have led to the acknowledgment that the competency needs of business leaders are changing, and that outdated standards for the training of managers are no longer effective.

As a result of the systematic and structural analysis of theoretical approaches to defining a manager's competencies according to international standards, it was determined that most theoretical models contain 4-12 competencies, which are often focused on ensuring the financial success of a company and team building, which are implemented through leadership, organizational skills and communication. Therefore, for a better perception and effective approach to the preparation of training programmes for future managers, competencies are grouped into managerial (organizational) competencies (necessary for the successful achievement of business goals), non-specialized supra-professional competencies – soft skills (important both in professional activities and in everyday life), and general professional competencies – hard skills (applied to a certain group of positions). The higher the professional level of the manager, the greater the role played by soft skills, and the further back general professional competencies are pushed into the background.

An analysis of (predicted) changes in basic managerial competencies in the near future in order to determine the priorities for the training of competitive managers showed that non-specialized cognitive competencies, i.e. thinking skills, creativity, innovativeness, providing effective adaptation to the turbulence of the external global environment, will come to the fore in the next 5 years. Critical thinking will become no less important because there is an urgent need for managers to make informed decisions on various aspects of company management. Complicated business processes require non-standard solutions, so creative thinking will be at the top of the key competencies of company managements. Today, many global corporations are investing in the development of the emotional intelligence of their employees, i.e. their ability to clearly understand and manage their emotions and the emotions of the interlocutor, and to build the actual communication process on this basis. Given these trends in the transition of the dominant role of cognitive skills over basic managerial skills, it is advisable to transform training programmes for future managers towards the formation of critical thinking skills, innovative technologies, self-development, creativity, emotional intelligence and leadership skills. It should be noted that general managerial competencies do not become irrelevant, they only serve as a basis for supra-professional competencies, especially cognitive ones.

A study of "post COVID-19" global analytical reports by top international consultancies and other organizations identifying new competencies in the model of a successful manager showed that digital and intercultural competencies help to overcome and adapt to the challenges of the global economy in today's environment. The digital competencies of a manager should be explained on the basis of three dimensions: a technological dimension through which managers are able to explore and solve problems with the help of digital technology in a new way; a cognitive dimension (the ability to read, analyse, select and interpret information on the basis of relevance, reliability and suitability); and a social dimension (the ability to teach and upskill employees in digital literacy, constructively and responsibly, and the ability to be a leader-motivator and generator of change and innovation based on digital technologies). In Ukraine, there exists an understanding at government level that company managers lack such competencies. It is within this context that the Government's project "Digital Education Diia[MVF1][OK2]" was created and implemented, which aims to increase the digital literacy of 6 million citizens. However, the system of training future managers in public universities is far from perfect. This is due to outdated human resources, the influence of the philosophy of the classical post-soviet education system, and the lack of sufficient funding and access to modern software. However, global analytical reviews have indicated a low level of digital competence in many countries around the world, not just in Ukraine.

Cultural self-awareness is a skill that helps managers prevent cultural conflicts and understand the impact of their own cultural beliefs and values on a person, thereby taking into account the practice of management in different cultures, as well as the impact of their own managerial and life experience. Acquiring cultural self-awareness skills is enhanced by intercultural understanding because managers can learn to recognize how their culturally implicit beliefs and values govern their behaviour in the workplace and they can see how their actions affect other workers from different cultures. Therefore, cultural self-awareness is primarily a tool for personal development that helps managers adapt their thinking and become more learning-oriented through increased interest in sources of cultural diversity, starting with their own beliefs, values and understanding of how culturally driven beliefs and values shape leadership behaviour. These two groups of skills in the system of competencies of a successful leader can be attributed to the supra-professional.

Within the framework of this scientific research, an attempt has been made to present a systematic vision of the model of a successful manager, thereby taking into account the impact of the latest globalization challenges. However, there is a need for further research in terms of empirical confirmation of the impact that the established set of managerial competencies has made on the financial and psychological well-being of managers and the financial success of employers.

References

1. N. I. Daliuk, Formuvannia profesiinykh kompetentsii suchasnoho menedzhera [*Formation of professional competence of a particular manager*]. *Visnyk Natsionalnoho universytetu vodnoho hospodarstva ta pryrodokorystuvannia.* **1**(57), 109–113 (2012)
2. T. O. Kozhan, Vyznachennia vydiv kompetentsii menedzhera z personal [Visiting the competence of the manager and the staff]. *Sotsialno-trudovi vidnosyny: teoriia ta praktyka.* **2**, 98–104 (2013)
3. I. V. Kryvoviaziuk, Skladnyky formuvannia liderskykh kompetentsii suchasnoho kerivnyka pidpryiemstva [Components of forming the leading competencies of a modern business manager]. *Problemy systemnoho pidkhodu v ekonomitsi.* **6**(68), 90–96 (2018)
4. R. Palšaitis, K. Čižiūnienė, K. Vaičiūtė, Improvement of Warehouse Operations Management by Considering Competencies of Human Resources. *10th International Scientific Conference Transbaltica 2017: Transportation Science and Technology. Procedia Engineering.* **187**, 604–613 (2017)
5. M. Tomastika, J. Strohmandlb, P. Cechc, Managerial competency of crisis managers. *Procedia – Social and Behavioral Sciences.* **174**, 3964–3969 (2015)
6. S. Blackburn, L. LaBerge, C. O'Toole, J. Schneider, *Digital Strategy in a Time of Crisis* [online]. Available at: https://www.mckinsey.com/business-functions/mckinsey-digital/our-insights/digital-strategy-in-a-time-of-crisis (2020)
7. Laycock, L. Gundry, *Case Study: Digital transformation in the post-COVID-19 world* [online]. Available at: https://www.smartindustry.com/articles/2020/case-study-digital-transformation-in-the-post-covid-19-world/ (2020)
8. G. Wilson, *Accenture: industrial manufacturing post COVID-19* [online]. Available at: https://www.manufacturingglobal.com/leadership/accenture-industrial-manufacturing-post-covid-19 (2020)
9. M. Javidan, D. Bowen, The "Global Mindset" of managers: What it is, why it matters, and how to develop it. *Organizational Dynamics.* **42**(2), 145–155 (2013)
10. Z. Nyikes, Contemporary Digital Competency Review. *Interdisciplinary Description of Complex Systems.* **16**(1), 124–131 (2018)
11. S. Zeike, K. Bradbury, L. Lindert, H. Pfaff, Digital Leadership Skills and Associations with Psychological Well-Being. *International Journal of Environmental Research and Public Health.* **16**(14) (2019)
12. Z. Ali, I. M. Zwetsloot, N. Nada, Operational capabilities to infuse organizational innovation in SMEs. *3rd World Conference on Technology, Innovation and Entrepreneurship (WOCTINE). Procedia Computer Science.* **158**, 260–269 (2019)
13. S. Kuroda, I. Yamamoto, Good boss, bad boss, workers' mental health and productivity: Evidence from Japan. *Japan and the World Economy.* **48**, 106–118 (2018)
14. J. Koponen, S. Julkunena, A.Asai, Sales communication competence in international B2B solution selling. *Industrial Marketing Management.* **82**, 238–252 (2019)

15. I. V. Ivanova, *Profesionalizatsiia menedzhmentu [Professionalization of management]*. Kyiv: Natsionalnyi torhovo-ekonomichnyi universytet (2006)

16. Z. Skorková, Competency models in public sector. *3rd International Conference on New Challenges in Management and Organization: Organization and Leadership, 2 May 2016, Dubai, UAE. Procedia – Social and Behavioral Sciences*. **230**, 226–234 (2016)

17. O. M. Arapova, V. P. Fridrif, V. A Modyrka. Kompetentnisnyi pidkhid yak chynnyk pidvyshchennia konkurentospromozhnosti pidpryiemstva [Competent approach as a contributor to the competitiveness of the company]. *Ekonomika: realii chasu*. **1**(6), 207–2011 (2013)

18. L. M. Spencer Jr., S. M. Spencer, *Competence at Work: Models for Superior Performance*. New York: John Wiley & Sons, Inc. (1993)

19. L. Bals, H. Schulze, S. Kelly, K. Stek, Purchasing and supply management (PSM) competencies: Current and future requirements. *Journal of Purchasing and Supply Management*. **25**, 2–14 (2019)

20. A. Maria, A.Ofei, Y. Paarima, T. Barnes, Exploring the management competencies of nurse managers in the Greater Accra Region, Ghana. *International Journal of Africa Nursing Sciences*. **13** (2020)

21. F. J. Lara, H. Mogorrón-Guerrero, S. Ribeiro-Navarrete, Knowledge of managerial competencies: cross-cultural analysis between American and European students. *Economic Research-Ekonomska Istraživanja*. **33**(1), 2059–2074 (2020)

22. P. F. P. Freitas, C. C. Odelius, Managerial competencies: an analysis of classification sin empirical studies. *Cadernos EBAPE.BR*. **16**(1), 37–49 (2018)

23. K. Dziekoński, Project Managers' Competencies Model for Construction Industry in Poland. *7th International Conference on Engineering, Project, and Production Management. Procedia Engineering*. **182**, 174–181 (2017)

24. J. W. S. P. Do Vale, B. Nunes, de M. M. Carvalho, Project Managers' Competences: What Do Job Advertisements and the Academic Literature Say? *Project Management Journal*. **49**(3), 82–97 (2018)

25. L. A. Toronii, V. M. Bondarenko, Udoskonalennia suchasnoi modeli kompetentsii menedzhera. *Sotsialno-ekonomichni problemy suchasnoho periodau Ukrainy: zbirnyk naukovykh prats*. **4**(126), 69–72 (2017)

26. World Economic Forum, *The Future of Jobs Report 2020* [online]. Available at: http://www3.weforum.org/docs/WEF_Future_of_Jobs_2020.pdf (2020)

27. Institute for the Future, *Future work skills* [online]. Available at: http://www.iftf.org/-uploads/media/SR-1382A_UPRI_future_work_skills_ sm.pdf (2015)

28. W. Scott, Cognitive complexity and cognitive flexibility. *American Sociological Association*. **25**, 405–414 (1962)

29. R. M. Steers, C. J. Sanchez-Runde, L. Nardon, *Management Across Cultures. Challenges and Strategies*. New York: Cambridge University Press (2010)

30. A. Czarnecka, M. Szymura-Tyc. The competencies of global managers in multinational corporations. In P. Urbanek (ed.) *Economy Today: An Interdisciplinary Approach to Contemporary Economic Challenges. The determinants of an organization's value*. Łódź: Wydawnictwo Uniwersytetu Łódzkiego, 221–235 (2016)

Measuring and Promoting Entrepreneurial Intentions and Entrepreneurial Competence – Implications for Education

Johann Faltermeier[1,*]

[1]ibi research an der Universität Regensburg GmbH, Galgenbergstraße 25, 93053 Regensburg, Germany

Abstract. The questionnaire "Entrepreneurial Competence" (developed as part of the author's dissertation) can be used in practice, but above all also for the measurement of entrepreneurial intentions and competence within (higher) education. Data were collected in two periods at both universities, OTH Regensburg and VŠTE Budweis, for the underlying descriptive evaluation in the field of higher education. This study provides information on the entrepreneurial environment, but also on the students' entrepreneurial competence. At the two universities, 153 students have already taken part in an entrepreneurship course. The environment at both institutes is rated as above-average. The entrepreneurial intention is also clearly visible at the institutes and is significantly higher in the second survey. In the area of entrepreneurial knowledge and skills, the second survey revealed higher levels than at the time of the first initial measurement. In order to be able to better adapt the teaching to the entrepreneurial intentions and entrepreneurial competence of students in the future, the increasing use of activating elements within entrepreneurship and intrapreneurship education is recommended. Planning, role-play, case studies and project work, for instance, contribute to the targeted, group-specific and differentiated competence-based promotion of potential entrepreneurs and intrapreneurs. To achieve this, the initial measurement of entrepreneurial intentions and entrepreneurial competence is therefore essential.
Keywords: entrepreneurial competence; entrepreneurial education; entrepreneurship; intrapreneurship

1 Introduction

At the heart of entrepreneurial intentions and entrepreneurial competence is the conflict between entrepreneurship and intrapreneurship. "Entrepreneurial Intention" describes the intention to create an (innovative) company, while "Intrapreneurial Intention" is defined as the intention to create a quasi-independent, innovation-capable collaboration within an existing organization. Intrapreneurship, as a field of research, has been gaining traction since the 1980s and is based on the work of Pinchot [1-2], Hisrich [3] and Bitzer [4]. The measurement of the entrepreneurial intentions and entrepreneurial competence of the students at the Ostbayerische Technische Hochschule (OTH) Regensburg and the Technische und Ökonomische Hochschule in Budweis (VŠTE), as described herein, was carried out with the help of a database generated using a strategic instrument. The questionnaire method was applied to determine the entrepreneurial intentions, as well as the (cognitive) psychological aspects of entrepreneurial competence. For this purpose, the questionnaire "Entrepreneurial Competence", which was developed as part of the dissertation "Der Unternehmer in uns – Entwicklung einer Diagnostik der unternehmerischen Absichten zur Förderung der unternehmerischen Kompetenz im Entre- und Intrapreneuership" ("The entrepreneur within us – development of a diagnostic tool for entrepreneurial intentions to promote entrepreneurial competence in entrepreneurship and intrapreneurship"), was used. This was included in a large-scale study. The following spot checks, as well as all the results, are taken from the pre-test for this study, which was conducted in March 2017. In addition, a second survey was carried out using the questionnaire in August 2019 in order to better assess the impact of entrepreneurship education at both institutes over a period of two years. This contribution is the first to examine and interpret the results of the two study periods with regards to entrepreneurship education.

To date, direct comparisons between entrepreneurial and intrapreneurial intentions with regards to entrepreneurial skills and knowledge have not been the focus of much scientific work. Rather, it is often assumed that "internal and external entrepreneurship [. . .] is not fundamentally [different], at least not in terms of their peculiarities and behavioral characteristics" [5, p. 23 f.]. What kinds of qualities, skills, abilities, as well as forms of knowledge are relevant for entrepreneurs and intrapreneurs? In order to answer this question, it is first worth looking at the definition and systematization of entrepreneurial competence.

Entrepreneurial competence is one of eight key competences in the European Framework of Reference [6]. This competence can be defined as "the ability of the individual to put ideas into practice. This requires creativity, innovation

* Corresponding author: johann.faltermeier@ibi.de

and risk-taking, as well as the ability to plan and implement projects to achieve specific aims. Entrepreneurial competence helps the individual not only in their daily life at home or in society, but also at work [...]; it is the basis for [...] a social or commercial activity" [6, p. 11]. According to this, entrepreneurial competence is an everyday competence that can be highly relevant in the private as well as professional context. Systemically, entrepreneurial competence can first be understood as a bundle of skills and methodological knowledge, job-specific abilities and skills. This is characterized by the fact that it can be learned, changed, trained or improved in a short period of time by means of initial and further training. Knowledge and skills are referred to in research as cognitive aspects of entrepreneurial competence. This bundle is extended by psychological conditional factors, more precisely by abilities and qualities (aptitudes), which have little or no influence on the quality of a person's life. These psychological conditional factors can be changed in the long term through experience and training, i.e. they are relatively stable in time and situation [7-9].

2 Methodology

The conceptual research model is based on the findings of Ajzen [10-14], as well as on other modified models of the theory of planned behavior [15-18]. In order to be able to derive implications for teaching, a model oriented towards entrepreneurship education is already recommended here. This model, which sees entrepreneurship education as, among other things, a direct influential variable of intentions, is provided by Rauch and Hulsink [19]. These intentions, in turn, serve as a regressor of entrepreneurial behavior (cf. *Figure 1*).

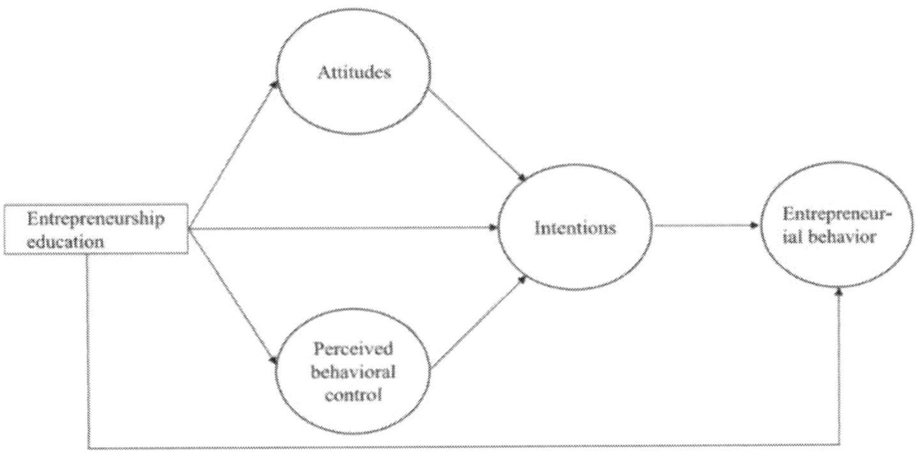

Figure 1. Theory of planned behavior

Source: Rauch & Hulsink, 2015, modified.

Intentions depend on personal aspects, e.g. attitudes towards entrepreneurial behavior. The determinant of perceived behavioral control, i.e. an individual's assessment of how easy or difficult it is to carry out intentional behavior, also provides evidence that entrepreneurial characteristics, but also, for example, entrepreneurial knowledge, have a significant influence on entrepreneurial behavior [11-13, 19]. However, the exact composition of the heterogeneous learning groups is often not known in teaching. In order to find out the learners' intentions and entrepreneurial skills, it is advisable to take an initial measurement of these aspects.

The "Entrepreneurial Competence" questionnaire consists of individual instruments, which serve to measure entrepreneurial competence, differentiated according to entrepreneurial intentions. To capture entrepreneurial intentions in a context-based manner, tools are selected for the desired occupation and types of intention (metric). These are partly transferred from existing surveys, such as the Global University Entrepreneurial Spirit Students' Survey (GUESSS), and are supplemented by own items, in particular for intrapreneurial intentions [20]. Tested and proved scales and items are used for recording cognitive and psychological aspects of entrepreneurial competence, e.g. for personality ("Big Five Inventory") [21]. Mostly nominal and interval scales (7-point Likert scales) are used. The data were collected as part of the project "Entrepreneurial skills on the Czech-Bavarian labour market at the Eastern Bavarian Institute of Technology (OTH) Regensburg and at the Institute of Technology and Business in Budweis (VŠTE)".

The spot check consists of 132 valid data sets for the first survey and 281 valid data sets for the second survey at both institutes. A total of 413 students – 224 students from OTH Regensburg and 189 students from VŠTE Budweis – participated successfully in the surveys. All students provided information on their age, with 412 students also providing information about their gender. For the first and second surveys, the spot check consisted of a total of 301 students in the age group 18 – 24 years, followed by 25 – 34 years (n = 77) and people older than 35 years (n = 35). 270 female and 142

https://doi.org/10.1051/shsconf/20219002003

male students took part in the survey. The following table summarizes the demographic data on age and gender according to institute and study period (cf. *Tables 1* and *2*).

Table 1. Age of participants

Institute	Time of data collection	18-24 years	25-34 years	35 years and older	Total
OTH Regensburg	First data collection	40	15	1	56
	Second data collection	122	43	3	168
VŠTE Budweis	First data collection	72	4	0	76
	Second data collection	67	15	31	113
	Total	301	77	35	413

Source: Author.

Table 2. Gender distribution of participants

Institute	Time of data collection	Female	Male	Total
OTH Regensburg	First data collection	40	16	56
	Second data collection	103	64	167
VŠTE Budweis	First data collection	48	28	76
	Second data collection	79	34	113
	Total	270	142	412

Source: Author.

3 Results and Implications

3.1 Results for institutional environment

In order to be able to interpret the results regarding the teaching of entrepreneurial competence at the two universities in a structured manner, the first question that arises is: How many of the students have already taken part in a course in the subject entrepreneurship before the data collection?

At OTH Regensburg, 15 students stated in the first survey that they took part in an entrepreneurship course, and 50 stated so in the second survey. At VŠTE Budweis, the numbers were 42 students and 46, respectively.

This gives a total number of 153 students who had attended a corresponding course even before the corresponding data collection. 237 students (N = 390) had not attended a relevant course previously. *Figure 2* summarizes these findings.

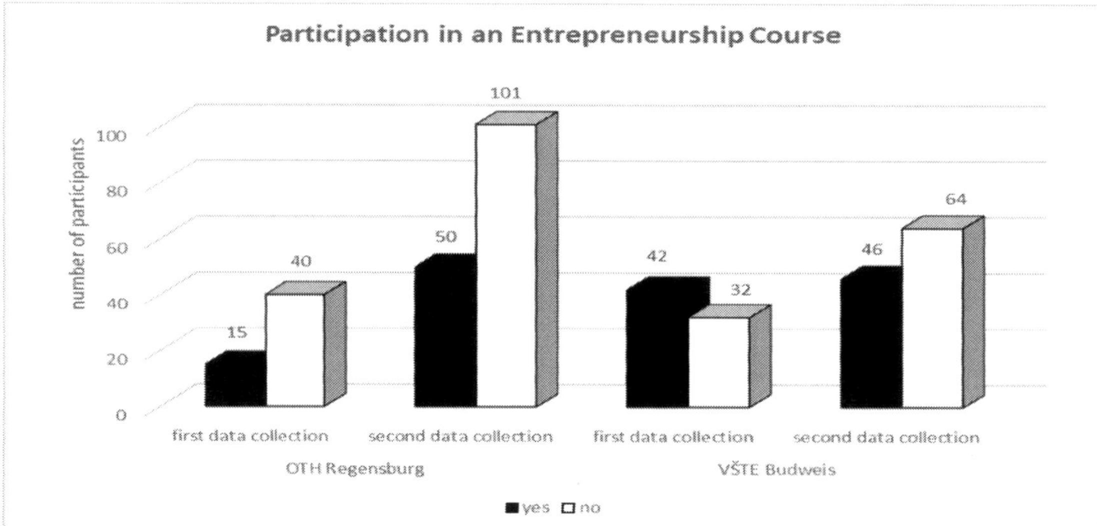

Figure 2. Participation in an entrepreneurship course

Source: Author.

 https://doi.org/10.1051/shsconf/20219002003

Figures 3 and *4* illustrate how the two universities are judged by the students with regard to the entrepreneurial environment and the promotion of entrepreneurial competence.

The answers could be given on a Likert scale from 1 to 7.

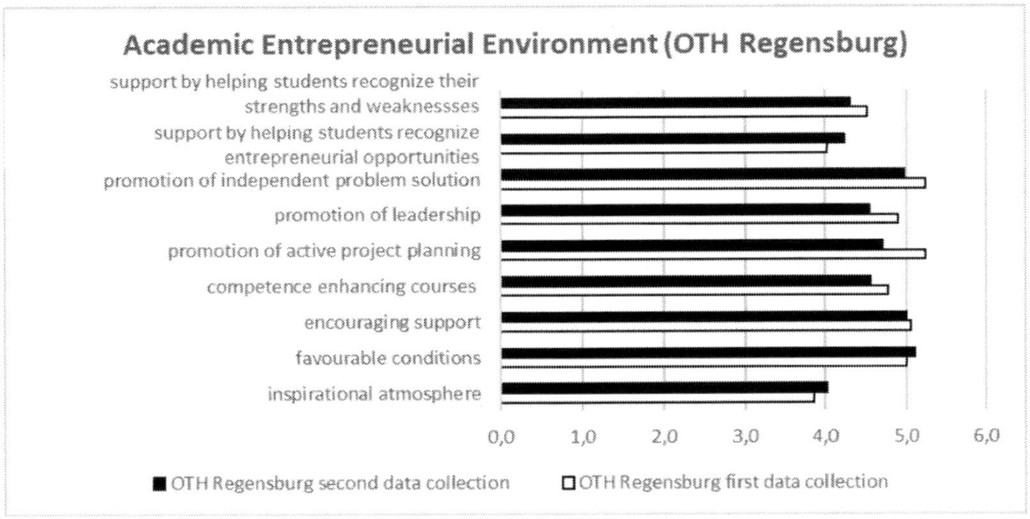

Figure 3. Academic entrepreneurial environment at OTH Regensburg

Source: Author.

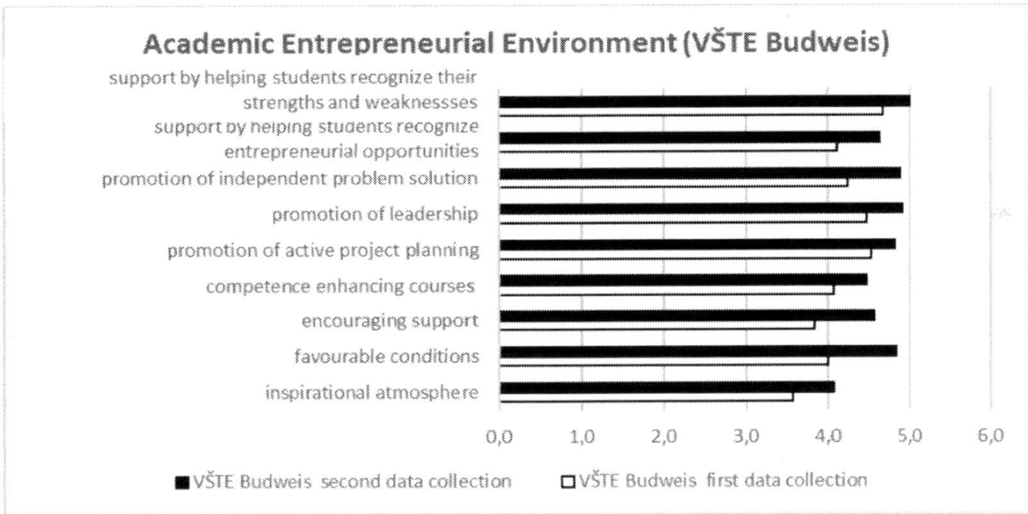

Figure 4. Academic entrepreneurial environment at VŠTE Budweis

Source: Author.

In order to promote the entrepreneurial environment after the first data collection at both institutes, an online course with teaching content on entrepreneurship was created. In addition, several conferences on entre- and intrapreneurship were organized and various professional articles and monographs published between the study periods. The teaching-learning methods were also further adapted. The following section on the implications for education seeks to explain in detail how even more intention-based teaching can contribute to the promotion of entrepreneurial competence in the future.

Firstly, the data obtained from the two investigation periods will be further interpreted. With regards to the entrepreneurial environment at the institutes, it is noticeable that in the second survey that all the factors surveyed at VŠTE Budweis were rated higher than at the beginning of the project in 2017. This is evidenced especially by the higher ratings for the factors "promotion of independent problem solving" ("The lectures/offers at my institution contribute to me working independently on problems"), "encouraging support" ("At my institution, students are encouraged to become entrepreneurs") and "favorable conditions" ("At my institution there are favourable conditions for becoming a founder and, as a consequence, an entrepreneur").

At OTH Regensburg, the changes are less noticeable. However, positive trends can be seen in the factors "favorable conditions" and "inspirational atmosphere" ("The atmosphere at my institution inspires me to develop business ideas for my own company"). The factor "promotion of active project planning" ("The lectures/offers at my institution promote my ability to actively plan and implement projects") was rated worse during the second survey. In this respect, however, it is important to point out the problem of two different spot checks (cf. methodology). During the second survey, different students were interviewed than in the initial measurement. Depending on the subjective perception, there may therefore be certain distortions when making a comparison.

3.2 Results for entrepreneurial intentions

Entrepreneurial intentions, both firm and seriously considered, were also recorded and evaluated with the help of the input measurement. The intention to found one's own company (entrepreneurial intention) allows for quite good estimations of the founding intentions across the two investigation periods (cf. *Figures 5* and *6*).

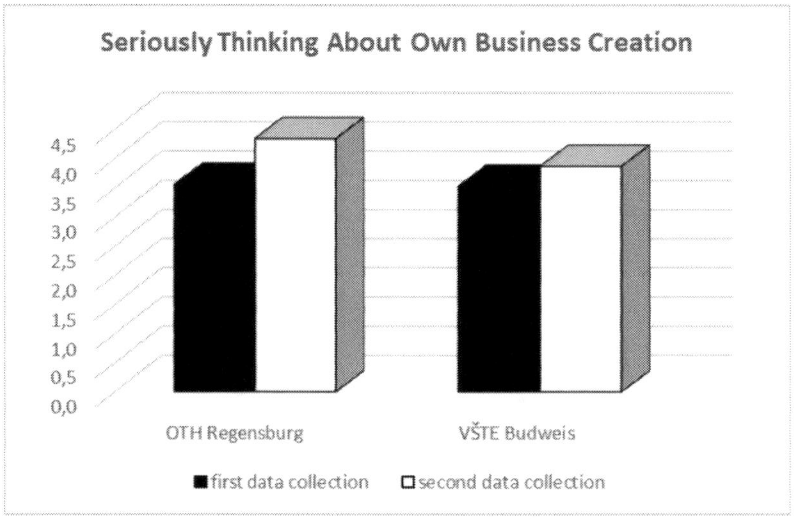

Figure 5. Seriously thinking about own business creation

Source: Author.

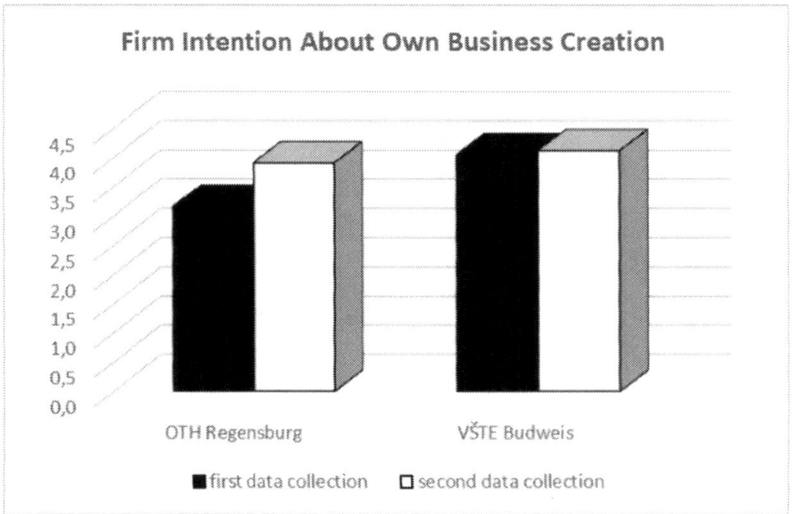

Figure. 6. Firm intention about own business creation

Source: Author.

 https://doi.org/10.1051/shsconf/20219002003

Looking at the two intention-based items "I have already seriously considered setting up or taking over a company" (cf. *Figure 4*) and "I have the firm intention of setting up or taking over a company one day", it is clear that the indicated intentions are higher at both institutes in the second survey. However, at OTH Regensburg, the entrepreneurial intention is significantly higher than at the VŠTE Budweis.

On the basis of the above findings, it can be concluded that the confidence in one's own entrepreneurial competence is more clearly visible among the students in the second survey than in the first.

3.3 Results for entrepreneurial competence

With regards to entrepreneurial competence, the students' knowledge and skills can be interpreted from the first and second surveys. The following two illustrations provide a corresponding group comparison of the two surveys at OTH Regensburg (cf. *Figure 7*) and at VŠTE Budweis (cf. *Figure 8*).

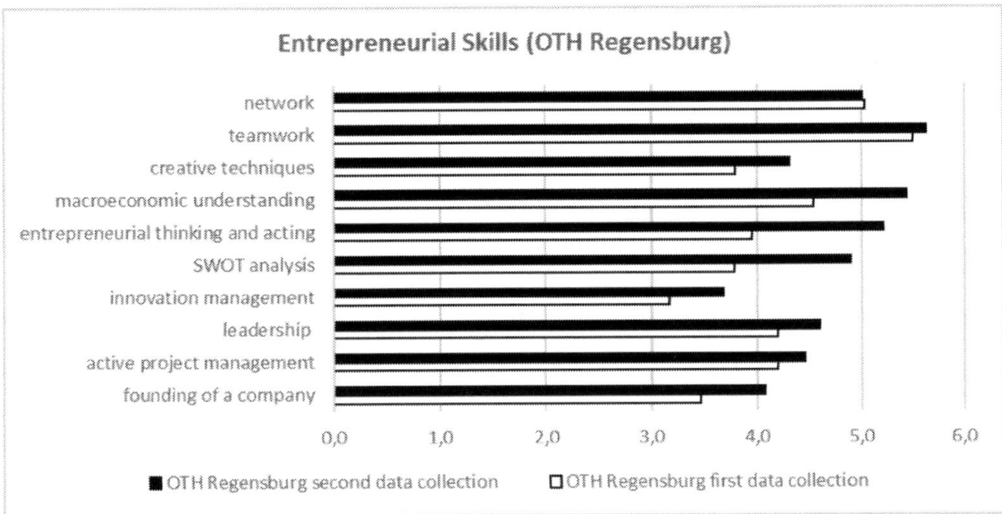

Figure 7. Entrepreneurial skills – OTH Regensburg

Source: Author.

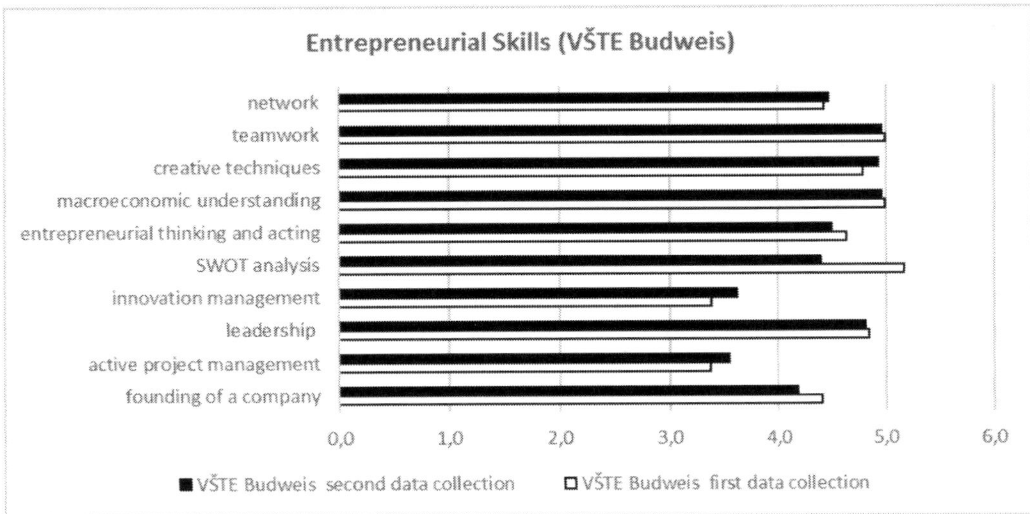

Figure 8. Entrepreneurial skills – VŠTE Budweis

Source: Author.

The students of OTH Regensburg assessed almost all factors related to their entrepreneurial knowledge and skills more highly in the second survey than in the first. This is particularly true in the case of "macroeconomic understanding", "entrepreneurial thinking and acting" and the implementation of the "SWOT analysis". At VŠTE Budweis, however, the differences between the two surveys are less clear. Knowledge of the "SWOT analysis" is much weaker, whereas that of "innovation management" and "active project management" are assessed better. However, the values are in the middle segment with an average value of 4 on a Likert scale of 1 to 7. In order to be able to promote entrepreneurial knowledge and (cognitive) skills, but also the (psychological) characteristics and skills even more specifically in the future, it seems logical to adapt the teaching to the entrepreneurial intentions of the students. The following section shows which activating elements can be used appropriately in entre- and intrapreneurship education.

3.4 Implications for education

Direct implications for entre- and intrapreneurship education can be deduced from the large-scale study (author's dissertation) and the descriptive analysis on which it is based. In order to be able to adapt the teaching to the students, two questions arise: Which entrepreneurial intentions, both psychological and cognitive aspects thereof, are important individually or within a group? Which learning environments should be chosen to promote entrepreneurial competence in a holistic way?

In order to answer the first question, the existing findings suggest an assessment of needs is required. This assessment compares the target values of the ideally typical expression of entrepreneurs and intrapreneurs to the actual values of the input measurement, while differentiating between entrepreneurial and intrapreneurial intentions. In order to answer the second question, a suitable teaching-learning setting must be chosen, taking into account the results of the target-actual comparison. This should address the identified needs. In addition, it is advisable to go as far as possible beyond the traditional transmission of knowledge. Although lectures through front-line teaching allow for the transmission of factual knowledge, they do not activate students. In addition to technical competence, it is also recommended to promote methodological and social competence within the framework of holistic entrepreneurial competence. Methodological competence can, for example, be introduced into teaching by means of problem-solving approaches and social competence by means of techniques of cooperation and communication [22-23]. *Figure 9* shows which elements of entrepreneurship and intrapreneurship teaching activate students. Since these elements are suitable in different ways for entrepreneurship or intrapreneurship education, different positions have been chosen in the space between the two fields of study.

Figure 9. Activating elements in entrepreneurship and intrapreneurship education

Source: Author.

These activating elements within entrepreneurship and intrapreneurship teaching, which can be expanded as desired, can serve to promote holistic entrepreneurial competence (entrepreneurial competence). Planning and role-playing games have a strong relationship to methodological and social competence, while the case study method and multimedia formats serve more to convey and apply factual knowledge. These three methods appear equally interesting for entre- and intrapreneurship education. On the other hand, project work is suited to the targeted promotion of social competence by means of interdisciplinary teams working communicatively together on a task. Initial findings suggest that social competence is in turn appropriate in intrapreneurship teaching. Entrepreneurship teaching, on the other hand, often requires methodological competence, since founders are often decision-makers and creators. Here, problem-solving approaches such as the use of creativity techniques, seem to be useful [24-25].

After completing entrepreneurship and/or intrapreneurship education, an evaluation measurement may be carried out to verify the success of the activating elements used. Within the context of the author's dissertation, an evaluation model for the efficacy of competence promotion was created, which takes into account the theory of planned behavior according to Ajzen, as well as the proposed modification according to Rauch and Hulsink. Such "post-evaluations" (after apprenticeship) makes it possible to compare current entrepreneurial competence and entrepreneurial intentions with the ideal-typical expression in entrepreneurship and intrapreneurship. Following this evaluation measurement, it may again be possible to integrate activating elements into further teaching and new action strategies to promote and instigate entrepreneurial intentions. In this way, students with a concrete business idea could be given targeted start-up advice in addition to their teaching. For potential intrapreneurs, an internship semester in innovation management would again be conceivable. The aim of all such interventions should be not only to promote entrepreneurial intentions and entrepreneurial skills, but also to ensure that the intentions ultimately translate into concrete entrepreneurial behavior.

In order to be able to assess the immediate success of teachers, it also seems logical to undertake an evaluation before ("pre-evaluation") and after the courses ("post-evaluation"). A recurring evaluation of potential entrepreneurs and intra-entrepreneurs could highlight even better the impact of entrepreneurship and intra-entrepreneurship teaching on entrepreneurial intentions and ultimately on entrepreneurial behavior. The teacher's success could then also be derived.

4 Conclusion

An initial measurement of the entrepreneurial intentions and the entrepreneurial competence seems necessary in order to be able to adapt the teaching even more to the needs of the students. Entrepreneurship and intrapreneurship intentions can be distinguished, but they have one thing in common: to enable students to put ideas into practice by means of creativity, risk-taking and other attributes. Entrepreneurial competence has matured into an everyday competence that can be helpful in any life situation. The theory of planned behavior of Ajzen [11] and current modifications in entrepreneurship [15-19] makes it possible to better understand entrepreneurial intentions. On this basis, entrepreneurial competence – entrepreneurial qualities, knowledge and skills – can be promoted holistically.

Using the questionnaire "Entrepreneurial Competence", which was developed as part of the author's dissertation, findings in the field of higher education at OTH Regensburg and VŠTE Budweis, as well as on entrepreneurial intentions and entrepreneurial competence were obtained from the two surveys conducted. At both institutes combined, 153 students have already taken part in an entrepreneurship course. Education in the entrepreneurial field was also rated as above-average at both institutes. At VŠTE Budweis, the factors were rated much higher in the second survey than in the first. At OTH Regensburg, only minor changes were observed between the surveys. In August 2019, OTH Regensburg's students assessed their entrepreneurial knowledge and skills even more highly than in March 2017. VŠTE Budweis shows less marked differences in the cognitive aspects of entrepreneurial competence.

These descriptive results make it clear that an even stronger alignment of teaching to the needs of students by means of intention-based teaching-learning settings seems useful. The latter require, in particular, activating elements that meet the needs of the two intention groups. Depending on the type of intention, different elements within teaching can be applied that go beyond the classical transmission of knowledge. In order to promote holistic competence in practice (entrepreneurial competence), a wide range of activating elements can be used within entrepreneurship and intrapreneurship teaching, which are related to technical, methodological and social competence. These can be integrated into the teaching in a variety of ways, depending on their character. Thus, for example, case studies and multimedia formats are more suited to the transmission of factual knowledge (specialist competence), whilst project work has a strong connection to social competence and creativity techniques are increasingly connected to methodological competence. While planning, role-play, multimedia formats and case studies should be integrated into entre- and intrapreneurship education, project work, for example, seems better suited to intrapreneurship education because of its social component, whilst the teaching of creative techniques for entrepreneurship education is a probate means of promoting entrepreneurial competence. However, these activating elements can be expanded as desired. Depending on the focus, various elements can be selected within entre- and intrapreneurship education. While entrepreneurship teaching should probably focus on problem-solving approaches in the sense of imparting methodological competence in addition to specialist competence, intrapreneurship teaching should probably focus more on social competence in the sense of social leadership of innovation projects in addition to factual knowledge. However, such a systematization should not obscure the fact that entrepreneurs should also have social competence and intra-entrepreneurs should also have methodological competence in the sense of a holistic competence to act.

Acknowledgement

This article contains a partial output of the dissertation „Der Unternehmer in uns – Entwicklung einer Diagnostik der unternehmerischen Absichten zur Förderung der unternehmerischen Kompetenz im Entre- und Intrapreneurship" ("The entrepreneur within us – development of a diagnostic tool for entrepreneurial intentions to promote entrepreneurial competence in entrepreneurship and intrapreneurship"). The dissertation is expected to be published in spring 2021.

References

1. G. Pinchot, *Intrapreneuring – Mitarbeiter als Unternehmer [Intrapreneuring - employees as entrepreneurs]*. Wiesbaden: Gabler Verlag (1988)
2. G. Pinchot, Intrapreneuring. *Longe Range Planning*. **29**(3), 405–411 (1996)
3. R. D. Hisrich, Entrepreneurship/Intrapreneurship. *American Psychologist*. **45**(2), 209–222 (1990)
4. M. Bitzer, *Intrapreneurship – Unternehmertum in der Unternehmung [Intrapreneurship - entrepreneurship in the company]*. Stuttgart, Zürich: Schäffer, Verlag für Wirtschaft und Steuern (1991)
5. R. Wunderer, H. Bruch, *Umsetzungskompetenz: Diagnose und Förderung in Theorie und Unternehmenspraxis [Implementation skills: Diagnosis and support in theory and business practice]*. München: Vahlen (2000)
6. Europäische Kommission, *Schlüsselkompetenzen für lebensbegleitendes Lernen – Ein europäischer Referenzrahmen [Key Competences for Lifelong Learning - A European Framework of Reference]*. Luxemburg: Amt für amtliche Veröffentlichungen der Europäischen Gemeinschaften (2007)
7. J. Erpenbeck, L. von Rosenstiel, *Handbuch Kompetenzmessung. Erkennen, verstehen und bewerten von Kompetenzen in der betrieblichen, pädagogischen und psychologischen Praxis [Competency measurement manual. Recognizing, understanding and evaluating competencies in operational, educational and psychological practice]*. Stuttgart: Schäffer-Poeschel Verlag (2007)
8. G. F. Müller, Unternehmerische Kompetenz [Entrepreneurial competence]. In W. Sarges. *Management-Diagnostik*. Hogrefe Verlag, Göttingen, 467–473 (2013)
9. G. F. Müller, M. Sauerland, G. Raab, *Wir alle sind Unternehmer! – Die Psychologie erfolgreicher Gründung und Führung von Unternehmen [We are all entrepreneurs! - The psychology of successful business creation and management]*. Hamburg: Windmühle Verlag (2016)
10. I. Ajzen, From intentions to actions. A theory of planned behavior. In J. Kuhl, J. Beckmann. *Action control: From cognition to behaviour*. New York: Springer-Verlag, 11–39 (1985)
11. I. Ajzen, The theory of planned behavior. *Organizational Behavior & Human Decision Processes*. **50**(2), 179–211 (1991)
12. I. Ajzen, Perceived behavioural control, self-efficacy, locus of control, and the theory of planned behaviour. *Journal of Applied Social Psychology*. **32**(4), 665–683 (2002)
13. I. Ajzen, *Attitudes, personality and behavior. Maidenhead*. Berkshire: Open University Press, McGraw-Hill Education (2005)
14. I. Ajzen, The theory of planned behaviour is alive and well, and not ready to retire: a commentary on Sniehotta, Presseau, and Araujo-Soares. *Health Psychology Review*. **9**(2), 131–137 (2015)
15. I. Ajzen, Reasoned Action in the Service of Goal Pursuit. *Psychological Review*. **126**(5), 774–786 (2019)
16. I. Ajzen, M. Fishbein, S. Lohmann, D. Albarracin, The Influence of Attitudes on Behavior. *Handbook of Attitudes*. **1**, 197–255 (2019)
17. F. La Barbera, I. Ajzen, Control Interactions in the Theory of Planned Behavior: Rethinking the Role of Subjective Norm. *Europes Journal of Psychology*. **16**(3), 401–417 (2020)
18. M. Bosnjak, I. Ajzen, The Theory of Planned Behavior: Selected Recent Advances and Applications. *Europes Journal of Psychology*. **16**(3), 352–356 (2020)
19. A. Rauch, W. Hulsink, Putting entrepreneurship education where the intention to act lies: An investigation into the impact of entrepreneurship education on entrepreneurial behavior. *Academy of Management Learning & Education*. **14**(2), 187–204 (2015)
20. H. Bergmann, S. Golla, *Unternehmertum an Hochschulen in Deutschland. Ergebnisse des Global University Entrepreneurial Spirit Students' Survey 2016 [Entrepreneurship at universities in Germany. Results of the Global University Entrepreneurial Spirit Students 'Survey 2016]*. St. Gallen, Fulda: Universität St. Gallen, Hochschule Fulda (2016)
21. B. Rammstedt, O. P. John, Kurzversion des Big Five Inventory (BFI-K): Entwicklung und Validierung eines ökonomischen Inventars zur Erfassung der fünf Faktoren der Persönlichkeit [Short version of the Big Five Inventory

(BFI-K): Development and validation of an economic inventory to record the five factors of personality]. *Diagnostica.* **51**(4), 195–206 (2005)

22. H. Grüner, Entrepreneurial Learing – Ist eine Ausbildung zum Unternehmertum möglich? [Entrepreneurial Learing - Is entrepreneurship training possible?]. *Zeitschrift für Berufs- und Wirtschaftspädagogik.* **5**, 485–509 (1993)
23. H. Klandt, Entrepreneurship spielend lernen: Erfahrungen beim Einsatz eines Computerplanspiels zur Vermittlung der mittelständischen Unternehmerrolle [Learning entrepreneurship through play: experience in using a computer simulation game to convey the role of medium-sized entrepreneur]. In G. Faltin, S. Ripsas, J. Zimmer. *Entrepreneurship: Wie aus Ideen Unternehmen werden.* München: C. H. Beck Verlag, 197–216 (1998)
24. H. Plattner, C. Meinel, U. Weinberg, *Design Thinking: Innovation lernen – Ideenwelten öffnen.* München: mi-Wirtschaftsbuch, FinanzBuch Verlag GmbH, (2009)
25. R. Schulte, H. Klandt, *Aus- und Weiterbildungsangebote für Unternehmensgründer und selbständige Unternehmer an deutschen Hochschulen [Training and further education offers for company founders and self-employed entrepreneurs at German universities]*. Bonn: Bundesministerium für Bildung, Wissenschaft, Forschung und Technologie, (1996)

Relationship Between Organizational Culture, Employee Satisfaction and Organizational Commitment

Sarantuya Jigjiddorj[1,*], *Altanchimeg* Zanabazar[1], *Tsolmon* Jambal[2], *Buyankhishig* Semjid[3]

[1]National University of Mongolia, Business School, Management Department, Mongolia
[2]Institute of Technology and Business, Faculty of Corporate Strategy, Department of Management, Nemanická 536/7, 370 01 Česke Budejovice, Czech Republic
[3]National University of Mongolia, Business School, Management Department, Mongolia

Abstract. Organizational culture is an important human resource management issue that affects the success and sustainability of a company. In organizational settings, culture describes what is unique about an organization, as expressed by the shared beliefs and values established by the founders and communicated through different ways. This shapes employees´ perceptions and behaviour, as well as the principles that apply to members of an organization.
Job satisfaction refers to an employee´s feelings about their job and is frequently studied in relation to organizational culture and other variables, such as job commitment, performance and organizational commitment.
Organizational commitment refers to whether an employee wishes to continue working for an organization or leave it. Employees tend to be attracted to those organizations with a culture that values their work and focuses on organizational wellbeing. There is a positive relationship between organizational culture, job satisfaction and the organizational commitment of employees.
Keywords: Organizational culture; job satisfaction; organizational commitment; case study

1 Introduction

An organization is a group of people united by common goals, with an open system that adapts to the ever changing external environment. Regardless of company size and the industry they operate in, organizations embrace three elements- goals, people and systems. People interact within a system, thereby contributing to the achievement of an organization´s goals. Every organization embraces its own culture, which includes its goals, values and interactions with the external environment. What makes an organization´s culture unique are the feelings, beliefs, standards, customs and rules that it is based on.

By nurturing a strong positive organizational culture, employees become more committed and responsible. It unites the ambitions of employees to become goal oriented. If an organization has loyal and committed employees, this may be a sign of job satisfaction, which is essential for achieving an organization's goals.

2 Background

Organizational culture is an important topic for researchers and practitioners of human resource management. In organizational settings, culture describes what is unique about an organization, as expressed by the shared beliefs and values established by the founders and communicated through different ways. This shapes employees´ perceptions and behaviour, as well as the principles that apply to members of an organization. The majority of researchers agree that organizational culture is a system of shared values, beliefs and behaviours of members of an organization [1].

According to the Schein [2], organizational culture can be observed in three forms, namely a) observable artefacts - specific attitudes, beliefs, values and other features like dress code, stories, rituals and special ceremonies that are specific to an organization; b) supported values – values widely promoted by the leaders and which are very likely reflected in the behaviour of employees. These values generally guide the behaviour of members of an organization in different situations; and c) basic assumptions or perceived situations – these regulate the collective actions of the members, which are the result of long-term practice and are valued by every organizational member, who practices them instinctively. In doing so, a sense of belonging to a community is generated.

* Corresponding author: sarantuyaj@num.edu.mn

Numerous studies have been conducted into types of organizational cultures, thereby referring to their format and functions [3], environmental requirements [4], and models [5] based on employee relations. They identified four completely different cultures, namely hierarchical, market, clan and adhocracy. Other researchers identified the existence of sub-cultures within an organization [6] based on individualistic and collective cultures related to employees' social groupings. Beyond this, organizational culture has also been studied in relation to workplace related issues, such as wellbeing or quality of work life. This concerns an organization's ability to promote and maintain the physical, psychological and social wellbeing of employees at all levels and for all jobs [7]. This includes the employees' physical and mental health, sense of happiness and social wellbeing, which are recognized in the term job satisfaction [8].

A widely studied topic closely linked to organizational culture is that of job satisfaction. Job satisfaction refers to an employee´s feelings about their job and is frequently studied in relation to organizational culture and other variables, such as job commitment, performance and organizational commitment. Job satisfaction is an important issue for all organizations because it is a crucial factor for organizational success and sustainability [9]. According to Lund [10], if job satisfaction is high, this reflects a strong organizational culture, and vice versa, if job satisfaction is low, the organizational culture is weak. Other studies show that satisfied employees tend to be more productive, creative and committed to their employers. According to [11], job satisfaction leads to higher productivity through greater motivation and the enhancement of the working capabilities of employees.

Barbara and Dormio [12] conducted research into the effect of organizational culture on job satisfaction. They showed that organizational culture had a positive effect on job satisfaction. The research conducted by Nayak and Barik [13] produced a similar result. Moreover, Park & Kim [14], Cronley & Kim [15] and Belias & Koustelios [16] conducted comprehensive studies into organizational culture and related variables, all of which also confirmed that organizational culture has a positive influence on job satisfaction.

Another widely observed and often researched topic is that of the relationship between organizational culture and organizational commitment. Organizational commitment refers to an individual's psychological attachment to the organization they work for. The relationship between organizational culture and organizational commitment is theoretically described by Peters and Waterman [17]. They suggest that organizational culture influences a variety of organizational and individual outcomes including productivity, performance, commitment and self-confidence.

Organizational culture is considered to have a strong relationship to organizational commitment. This is reflected in how members of an organization follow the provisions made by an organization. If an organization is well-organized, whereby the leader takes their authority and responsibility seriously, then its members will certainly be committed to the organization [18]. Another study suggests that organizational culture in the form of good communication, trust and an innovative culture, positively influences organizational commitment [19].

Organizational cultures characterized as being collaborative, a brotherhood, participatory, consensual, constructive and supportive are more likely to benefit from higher levels of organizational commitment. Abdullah, Shamsuddin and Wahab [20] showed this to be true. Organizational culture has also been studied in relation to different variables, such as leadership and employee performance [21], and the impact of it on leadership style, organizational commitment and its relationship to job satisfaction [22].

2.1 Problem statement

Numerous studies have previously been conducted on the relationship between organizational culture, job satisfaction and employee commitment in universities in Pakistan [23]. Denisten, Sarros and Gray [24] attempted measuring the levels of employee satisfaction in organizational cultures against the job satisfaction of marketing staff. Similar research was conducted in the fertilizer industry in Pakistan. These attempts to measure the correlation between organizational culture, job satisfaction and organizational commitment serve as examples for extending the research on the selected topic.

When taking onto consideration past research and the opportunities to apply such concepts to the analysis of the current state of affairs within a local context, the following problems statements were formulated:

- The examined organizational culture has an impact on job satisfaction;
- The examined organizational culture has an impact on organizational commitment.

2.2 Scope of the study

The purpose of the study presented here is to explore the impact of the organizational culture of an insurance company on the job satisfaction and organizational commitment of its employees. The study´s aims are, based on the results of an analysis, to enhance human resource management practices, including potential changes to the workplace environment, to improve job satisfaction and organizational commitment.

Research design: Based on previous studies, the following research model was developed to achieve the proposed aims of the study

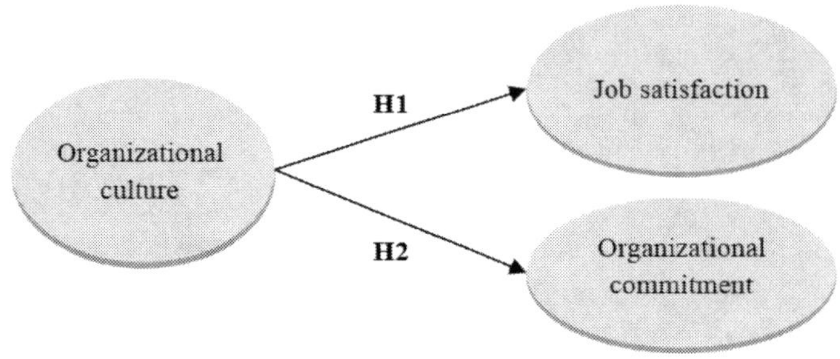

Figure 1. Research model

Source: Authors.

Research hypotheses:

H1: Organizational culture has an impact on job satisfaction.

H2: Organizational culture has an impact on organizational commitment

3 Methodology

The beneficiary of the conducted socio-metric survey and analysis is a major insurance company operating in the local market. SPSS 23 software was used for the data analysis. Reliability tests, and correlation and regression analyses were conducted to determine the relationship between the independent and dependent variables.

The survey was conducted between 4th and 20th November 2019. Out of the 180 questionnaires distributed to employees, 152 were returned (a response rate of 84.4%). After cleaning the data, it was determined that the data from 151 questionnaires could be used for the analysis.

The questionnaire consisted of 61 questions, each with a 5-point Likert scale, grouped into 3 categories. The background data of the respondents is summarized in Table 1.

Table 1. Background of the respondents

Gender		**Age level**		**Education background**		**Years of service in current position**	
Gender	**%**	**Age (in yrs)**	**%**	**Education level**	**%**	**Tenure**	**%**
Male	35.8	≤ 20	0.7	Master's degree	15.9	≤ 1 year	31.8
		21-25	7.3			1-3 years	23.3
		26-30	24.5	Undergraduate degree	76.8	4-10 years	33.1
Female	64.2	31-35	25.8	Professional education	4.6	11-15 years	3.3
		36-40	15.9			16-20 years	4.6
		41-50	12.6	Incomplete secondary education	1.3	> 21 years	3.3
		> 55	11.3				

Source: Authors.

By demographics: 64.2% of the respondents were female; 58.3% were younger than 35 years old; and 55.1% had worked in their current positions for up to 3 years.

By the positions held by the respondents: 68.2% were employees; 9.9% senior executives; and the remaining 16.6% were department or branch managers (see Figure 2).

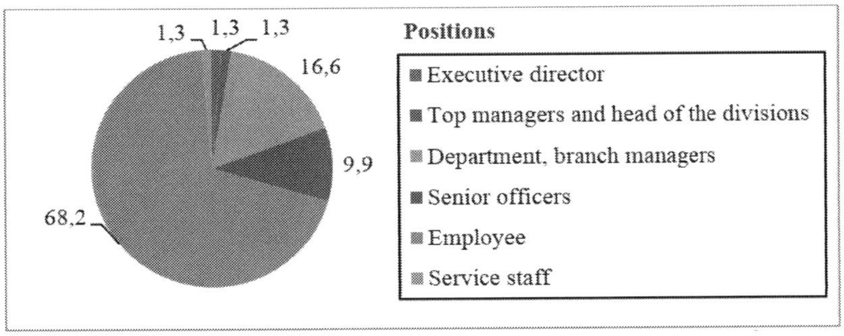

Figure 2. Positions held by the respondents

Source: Authors.

3.1 Sample size

To ensure the adequacy of the sampling, Kaiser-Meyer-Olkin (KMO) and Bartlett's tests were performed. In general, the higher the KMO value, the greater the adequacy of the sample size and the less likely remedial measures need to be taken. The significance test (Sig.) checks the variances across samples. A KMO value less than 0.5 indicates that the sample size is inadequate. The results of the Kaiser-Meyer-Olkin (KMO) and Bartlett´s tests are summarized in Table 2.

Table 2. Summary of sampling adequacy tests

KMO and Bartlett's tests		
Kaiser-Meyer-Olkin measure of sampling adequacy		0.725
Bartlett's test of sphericity	Approx. Chi-Square	17293.155
	df	6328
	Sig.	.000

Source: Authors.

According to the test results, the KMO value is 0.725, which means that the sample size is adequate. The significance value (Sig. = 0.000) of Bartlett's test of sphericity means that the variances across the samples are equal and that the data are reliable for conducting further analysis.

3.2 Reliability analysis of variable

Data reliability was tested to ensure the consistency of the set of variables. For the measurement of the closeness, or scale reliability of the related set of items, Cronbach α was estimated. According to Nunnally [25], a Cronbach α value of 0.7 is an acceptable value for measurement. On the basis of the results (see Table 3), it can therefore be said that the design of the developed questionnaire was appropriate for the purpose of the survey, which was found to be capable of measuring the proposed issues. The results of the test are summarized in the table below.

Table 3. Reliability test results

Set of items	Number of questions	Cronbach's α
Organizational culture	15	.912
Organizational commitment	15	.843
Job satisfaction	34	.969

Source: Authors.

3.3 Correlation analysis

The results of the correlation analysis are summarized in Table 4.

Table 4. Correlation Analysis

Correlations				
		OC	OCOM	JS
OC	Pearson correlation Sig. (2-tailed) N	1		
OCOM	Pearson correlation Sig. (2-tailed) N	.558** .000 151	1	
JS	Pearson correlation Sig. (2-tailed) N	.718** .000 151	.450** .000 151	1

** correlation is significant at the 0.01 level (2-tailed).

Source: Authors.

Although the correlation between organizational culture [r=.718; N=151; p<0.01] and job satisfaction [r=.522; N=405; p<0.01] is strong, the correlation with organizational commitment [r=.558, N=151, p<0.01] is only slightly positive.

3.4 Regression analysis

To test the hypotheses of the study, regression analyses were performed, the results of which are presented in Table 5.

Table 5. Regression analysis

Model summary

Model	R	R Square	Adjusted R Square	Std. error of the estimate
1	.718a	.515	.512	11.6222
2	.558a	.311	.306	2.762

a. Predictors: (Constant), OC

Source: Authors.

ANOVA[a]

Model		Sum of Squares	df	Mean Square	F	Sig.
1	Regression	49181.427	1	49181.427	158.373	.000b
	Residual	46270.745	149	310.542		
	Total	95452.172	150			
2	Regression	512.738	1	512.738	67.235	.000b
	Residual	1136.282	149	7.626		
	Total	1649.020	150			

a. Dependent variable: JS, OCOM

b. Predictors: (Constant), OC

Source: Authors.

Coefficients[a]

Model		Unstandardized coefficients		Standardized coefficients	t	Sig.
		B	Std. Error	Beta		
1	JS	.944	.154	.718	12.585	.000
2	OCOM	.199	.024	.558	8.200	.000

a. Dependent variable: JS, OCOM

Source: Authors.

According to the test results, there is a positive correlation between organizational culture and job satisfaction (51.2%; R Square=.512) and a slightly positive one with organizational commitment (30.6%; R Square=.306). In the case of the surveyed organization, the organizational culture of the organization has a positive impact on the job satisfaction of its employees. An increase of one unit in organizational culture results in $\beta = .718$, or in other words, an increase in job satisfaction [$\beta = .718$; $p < 0.01$]. However, the results also show that organizational culture has only a slightly positive correlation with organizational commitment. Clearly, an increase of one unit in organizational culture only increases organizational commitment by $\beta = 0.558$ unit or [$\beta = .558$, $p < 0.01$].

The aforementioned results are graphically presented in Figure 3.

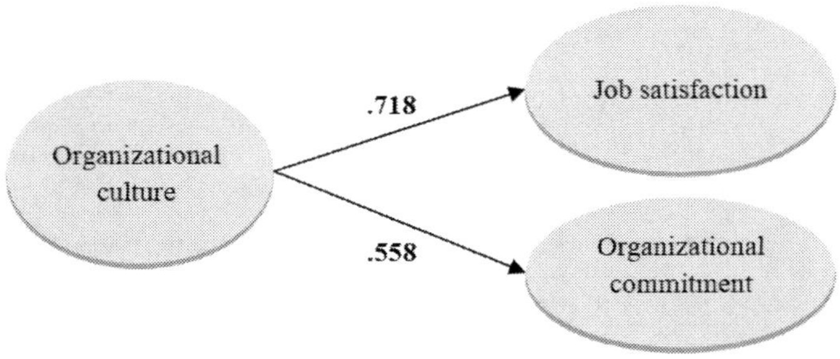

Figure 3. Results of the study

Source: Authors.

The strength of the independent variable in relation to the dependent variables shows a positive effect (high Beta coefficient value) and demonstrates strong statistical significance ($p < 0.01$). This enables us to confirm both hypotheses.

As was the case in other studies [23], the survey results demonstrate that there is a strong correlation between organizational culture and job satisfaction, and less so to organizational commitment. It should be noted that there is room for improvement, such as covering different industry representatives and including more respondents, in order to better observe the correlation between organizational culture and organizational commitment.

4 Conclusion

Organizational culture is a strong variable that ensures the long-term sustainability of an organization by enabling it to distinguish itself from its competitors. Organizations with a strong culture are better able to retain committed employees who feel satisfied about their jobs, which reduces staff turnover.

Job satisfaction is a challenging issue for the management of any organization. Through organizational culture, it is possible to increase job satisfaction, which in turn can nurture an employee´s commitment and willingness to stay for a longer period of time.

The presented study was based on the evidence of similar previous studies conducted in different countries among diverse communities into the relationship between organizational culture and job satisfaction, and organizational culture and organizational commitment. The principles applied in those earlier studies were applied to a local case, namely a local insurance company. The survey results showed that there is a positive relationship between the selected variables, thereby confirming the results of the previous studies.

According to our study, an increase in the strength of organizational culture by one unit, leads to an increase in job satisfaction. Similarly, an increase in the strength of organizational culture by one unit leads to an increase in organizational commitment ($\beta = 0.558$ unit), although to a lesser degree than the former.

We can conclude that organizational culture has a strong impact on employee job satisfaction and less so on organizational commitment.

On the assumption that the principles of the presented study are valuable and reliable, it may be useful to add more dependent variables in relation to organizational culture in order to improve human resources management within an organization and to encourage people to become more goal oriented and committed to achieving organizational goals.

References

1. D. Ravasi, M. Schultz, Responding to Organizational Identity Threats: Exploring the Role of Organizational Culture. *Academy of Management Journal*. **49**(3), 433-458 (2006)
2. E. H. Schein, *Organizational culture and Leadership*. 3rd ed. San Francisco: Jossey-Bass (2004)

3. T. Deal, A. Kennedy, *Corporate Cultures: The Rites and Rituals of Corporate Life.* Boston: Addison-Wesley Publishing (1982)

4. R. Daft, *Essentials of Organization Theory & Design.* Cincinnati: South-Western College Publishing (2001)

5. K. S. Cameron , R. E. Quinn, *Diagnosing and Changing Organizational Culture Based on the Competing Values Framework. The Jossey-Bass Business & Management Series.* San Francisco: Jossey-Bass, A Wiley Imprint (2006)

6. D. Denison, C. Lief, J. L. Ward, Culture in Family-Owned Enterprises: Recognizing and Leveraging Unique Strengths. *Family Business Review.* **17**(1), 61–70 (2004)

7. P. Torri, E. Toniolo, Organizational well being: Challenge and future foundation. *Giornale italiano di medicina del lavoro ed ergonomia.* **32**(3), 363–367 (2010)

8. A. M. Grant, M. K. Christianson, R. H. Price, Happiness, Health, or Relationships? Managerial Practices and Employee Well-Being Tradeoffs. *Academy of management perspectives.* **21**(3), 51–63 (2007)

9. S. P. Robbins, T. A. Judge, *Organisational Behavior.* 15th ed. Boston: Pearson (2013)

10. D. Lund, Organizational culture and job satisfaction. *Journal of Business & Industrial Marketing.* **18**(3), 219–236 (2003)

11. K. Bhatti, T. Qureshi, Impact of employee participation on job satisfaction, employee commitment and employee productivity. *International Review of Business Research Papers.* **3**(2), 54–68 (2007)

12. B. Bigliardi, A. I. Dormio, F. Galati, G. Schiuma, The impact of organizational culture on the job satisfaction of knowledge workers. *VINE.* **42**(1), 36–51 (2012)

13. B. Nayak, A. Barik, Assessment of the link between Organizational culture and job satisfaction (Study of an Indian Public Sector). *International Journal of Advanced System and Social Engineering Research.* **13**(3), 78–87 (2013)

14. J. S. Park, T. H. Kim, Do types of organizational culture matter in nurse job satisfaction and turnover intention? *Leadership in Health Services.* **22**(1), 20–38 (2009)

15. C. Cronley, Y. K. Kim, Intentions to turnover: Testing the moderated effects of organizational culture, as mediated by job satisfaction, within the Salvation Army. *Leadership and Organization Development Journal.* **38**(2), 194–209 (2017)

16. D. Belias, A. Koustelios, Organizational Culture and Job Satisfaction: A Review. *International Review of Management and Marketing.* **4**(2), 132–149 (2014)

17. T. J. Peters, R. H. Waterman, *In Search of Excellence: Lessons from America's Best-run Companies, Reprint edition.* New York: Harper Business (2006)

18. T. Lim, Relationships among Organizational Commitment, Job Satisfaction, and Learning Organization Culture in one Korean Private Organization. *Asia Pacific Education Review.* **11**(3), 311–320 (2003)

19. M. Muliaty, M. Basri, J. Jasruddin, Effects of organizational transformation and culture on employees performance. *International Journal of Economic Perspectives.* **11**(3), 1287–1292 (2017)

20. N. H. Abdullah, A. Shamsuddin, E. Wahab, Does organizational culture mediate the relationship between transformational leadership and organizational commitment? *International Journal of Organizational Leadership.* **4**, 18–32 (2015)

21. A. Xenukou, M. Simosi, Organizational culture and transformational leadeship as predictors of buisness unit performance. *Journal of Managerial Psychology.* **21**(6), 566–579 (2006)

22. L. H. Yiing, K. B. Ahmad, The moderating effects of organizational culture on the relationships between leadership behavior and organizational commitment and between organizational commitment and job satisfaction and performance. *Leadership and Organizational Develoment.* **30**(1), 53–86 (2009)

23. M. R. Awan, K. Mahmood, Relationship among leadership style, organizational culture and employee commitment in university libraries. *Library Management Journal.* **31**(4), 253–266 (2010)

24. I. L. Densten, J. C. Sarros, J. Gray, Leadership and its impact on organizational culture. *International journal of business studies.* **10**(2), 1–26 (2002)

25. J. C. Nunnally, *Psychometric Theory*. New York: McGraw-Hill (1978)

Academic Entrepreneurship Education: Does Gender Matter?

Xenia Justus[1,*]

[1]Ostbayerische Technische Hochschule Regensburg, Human Resources Department, Department of Personnel Development, Prüfeninger Str. 58, 93049 Regensburg, Germany

Abstract. Academic entrepreneurship education aims to train and raise awareness amongst students about entrepreneurial activity, thereby increasing the number of academic start-ups and spin-offs. However, as yet, there are very few findings on the gender-specific differences amongst potential entrepreneurs. Building on the current state of research, gender-specific differences in the form of various different components of entrepreneurial competence are examined. This study focuses in particular on (1) entrepreneurial knowledge, (2) domain-specific interest in entrepreneurship, (3) interest in leadership roles, and (4) entrepreneurial and (5) intrapreneurial intention. Based on an online survey of 281 students of a German and Czech university, the mean differences show that there are statistically significant lower values for female students than for male students for all the variables investigated, that is with exception to intrapreneurial intention. These findings underline the need for targeted promotion of female entrepreneurship within the context of academic entrepreneurship education.
Keywords: female entrepreneurship; entrepreneurship education; gender research; entrepreneurial competence; interest in entrepreneurship; leadership role

1 Introduction

"The incidence of female entrepreneurs is lower than that of male entrepreneurs in almost every country in the world" [1, p. 261]. Germany is no exception: here, the majority of companies are founded by men [2]. In 2017, the proportion of women in various start-up activities in Germany was 37% [3].

Although an entrepreneur is often associated with the founding of a company, there is more to entrepreneurship than just this. Bechter et al [4] emphasise that an entrepreneur is constantly on the lookout for new opportunities and does not shy away from taking risks. Entrepreneurs also seek to implement innovations in a marketable way and make them available [4, 5].

In the last few years, female entrepreneurship has attracted more attention as a field of research. In current research discussions, several reasons have been put forward as to why the field of entrepreneurship is dominated by men [2, 6, 7, 8, 9]. However, the findings concerning this complex issue are not particularly conclusive. Within this context, Bechter et al [4] refer to gender-specific socialisation, gender-role stereotypes, negative self-perception, financial framework conditions and the very patriarchal structures that still frequently exist in society [10]. "In addition, fewer role models are available to female than male entrepreneurs" [9, p. 3]. Since role models are of particular relevance for entrepreneurship, an excessively male-dominated entrepreneurial image could have a negative influence on the entrepreneurial propensity of women [2].

Until now, research into the gender-specific differences amongst potential graduate entrepreneurs has received little attention. The focus of empirical studies often lies in the differences in the case of fully established start-ups, whereby gender is merely a control variable [11,12]. However, the empirically proven gender gap has prompted a growing number of researchers into start-up entrepreneurship to subject their models to gender-specific tests [12], since female-specific behaviour is overlooked in most theories [13]. "Concepts in entrepreneurship theory and practice have long been either dominated by a supposedly gender-neutral perspective, or mainly eulogized as part of a dominant male discourse lacking the complexity of theories on gender aspects" [6, p. 4]. In doing so, motivational components of entrepreneurial competence, for example domain-specific interest in entrepreneurship, interest in leadership roles and entrepreneurial intention, but also cognitive aspects, such as entrepreneurial knowledge, which can be positively influenced in the context of academic entrepreneurship education, are frequently neglected [14]. Here, competence can be understood as all cognitive skills and capacity to solve certain problems, as well as the associated motivational, volitional and social preparedness and skills to be able to use the solutions successfully and responsibly in various situations [15].

* Corresponding author: xenia.justus@oth-regensburg.de

This paper attempts to close some of these research gaps. This study on gender-specific differences in academic entrepreneurship concentrates on the gender of the students as an independent variable and (1) entrepreneurial knowledge, (2) domain-specific interest in entrepreneurship, (3) interest in leadership roles, (4) entrepreneurial and (5) intrapreneurial intention as dependent variables.

1.1 Entrepreneurial knowledge

In cognitive psychology, a distinction is made between declarative and procedural knowledge [16, 17]. Declarative knowledge denotes static knowledge of domain-specific facts and concepts [18], for example knowledge of the financing options for technology-based start-ups. Procedural knowledge is understood to be practical knowledge [18], for example the development of business ideas. In this, the knowledge is not taught directly but rather is constructed individually by the students [17]. In academic entrepreneurship education, both declarative and procedural knowledge about entrepreneurship, business management, innovation, business planning and control is taught. Without entrepreneurial knowledge it would hardly be possible to establish a company, even if all other motivational conditions were fulfilled, since the rational problem solving cannot take place without the specialist knowledge this requires [19]. In this study, entrepreneurial knowledge was operationalised through knowledge about founding a company, active project management, staff management and motivation, innovation management, general economic understanding, creativity techniques and networks [20].

1.2 Domain-specific interest in entrepreneurship

Current approaches in interest research understand interest to be a specific relationship between a person and an object [21]. The interest in an object, a situation, an action or content normally forms gradually over previous experiences and has a certain stability [22]. In order to support the development of new interest in a certain topic or area (e.g. entrepreneurship), there first has to be initial contact which arouses curiosity in the corresponding content: thus, an instable, situational interest arises. The stabilisation of this interest takes time and is highly dependent on external influences. Aided by internalisation and identification, the transition from situational to stable, individual interest then ultimately takes place [21]. Against this background, the interest in entrepreneurship can be seen as a significant condition for the successful commencement and implementation of action (here: the establishment of one's own company).

1.3 Interest in leadership roles

Leadership means proactively thinking ahead and acting, anticipating and precipitating changes, and discovering new opportunities [4, 23]. Both concepts, entrepreneurship and leadership, are held responsible these days, to the same extent in practice and in theory, for the long-term success of entrepreneurship. However, the relationship between entrepreneurship and leadership is often overlooked. Rather, the impression is given that, despite the coexistence of both in successful entrepreneurship, it is usually only one of the two that is responsible for the success of a company [24].

1.4 Entrepreneurial and intrapreneurial intention

Entrepreneurial intention relates to the subjectively evaluated likelihood of establishing one's own company. In meta-analyses, intentions account for up to 31% of the actual, self-reported behaviour [25,26]. With respect to gender-specific differences, the majority of studies thus far have pointed to a greater likelihood of founding a company amongst men [12]. This could be, for example, because an entrepreneurial scene dominated by men offers few reference points for women with which they can identify and seek access [12]. Furthermore, the results of an empirical investigation by Walter and Walter [26] showed that an application-oriented entrepreneurial education effects a deeper interest in entrepreneurship amongst male students than amongst female students. This difference may indicate that the educational offerings do not appeal to female students in their current form. Intrapreneurship can be differentiated quite clearly from entrepreneurship. Moriano et al [23] define intrapreneurship as "(...) entrepreneurial behaviour within an existing organization, (...) which is comprised of three key dimensions: risk taking, innovation and proactiveness" (p. 105). An intrapreneur is also proactive and develops innovative ideas for new projects; in contrast to an entrepreneur, however, they do not bear the full financial and social risk [27].

Building on the analysis of the theoretical basis and the actual state of research, the hypotheses of this empirical study are explained in detail below.

- Hypothesis 1a: There are statistically significant positive correlations between entrepreneurial knowledge, interest in entrepreneurship, interest in leadership roles and entrepreneurial intention.
- Hypothesis 1b: There are no statistically significant correlations between entrepreneurial and intrapreneurial intention.
- Hypothesis 2a: Female students evaluate their entrepreneurial knowledge as statistically significantly lower than male students.

- Hypothesis 2b: Female students demonstrate statistically significantly lower expressions of domain-specific interest in entrepreneurship, interest in leadership roles and entrepreneurial intention.
- Hypothesis 2c: There are no statistically significant differences in intrapreneurial intention between female and male students.

In the following section, the hypotheses formulated here are examined statistically.

2 Methodology

The already existing data from the EU project "Entrepreneurial skills on the Czech/Bavarian labour market" (goal ETZ 2014-2020, INTERREG V) were used to examine the hypotheses formulated above. The data were collected within the context of a non-experimental, cross-functional field study. The web-based student survey is looked at in detail below.

2.1 Description of the random sample

As already mentioned, the data were collected within the framework of an EU project (see above) on the basis of an online questionnaire survey conducted during the summer semester 2019 at the Ostbayerische Technische Hochschule Regensburg (OTHR) and the Institute of Technology and Business in České Budějovice (VŠTE). In total, 281 students participated. At the time of the data acquisition, 59.8% of the respondents were enrolled at OTHR and 40.2% at VŠTE (N = 281). Of the participating students, 65.0% were female and 35.0% male. Most of the respondents (67.3%) were aged 18 to 24 years. 86.4% of those surveyed were enrolled in a Bachelor course and 93.5% were studying business administration.

2.2 Survey materials

To identify the gender-specific difference, relevant personal characteristics were collected by means of an online questionnaire survey. In order to measure the subject-specific interests of students in entrepreneurship, an adapted and abbreviated questionnaire regarding the interest of study [22] was used (a seven point Likert scale from 1 "does not apply at all" to 7 "completely applies"). The scale consisting of three items takes into consideration personal value-oriented valences as well as the intrinsic character of interest (α = .78; exemplary item: "It is very important to me that I can attend entrepreneurship classes"). In order to measure the interest in leadership roles, the leadership motivation scale by Nicholson and Arnolds [28] was used. The scale consists of five items measured on a seven point Likert scale for interest in assuming management responsibility (α = .87; exemplary item: "I want to lead employees"). The entrepreneurial and intrapreneurial intention was gauged with one question in each case (exemplary item for entrepreneurial intention: "I have already thought seriously about establishing my own company or taking one over"; exemplary item for intrapreneurial intention: "It is my professional goal to become an innovation-capable employee (intrapreneur)"; a seven point Likert scale). The entrepreneurial knowledge of students was operationalized by the use of a scale (α = .88) consisting of 10 items (exemplary item: "Please assess your previously acquired competencies in founding a business"; a seven point Likert scale). On this occasion, the corresponding scale of the Global University Entrepreneurial Spirit Students' Survey [20] was abbreviated and adapted.

2.3 Analyses

The statistics programme IBM SPSS 25 was used for the analyses. First of all, fundamental analyses of scales for the control of distributional assumptions, as well as for the verification of the reliability and validity thereof, were carried out. Subsequently, various descriptive and correlation analyses were performed and the assessment of the differences in mean value examined using the t-test.

3 Results

After the fundamental analysis of scales, the descriptive and correlation analyses were performed. The mean values, standard deviations and correlative findings are presented in Table 1.

Table 1. Mean values, standard deviations and intercorrelations for the investigated components of entrepreneurial competence

Variable	M (SD)	(1)	(2)	(3)	(4)	(5)
(1) entrepreneurial knowledge	4.60 (0.92)	-				
(2) interest in entrepreneurship	4.77 (1.39)	.25**	-			
(3) interest in leadership roles	5.33 (1.07)	.45**	.21**	-		
(4) entrepreneurial intention	4.13 (2.23)	.34**	.40**	.36**	-	
(5) intrapreneurial intention	3.97 (1.78)	.19**	.20**	.21**	.13ns	-

Source: Author.

*p < .05; **p < .01

As Table 1 shows, the mean values for entrepreneurial knowledge, interest in entrepreneurship and interest in leadership roles lie significantly above the average rating. The mean values for entrepreneurial and intrapreneurial intention on the other hand lie only slightly above the average rating. The standard deviations of the ratings investigated are relatively homogenous for entrepreneurial knowledge, interest in entrepreneurship and interest in leadership roles and should therefore be described as marginal. In the case of entrepreneurial and intrapreneurial intention, the values of the standard deviations indicate a rather heterogeneous distribution of attributes. This first finding points to the students in the business studies course showing a marked interest in the field of entrepreneurship. As expected, average, statistically significant positive correlations could be determined for correlations between the variable examined ($r = .19\text{-}.45$, $p < .05$). Furthermore, the conjecture that there is no statistically significant correlation between entrepreneurial and intrapreneurial intentions is confirmed. Both constructs can be clearly, empirically distinguished from one another ($r = .13$, $p > .05$).

In order to examine the hypotheses and to identify the gender-specific difference in the expression of different components of entrepreneurial competence, the mean differences were subsequently examined using the t-test. The results are presented in Table 2.

Table 2. Mean values, standard deviations, sample size, t-values and effect size of the gender-specific group comparison

Variable	Female students M (SD, N)	Male students M (SD, N)	t-values	d_{Cohen}
entrepreneurial knowledge	4.52 (0.92, 163)	4.78 (0.90, 79)	$t_{(240)} = 2.09$, $p < .05$	0.29
interest in entrepreneurship	4.58 (1.40, 150)	5.13 (1.31, 75)	$t_{(223)} = 2.83$, $p < .05$	0.40
interest in leadership roles	5.12 (1.00, 149)	5.77 (1.08, 75)	$t_{(222)} = 4.47$, $p < .005$	0.63
entrepreneurial intention	3.77 (2.22, 162)	4.91 (2.06, 79)	$t_{(239)} = 3.85$, $p < .005$	0.53
intrapreneurial intention	3.98 (1.80, 162)	3.95 (1.77, 79)	$t_{(239)} = 0.13$, $p > .05$	0.02

Source: Author.

The t-tests carried out for the independent samples showed statistically significant mean differences between female and male students for all the variables investigated. The male students consistently produced higher statistically significant mean values than their female counterparts. It was only in the expression of intrapreneurial intention that, as expected, no statistically significant differences between the two comparative groups could be determined. As a consequence, the hypotheses formulated above were confirmed.

4 Discussion

Table 1 illustrates that the students surveyed evaluate their entrepreneurial knowledge as better than average. In addition, the findings indicate that students have a marked interest in entrepreneurship as well as an interest in leadership roles. Since students generally select their course of study based on their interests, it is reasonable to assume that they would have particular interest in certain aspects of economic sciences. In their studies, they are also confronted with many different areas of economic sciences, e.g. with entrepreneurship, founding start-ups, and therefore have the opportunity to develop an interest, even if this was absent at the beginning. The values for entrepreneurial and intrapreneurial intentions also conform to the hypotheses. Some of the students surveyed have already given serious consideration to establishing their own companies, some have not considered this at all, which also confirms the heterogeneous standard deviation. These findings are in line with founding activities in general. Serious consideration had also not been given to intrapreneurship. This is probably something that would generally change with the first major internship or with the first job, since innovations are context related.

The correlative findings underline that persons with high values in interest in entrepreneurship also demonstrate greater entrepreneurial knowledge and interest in leadership. Particularly notable is the comparatively high positive correlation between interest in entrepreneurship and entrepreneurial intention, which also conforms to the hypotheses. Within the framework of the aforementioned EU project, it was investigated as to which factors contribute most strongly to the formation of entrepreneurial intention. In this, domain-specific interest in entrepreneurship ($\beta = .35$, $p < .001$), interest in leadership roles ($\beta = .23$, $p < .01$) and entrepreneurial knowledge ($\beta = .21$, $p < .01$) were identified as statistically significant predictors. These variables go a long way to explain the variance for entrepreneurial intention ($R^2 = .35$, $F(1,124) = 22.08$, $p < .001$) [29]. The constructs entrepreneurial and intrapreneurial intention can be clearly differentiated from one another, both theoretically and empirically, which confirms the findings of this study. It is clear that not every person who thinks and acts innovatively has the intention of setting up their own business.

The findings of this study show that male students evaluate their own entrepreneurial knowledge as being much higher than do their female counterparts. Whether or not male students actually do have more comprehensive entrepreneurial knowledge than female students or simply have more confidence in themselves can only be verified with further validity criteria (e.g. comparison of examination grades). In addition, they display a significantly stronger domain-specific interest in entrepreneurship, interest in leadership roles and entrepreneurial intentions. Against the theoretical background discussed above and the findings of empirical studies, these results are plausible, confirm the hypotheses of this investigation and, furthermore, replicate actual founding activity in Europe. Within this context, female students may have lesser entrepreneurial self-efficacy and risk tolerance, and/or the current offerings in academic entrepreneurship education do not appeal to female students.

The following limitations should be noted for this study: (a) students from two universities were surveyed. The robustness of the data could be increased through the inclusion of further universities or colleges as well as of other faculties; (b) the validity of intention as a behavioural predictor has been proven thus far only in a general context and not in the context of establishing a business [12, 25]. As a result, how many female and male students actually do realise their entrepreneurial plans cannot be quantified.

This paper has specific implications for the research and promotion of business establishment. Since the effect of entrepreneurship education would appear to be dependent on gender, the first step should be to verify whether or not the current promotional measures are sufficiently tailored to the needs of potential female entrepreneurs. The second step should be the development, testing and implementation of targeted offerings for female students. Further research into female entrepreneurship can contribute to the better conception of academic entrepreneurship education.

5 Conclusion

Although the proportion of women establishing businesses has increased across Europe in recent years, as has the number of women who see establishing a company as an alternative to a conventional career path [3], the still low level of female entrepreneurship raises the question of how these numbers can be raised. Against this background, the following recommendations can be derived for targeting women within the context of entrepreneurship education [4, 30, 31]:

- The offers already available should be adjusted for female entrepreneurs. New formats should be conceived, e.g. coaching for female students to illustrate the opportunities and risks of realising their own ideas, or women-to-women mentoring programmes.
- For the entrepreneurial motivation of female students, highlighting real role models can be of great importance. To this end, guest lectures by successful female entrepreneurs could be employed to a greater extent. By showing female students that successful female entrepreneurs are not superwomen, the entrepreneurial self-efficacy of the female students will also be bolstered.

However, in order to improve the theoretical and empirical basis as well as the quality and the effectiveness of the targeted measures, further empirical gender studies in the area of female entrepreneurship are needed [6]. In addition, the corresponding changes in family and economic policy, in the structure of the labour market, and in societal norms are essential [4].

References

1. K. Terrell, M. Troilo, Values and female entrepreneurship. *International Journal of Gender and Entrepreneurship*. **2**(3), 260–286 (2010)
2. B. Schulte, *Unternehmensnachfolgen und Entrepreneurship Education*. Wiesbaden: Springer (2019)
3. G. Metzger, *KfW Start-up Monitor 2018* [online]. Available at: https://www.kfw.de/PDF/Download-Center/Konzernthemen/ Research/PDF-Dokumente-Gründungsmonitor/Gründungsmonitor-englische-Dateien/KfW-Start-Up-Monitor-2018_EN.pdf (2018)
4. M. Bechter, M. Raich, P. Schober, Auf dem Weg zur erfolgreichen Unternehmerin – Vom Entrepreneurship zum Leadership [On the way to becoming a successful entrepreneur - from entrepreneurship to leadership]. In M. Raich,

H. Pechlaner, H. H. Hinterhuber, *Entrepreneurial Leadership.* Wiesbaden: Deutscher Univesitäts-Verlag, 131–144 (2007)

5. D. F. Kuratko, *Entrepreneurship. Theory, process, practice.* Boston: Cengage Learning (2017)
6. S. Birkner, K. Ettl, F. Welter, I. Ebbers, Women's Entrepreneurship in Europe: Research Facets and Educational Foci. In S. Birkner, K. Ettl, F. Welter, I. Ebbers, *Women's Entrepreneurship in Europe.* Cham: Springer Nature Switzerland, 3–14 (2018)
7. T. L. Michaelis, J. M. Pollack, P. Mulvey, Gender Bias and Venture Funding: Discussing Bias in the Entrepreneurship Classroom. *Entrepreneurship Education and Pedagogy.* **3**(2), 154–181 (2019)
8. J. Wu, Y. Li, D. Zhang, Identifying women's entrepreneurial barriers and empowering female entrepreneurship worldwide: a fuzzy-set QCA approach. *International Entrepreneurship and Management Journal.* **15**(3), 905–928 (2019)
9. C. S. Zerwas, *Work-Life-Balance and Women's Entrepreneurship. An Exploration of Influencing Factors.* Cham: Springer Nature Switzerland (2019)
10. L. Balachandra, T. Briggs, K. Eddleston, C. Brush, Don't Pitch Like a Girl!: How Gender Stereotypes Influence Investor Decisions. *Entrepreneurship: Theory and Practice.* **43**(1), 116–137 (2019)
11. V. Iffländer, A. Sinell, M. Schraudner, Does Gender Make a Difference? Gender Differences in the Motivations and Strategies of Female and Male Academic Entrepreneurs. In S. Birkner, K. Ettl, F. Welter, I. Ebbers, *Women's Entrepreneurship in Europe.* Cham: Springer Nature Switzerland, 65–82 (2018)
12. S. G. Walter, *Gründungsintention von Akademikern* [*Founding intention of academics*]. Wiesbaden: Gabler (2008)
13. C. Henry, L. Foss, H. Ahl, Gender and entrepreneurship research: A review of methodological approaches. *International Small Business Journal.* **34**(3), 217–241 (2016)
14. A. Tittel, O. Terzidis, Entrepreneurial competences revised: Developing a consolidated and categorized list of entrepreneurial competences. *Entrepreneurship Education.* **3**, 1–35 (2020)
15. F. E. Weinert, A Concept of Competence: A Conceptual Clarification. In D. S. Rychen, L. H. Salganik, *Defining and Selecting Key Competencies.* Seattle: Hogrefe & Huber Publishers, 45–65 (2001)
16. J. R. Anderson, Skill acquisition: compilation of weak-method problem solutions. *Psychological Review.* **94**(2), 192–210 (1987)
17. A. Renkl, Wissenserwerb [Knowledge acquisition]. In: E. Wild, E., J. Möller, *Pädagogische Psychologie.* Berlin, Heidelberg: Springer, 3–26 (2009)
18. C. Brühwiler, L. Hollenstein, B. Affolter, H. Biedermann, F. Oser, Welches Wissen ist unterrichtsrelevant? *Zeitschrift für Bildungsforschung.* **7**, 209–228 (2017)
19. Z. Caha, S. P. Sassmannshausen, J. F. Faltermeier X. Justus, *Unternehmerische Kompetenzen an Hochschulen für angewandte Wissenschaften am Beispiel einer Hochschule in Ostbayern und Südböhmen* [*Entrepreneurial skills at universities of applied sciences using the example of a university in Eastern Bavaria and South Bohemia*]. Lüdenscheid: RAM-Verlag (2017)
20. H. Bergmann, S. Golla, *Unternehmertum an Hochschulen in Deutschland* [*Entrepreneurship at universities in Germany*] [online]. St. Gallen/Fulda: Schweizerisches Institut für Klein- und Mittelunternehmen. Available at: http://www.guesssurvey.org/resources/nat_2016/GUESSS-Bericht-Deutschland-2016_30Sept2016.pdf (2016)
21. U. Schiefele A. Krapp, K.-P. Wild, A. Winteler, Der "Fragebogen zum Studieninteresse" (FSI) [The "Questionnaire on interest in studying"]. *Diagnostica.* **39**(4), 335–351 (1993)
22. A. Krapp, T. Hascher, Theorien der Lern- und Leistungsmotivation [Theories of learning and achievement motivation]. In L. Ahnert *Theorien in der Entwicklungspsychologie Theories in Developmental. Psychology.* Berlin, Heidelberg: Springer, 252–281 (2014)
23. J. A. Moriano, F. Molero, G. Topa, J.-P. Lévy Mangin, The influence of transformational leadership and organizational identification on intrapreneurship. *International Entrepreneurship and Management Journal.* **10**(1), 103–119 (2014)
24. C. Müller, Entrepreneurship und Leadership – Erfolgsfaktoren am Beispiel eines Familienunternehmens. In: M. Raich, H. Pechlaner, H. H. Hinterhuber, *Entrepreneurial Leadership.* Wiesbaden: Deutscher Univesitäts-Verlag, 117–130 (2007)
25. J. Heckhausen, H. Heckhausen, *Motivation und Handeln* [*Motivation and Action*]. Berlin, Heidelberg: Springer (2010)

26. S. G. Walter, A. Walter, Deutsche Universitäten als Gründungsinkubatoren: Der Beitrag der Gründungsausbildung zur Gründungsintention von Studierenden [German universities as start-up incubators: The contribution of start-up training to the start-up intention of students]. *Schmalenbach Journal of Business Research.* **60**(6), 542–569 (2008)
27. S. Weber, M. Wiethe-Körprich, S. Bley, C. Weiß, F. Achtenhagen, Intrapreneurship-Verhalten an kaufmännischen Arbeitsplätzen [Intrapreneurship behavior in commercial workplaces]. *Empirische Pädagogik.* **29**(1), 84–105 (2015)
28. N. Nicholson, J. Arnold, From Expectations to experience: Graduates entering large corporations. *Journal of Organizational Behavior.* **12**(5), 413–429 (1991)
29. X. Justus, J. F. Faltermeier, S. P. Saßmannshausen, Denken und Handeln grenzüberschreitend fördern [Promote thinking and acting across borders]. In: *Forschung 2018* [online]. Available at: https://www.oth-regensburg.de/fileadmin/media/forschung/Dateien_2018/Forschungsbericht-2018.pdf (2018)
30. A. Renkl, Drei Dogmen guten Lernens und Lehrens: Warum sie falsch sind [Three Dogmas of Good Learning and Teaching: Why They Are Wrong]. *Psychologische Rundschau.* **66**(4), 211–220 (2015)
31. K. Wilson, Entrepreneurship Education in Europe. In J. Potter, *Entrepreneurship and Higher Education.* Paris: OECD, 119–138 (2008)

 https://doi.org/10.1051/shsconf/20219002006

Evaluation of the Level of Use of Motivational Factors in SMEs According to Selected Criteria

Iveta Kmecová[1,*], *Michal* Tlustý[1]

[1]Institute of Technology and Business, Faculty of Corporate Strategy, Okružní 517/10, 370 01 České Budějovice, Czech Republic

Abstract. The research presented in this contribution is an analysis of the level of use of motivational factors in small and medium-sized enterprises (SME) in terms of three criteria, as selected by the authors. The primary goal was to draw conclusions on the basis of the formulated hypotheses concerning the degree of use of motivational factors by SMEs. The literature review outlines the issue of work motivation and motivational factors. The practical part of the contribution presents the course of the specific calculations and the evaluation of the hypotheses. The primary data were obtained using the SAAPM questionnaire, which was created as part of a project by the Slovak Academic Association for Personnel Management (SAAPM). The data are evaluated using the one-sample t-test statistical method. The research sample consists of a total of 610 small and medium-sized enterprises (SMEs) in the Czech Republic.
Keywords: motivation; work motivation; motivational tools; small and medium enterprises (SMEs); one-sample t-test

1 Introduction

The management of people´s motivation and efficacy are strategic issues for all organizations that use human resources. Influencing work motivation is not only a matter for the human resources department of a company, but also for managers at all company management levels. Given the rising costs of labour, a successful company must fully use the potential of its employees and motivate them to deliver the best work performance possible. As the authors believe, the responsibility (read: mission) of every company should therefore be to provide its employees with a suitable incentive system that satisfies their employees and contributes to improving their performance and the achievement of the set goals.

2 Theoretical background

Currently, there are many motivational tools available to companies. In recent years, under the current economic conditions, companies have been forced to make changes within the business environment. Most companies have come to realize that the human factor is an important prerequisite for success and that reducing risks related to the human factor is important. This means that companies are increasingly striving to create a work environment that enables performance maximization and loss minimization [1].

It can be said that the issue of employee motivation, motivational tools, and their application in practice is a very widely discussed topic. According to [2], employee motivation should be seen as the mission of a company towards its stakeholders (i.e. towards its employees). Employees have different motivations for finding and keeping jobs [3]. This opinion is shared by [4], who claim that managers should know what motivational tool to apply towards employees because each employee is motivated by a different motivational factor. The success of an organization depends on motivated employees. To be able to perform everyday tasks, employees must feel motivated and satisfied [5]. It is therefore necessary, on the part of management, to create favourable conditions in companies and to have a suitable system of motivational tools in order to encourage work motivation and the engagement of employees. Internal benefits and motivational factors are used by multinational companies to establish long-term relationships [6]. According to [7], motivational factors are normally divided into five categories, whereby the best-known and most commonly used factors are financial or work-related. Within this context, companies try to create a pleasant working environment, make the work as interesting as possible, explain the added value of the work, etc.; these are mainly incentives. The third category covers social motivational factors. The fourth category consists of relational motivational factors, such as the atmosphere in the workplace, which is often reflected in the performance of teams of workers. The remaining category covers career motivational factors. According to [8, 9], adopting the right motivational strategy enables companies to benefit from the

* Corresponding author: kmecova @mail.vstecb.cz

improved performance of a project team, which results in high customer satisfaction. [10] found that motivational factors differ according to position, age and gender. Managers should take the relationship between motivation and the age of employees into account and motivate their employees in different ways [11,12].

This contribution evaluates the level of use of motivational factors in small and medium-sized enterprises in the Czech Republic.

3 Methodology

The aim of this contribution is to use the one-sample t-test statistical method to compare the results of two parts of each established hypothesis and then to reject or confirm the whole hypothesis. The hypotheses relate to the level of use of motivational factors in companies.

3.1 Data collection methods

The data for this contribution were obtained via a questionnaire survey conducted among SMEs in the Czech Republic and from research activities (Industry 4.0) conducted in cooperation with the Slovak Academic Association for Personnel Management (SAAPM). In total, 615 companies participated in the research. Data collection took place in the first half of 2020.

For the purposes of this contribution, the research sample was divided according to three criteria, namely: domestic or foreign ownership; industrial sector - manufacturing or logistics (these sectors were merged for the purposes of this contribution) or services; company size according to the number of employees (49 and fewer; 50 and more).

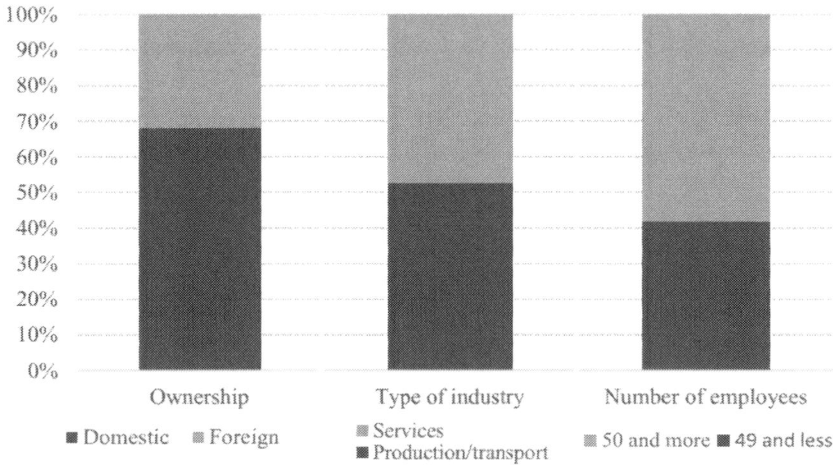

Figure 1. Division of companies

Source: Authors.

3.2 Data evaluation methods

To calculate the final value of the t-test, it is first necessary to develop auxiliary quantities. The first is the value of the arithmetic mean [13].

$$\bar{x} = \frac{1}{n} * (x_1 + x_2 + x_3 + \cdots + x_n) \quad (1)$$

Where:

x̄ = arithmetic mean

n = number of sample values

x = individual file values

Another indicator is the sample variance [14].

$$s^2 = \frac{1}{n} \sum_{i=1}^{n} (x_i - \bar{x})^2 \quad (2)$$

Where:

s^2 = sampling variance

x_i = individual values of the file

$\bar{x}$ = arithmetic mean

The standard deviation can now be calculated [14].

$$\sigma = \sqrt{s^2} \tag{3}$$

Where:

σ = standard deviation

s^2 = file variance

If all the above-mentioned quantities are known, a one-sample t-test can be calculated.

$$t = \frac{\bar{x} - \mu_0}{\sigma} \sqrt{n} \tag{4}$$

Where:

t = value of one-sample t-test

$\bar{x}$ = arithmetic mean

μ_0 = established null hypothesis

σ = standard deviation

n = number of sample values

Finally, the resulting values of the t-test must be compared with the critical values from the tables available on the internet.

For the calculations, it is important to mention that the companies answered the questions using a scale of 1-5, where: 1 = 0%; 2 = 25%; 3 = 50%; 4 = 75%; 5 = 100%. The percentage results are rounded to whole percentages.

4 Results

H1: Zero hypothesis: The level of use of motivational factors in both Czech-owned and foreign-owned companies is 75% (comparison).

$$\bar{x}_1 = \frac{1}{2080}(7836) = 3.77 = 9\ \% \tag{1}$$

$$\bar{x}_2 = \frac{1}{970}(3755) = 3.87 = 72\ \% \tag{1}$$

$$s_1^2 = \frac{1}{2080} * (x_1 - 3.77)^2 + (x_2 - 3.77)^2 + \cdots (x_n - 3.77)^2 = 1.1 \tag{2}$$

$$s_2^2 = \frac{1}{970} * (x_1 - 3.87)^2 + (x_2 - 3.87)^2 + \cdots (x_n - 3.87)^2 = 0.9 \tag{2}$$

$$\sigma_1 = \sqrt{1.1} = 1.04 \tag{3}$$

$$\sigma_2 = \sqrt{0.9} = 0.95 \tag{3}$$

$$t_{1.1} = \frac{3.77 - 4}{1.04} \sqrt{2080} = |-10.09| = 10.09 \tag{4}$$

$$t_{1.2} = \frac{3.87 - 4}{0.95} \sqrt{970} = |-4.26| = 4.26 \tag{4}$$

The critical values can be obtained from the table of critical values in [16]. The critical value for both parts of the formulated hypothesis is 1.96. Both t-test values are greater than the critical value set. As a result, both parts of the hypothesis can be rejected.

$p\text{–value}_1 = <2.2e\text{-}16$

$p\text{–value}_2 = 2.684e\text{-}05$

Both p-values are clearly lower than 5%, which also confirms the rejection of the hypothesis. The 95% interval for the first interval is 3.72 - 3.81, while the interval for the second selection is 3.81 - 3.93. The value for 75% (4 points) is not at this interval; this also confirms the rejection of the hypothesis. The results for both parts of the hypothesis are illustrated by means of the Gaussian curve (see Figures 2 and 3 below).

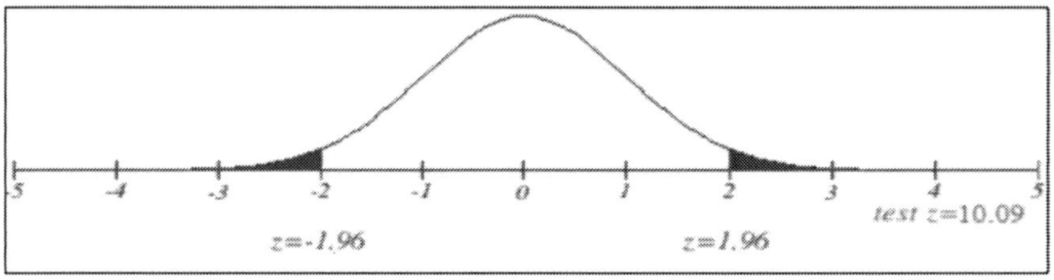

Figure 2. Gaussian curve of normal division – $H_{1.1}$

Source: Authors, [15].

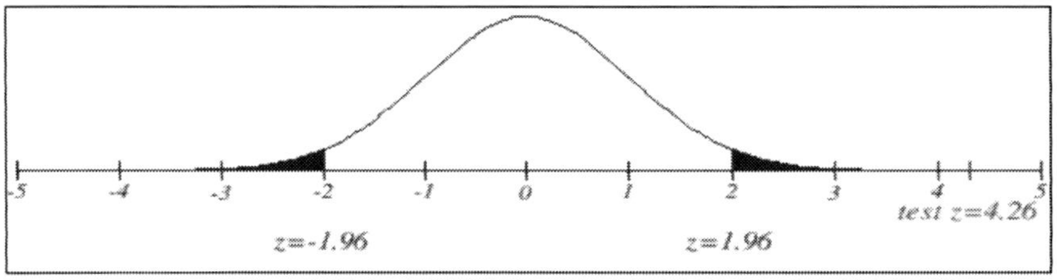

Figure 3. Gaussian curve of normal division – $H_{1.2}$

Source: Authors, [15].

H2: Zero hypothesis: The level of use of motivational factors in companies operating in the manufacturing or logistics sector is 75%, and for those operating in the services sector is 75% (comparison).

All the aforementioned calculations were carried out. It was subsequently possible to calculate the relevant t-test.

$$t_{2.1} = \frac{3.78 - 4}{1.03} \sqrt{1435} = |-8.09| = 8.09 \quad (4)$$

$$t_{2.2} = \frac{3.82 - 4}{0.99} \sqrt{1295} = |-6.54| = 6.54 \quad (4)$$

The table of critical values for the t-distribution in [16] shows that the critical value for both groups of companies is 1.96. The t-test values are clearly greater than the critical value. As a result, the hypothesis cannot be confirmed at a 95% significance level. The hypothesis is therefore rejected.

$\text{p-value}_{2.1} = 4.663\text{e-}15$

$\text{p-value}_{2.2} = 5.516\text{e-}11$

The comparison of both p-values confirmed the rejection of this hypothesis. Both p-values are clearly lower than 5%; the hypothesis is therefore rejected. This statement can be confirmed also by means of a 95% quartile. The quartile for the first part of the hypothesis, i.e. for those companies in the manufacturing and logistics sectors, is 3.73 - 3.84, and for the second part of the hypothesis, i.e. for those companies in the services sector, is 3.76 - 3.87. The obtained value (4) is clearly not in either of the quartiles; rejection of the hypothesis is therefore confirmed.

H3: Zero hypothesis: The level of use of motivational factors in small companies (49 or fewer employees) is 75%, and in larger companies (50 or more employees) is 75% (comparison).

$$t_{3.1} = \frac{3.74 - 4}{1.04} \sqrt{1275} = |-8.93| = 8.93 \quad (4)$$

$$t_{3.2} = \frac{3.84 - 4}{1} \sqrt{1776} = |-6.74| = 6.74 \quad (4)$$

Even in this case, the relevant critical value [16] is 1.96. Both t-test values are clearly higher than the critical value specified above; it is therefore recommended to reject this hypothesis.

$\text{p-value}_{3.1} = <2.2\text{e-}16$

$\text{p-value}_{3.2} = 8.325\text{e-}11$

Both p-values are lower than 5%, which confirms the rejection of the hypothesis. The 95% quartile for the first part of the hypothesis, i.e. for those companies with 49 or fewer employees, is 3.68 - 3.8, and for the second part of the hypothesis, i.e. for those companies with 50 or more employees, is 3.8 – 3.89. The value of 4 (75%) does not fall within any of the quartiles; rejection of the hypothesis was therefore correct.

H4: Zero hypothesis: The level of use of motivational factors now and planned in the future is 75% (comparison).

$$t_{4.1} = \frac{3.8 - 4}{1.02}\sqrt{3050} = |-10.83| = 10.83 \quad (4)$$

$$t_{4.2} = \frac{4.36 - 4}{0.84}\sqrt{3050} = 23.67 \quad (4)$$

The critical values [16] for both categories are identical - 1.96. Both resulting values of the t-tests are higher than the critical values; both parts of the hypothesis can therefore be rejected.

$$\text{p-value}_{4.1} = <2.2e\text{-}16$$

$$\text{p-value}_{4.2} = <2.2e\text{-}16$$

Both p-values are lower than 5%, which also contributes to the rejection of the hypothesis. This is confirmed by the analysis of the 95% quartiles. The 95% quartile for the current use of motivational factors is 3.76 - 3.84, while for planned use in the future is 4.33 - 4.39. The tested value of 4 does not fall within the quartiles; the hypothesis can therefore be decidedly rejected.

5 Discussion

The results of all hypotheses are presented in Table 1.

Table 1. Summary of results

Hypothesis	t-test values	Critical values	Result
H_1	10.09; 4.26	1.96; 1.96	Hypothesis rejected
H_2	8.09; 6.54	1.96; 1.96	Hypothesis rejected
H_3	8.93; 6.74	1.96; 1.96	Hypothesis rejected
H_4	10.83; 23.67	1.96; 1.96	Hypothesis rejected

Source: Authors.

H1: Zero hypothesis: The level of use of motivational factors in both Czech-owned and foreign-owned companies is 75% (comparison).

The first hypothesis examined whether the level of use of motivational factors in companies is affected by the fact that a company is Czech- or foreign-owned, and whether the level is about 75%. The results show that neither of the parts of the hypothesis confirm that the level of use of motivational factors is about 75%. Specifically, for Czech-owned this was about 69, and for foreign-owned companies about 72%. It can therefore be said that although the hypothesis was rejected, the use of motivational factors is more common in foreign-owned companies; however, the difference is not significant.

H2: Zero hypothesis: The level of use of motivational factors in companies operating in the manufacturing or logistics sector is 75%, and for those operating in the services sector is 75% (comparison)

For the verification of the second hypothesis, the companies were divided according to the industrial sector in which they operate. The first group includes those companies operating in the manufacturing or logistics sectors, the second group includes those companies operating in the services sector. Moreover, it was examined whether the level of use of motivational factors is close to 75%. The results indicate that there are no significant differences in the level of use of motivational factors according to the indicated industrial sectors. The values for both categories was about 70%. The hypothesis was therefore rejected.

H3: Zero hypothesis: The level of use of motivational factors in small companies (49 or fewer employees) is 75%, and in larger companies (50 or more employees) is 75% (comparison).

This hypothesis dealt with the influence of the number of employees on the level of use of motivational factors in companies, specifically that these values achieved 75%. The companies were divided into two categories: companies with 49 or fewer employees; and companies with 50 or more employees. It was found that the values for both categories were about 70%; the hypothesis was therefore rejected.

H4: Zero hypothesis: The level of use of motivational factors now and planned in the future is 75% (comparison).

The last formulated hypothesis examined the satisfaction of the companies with their current level of use of motivational factors, whether they plan to increase or reduce their use thereof, and whether the level of use is close to 75%. The first part of the hypothesis, concerning the current situation, was rejected; it was found that the level of use of motivational factors in companies is currently about 70%. The second part of the hypothesis was also rejected; however, in this case, the results indicated a level of 84%. This indicates that the companies are not satisfied with their current level of use of motivational factors and that they are planning significant improvements for the future.

6 Conclusion

In conclusion, the companies were subject to statistical analysis by means of a one-sample t-test. Firstly, the companies were divided into several categories according to predetermined criteria, namely: domestic or foreign ownership; industrial sector - manufacturing or logistics (these sectors were merged for the purposes of this contribution) or services; company size according to the number of employees (49 and fewer; 50 and more), on the basis of which hypotheses were subsequently formulated. The fourth hypothesis examined whether there are any differences in the current and planned level of use motivational factors. Secondly, auxiliary variables were calculated, such as arithmetic mean, variance, standard deviation, etc., for each hypothesis. Thirdly, specific t-test values were calculated and, on the basis of tables of critical values, a decision made as to whether to confirm or reject the hypothesis. The decision to reject a hypothesis was confirmed by means of calculated p-values and 95% quartiles. It was found that the values for the level of use of motivational factors did not achieve 75%, whereby neither differences in ownership, industrial sector nor company size proving to have a significant impact on these values. However, the last hypothesis provided interesting results. It was found that the current level of use of motivational factors in companies is about 70%, with companies appearing to be dissatisfied with this value and planning significant improvements (up to 84%) for the future.

Acknowledgement

This contribution is an output created within the project 8110-002 of the Internal Grant Competition called Innovation for the study "Human Resource Management".

The contribution uses data obtained from the research activities of the Slovak Academic Association for Personnel Management (SAAPM).

References

1. F. L. Cooke, R. Schuler, A. Varma, Human resource management research and practice in Asia: Past, present and future. *Human Resource Management Review*. **30**(4) (2020)
2. M. Bastons, M. Mas, C. Rey, Pro-stakeholders motivation: Uncovering a new source of motivation for business companies. *Journal of Management and Organization*. **23**, 621–632 (2017)
3. N. A. Muslim, D. Dean, D. Cohen, Employee Job Search Motivation Factors: An evidence from Electricity Provider Company in Malaysia. *Procedia Economics and Finance*. **35**, 532–540 (2016)
4. J. Faletar, D. Jelačić, L. Oblak, Comparison of motivation factors in two wood processing companies. *Acta Silvae et Ligni*. **111**, 1–11 (2016)
5. R. Alamelu, R. Amudha, R. Nalini, L. C. S. Motha, I. Bahavathi, Measure of motivation practices – An employee's insight. *International Journal of Economic Research*. **14**, 409–415 (2017)
6. S. Bhattacheryay, Multinational enterprises motivational factors in capitalizing emerging market opportunities and preparedness of India. *Journal of Financial Economic Policy*. In print (2020)
7. M. Yunus, E. Yusuf Iis, M. Adam, H. Sofyan, Does motivation mediate the effects of employee staff empowerment, talent, working environment, and career development on staff performance? *Quality – Access to Success*. **21**, 91–96 (2020)
8. M. M. Uthappa, A. N. Santosh Kumar, Impact of motivational factors on project team performance. *Test Engineering and Management*. **83**, 5168–5176 (2020)
9. K. C. Diwas, Worker productivity in operations management. *Foundations and Trends in Technology, Information and Operations Management*. **13**, 151–249 (2020)
10. S. Lorincová, P. Štarchoň, D. Weberová, M. Hitka, M. Lipoldová, Employee motivation as a tool to achieve sustainability of business processes. *Sustainability*. **11** (2019)
11. L. Ližbetinová, M. Hitka, C. Li, Z. Caha. Motivation of employees of transport and logistics companies in the Czech Republic and in a Selected Region of the PRC. *MATEC Web of Conferences*. **134** (2017)

12. S. Lorincová, M. Hitka, P. Štarchoň, K. Stachová, Strategic instrument for sustainability of human resource management in small and medium-sized enterprises using management data. *Sustainability*. **10** (2018)
13. M. Budíková, M. Králová, B. Maroš, *Průvodce základními statistickými metodami* [*Guide to basic statistical methods]*. Praha: Grada Publishing. (2010)
14. J. Neubaeur, M. Sedlačík, O. Kříž, *Základy statistiky – Aplikace v technických a ekonomických oborech* [*Basics of statistics – Applications in technical and economic fields*]. Praha: Grada Publishing (2016)
15. Imathas. *Gaussian curve creation tool*. [online]. Available at: http://www.imathas.com/stattools/norm.html (2020)
16. Lekopis. *Table of critical values of t-distribution*. [online]. Available at: http://www.lekopis.cz/Kap_5_3_8_2.htm (2020)

Comparative Analysis of the Level of Development of Entrepreneurial Competencies among Students in Russia and the Czech Republic

Anastasiya Pesha[1], *Marina* Shavrovskaya[1,*], *Zdeněk* Caha[2]

[1] Ural State University of Economics, Department of Labour Economics and HR Management, 8 Marta 62, 620144 Yekaterinburg, Russia

[2] Institute of Technology and Business, Faculty of Corporate Strategy, Nemanická 7, 37010 České Budějovice, Czech Republic

Abstract. The role of universities in the development of entrepreneurship is increasing. Within this context, universities are introducing specialized programs and disciplines aimed at developing students' innovative and entrepreneurial competencies. This article presents the results of a comparative analysis of the level of development of students' entrepreneurial competencies in Russia and the Czech Republic. To implement the key task, the authors analysed approaches to the classification of entrepreneurial competencies. On the basis of relevance and objectivity, the authors selected the EntreComp classification from 2016, which subsequently formed the basis for the developed methodological tools for the empirical research. The method was tested in three universities in the Russian Federation and one in the Czech Republic. The total number of participants was 436. The results of the analysis of the entrepreneurial competencies revealed significant statistical differences between the respondents from the two countries for 10 out of 15 evaluated competencies. At the same time, 68% of respondents in the Russian Federation and 64% in the Czech Republic were found to have an advanced level of development of one entrepreneurial competence. This fact is indicative of the similarity of the vast majority of the final results for the respondents from both countries and confirms the hypothesis that most students participating in the test did not have a lower than average level of competence development. The results of the study may be of interest to scientists and practitioners who study the assessment and development of entrepreneurial competencies in students and specialists.

Keywords: entrepreneurial competence; higher education; students; supra-professional competencies

1 Introduction

At present, many scientists around the world are focusing on the development and assessment of entrepreneurial competencies. Effective entrepreneurs are important for the development of a society because they create jobs, improve the quality of life of the population, contribute to economic growth and increase the welfare of a country. The UNESCO Agenda for the period up to 2030 is based on 17 sustainable development goals, which describe the main problems of human development [1]. During this period of socio-economic and political upheaval and instability, and within the context of global environmental risks and catastrophes, the world's scientific and political communities are focusing on the development of competencies in the field of sustainable development [2, 3, 4]. Cutting-edge research in the field of sustainability competencies has been conducted by scientists such as [2, 3] and others. It is on the basis of such research that the UNESCO document "Education for sustainable development goals: learning objectives" contained a list of key sustainable development competencies for education in 2017 [1, p.10]. Key competencies are "a set of knowledge, skills, and relationships that enable successful tasks and problem solving in relation to real world sustainability issues, challenges, and opportunities" [3, p. 204].

The focus on sustainable development competencies in the field of entrepreneurship is reflected in the works of researchers such as Ploum et al., etc. [5 ,6].

Research is also focusing on "internal entrepreneurship". This not only implies the presence of creative abilities to search for new ideas, but also the willingness to implement these ideas and to take responsibility for their implementation from a wide range of company specialists [3, 7, 8, 9].

In this regard, the role of universities in the development of students' entrepreneurial competencies in different fields is growing [10, 11, 12, 13]. Russian scientist Chepurenko [14] notes that the inclusion of the development of

* Corresponding author: marina_bel@list.ru

this range of competencies in the professional educational "menu" will "provide the country's economy with qualified young personnel who are competent in creating and running a business". Analysis of the exchange of 20 European best practices shows that long-term relationships with stakeholders and a supportive entrepreneurial ecosystem of educational institutions are key factors in the success of entrepreneurship training [15].

The goal of this article was to conduct and present an assessment and comparative analysis of the entrepreneurial competencies of university students in Russia and the Czech Republic.

To achieve this goal, the following tasks were implemented within the framework of the study:

1. Analysis of the approaches to the classification of entrepreneurial competencies for sustainable development.

2. Development and testing of a methodology for assessing the development of the entrepreneurial competencies of students in Russia and the Czech Republic.

3. Presentation of the results of a comparative analysis of the level of development of entrepreneurial competencies of university students in Russia and the Czech Republic based on the results of the tested methodology.

The methods used included content analysis of publications on the research topic, modelling of test tasks based on competence indicators, online testing, comparative analysis and methods for visualizing the results of the research.

The assumption was made that there are no statistically significant differences in the level of development of entrepreneurial competencies of the university students from both countries. In addition, the assumption was made that most of the students participating in the test would not have a lower than average level of competence development.

Approaches to the classification of entrepreneurial competencies (literature review)

With regards to the issue of the classification of entrepreneurial competencies as a basis for their development and evaluation during university studies, there is still no well-established and generally accepted list of competencies.

The first attempt to study entrepreneurial competencies was made in a large-scale cross-cultural study conducted by the United States Agency for International Development (USAID) [16]. As a result, a set of competencies was compiled on the basis of highly successful entrepreneurs. These included: achievement competencies (initiative, sees and uses opportunities, concern for high quality of work, obligations under contract, focus on efficiency); thinking and problem solving competencies (systematic planning, problem solving); directivity and control (perseverance, control); orientation to others (recognition of the importance of business relationships) [16].

The competence model presented by Wei-Wen Wu [17] includes a list of 23 competencies, and the Kaur and Bains [18] model includes 12 competencies. Inyang and Enuoh [19] note that entrepreneurial competencies include the knowledge, attitudes, and skills that an entrepreneur must acquire through management training and development to contribute to the creation of outstanding products. Bird [20] combines entrepreneurial competencies into two groups: basic or threshold competencies (important for the implementation of a business idea), and high-level competencies that allow entrepreneurs to act effectively or successfully beyond the launch of a business, ensuring the survival and growth of the organization.

Within the context of internal entrepreneurship, Hayton and Kelly [21] propose four competencies: innovation (recognizing opportunities and further exploiting these opportunities); mediation (accessing and combining new sources of knowledge and information); advocacy (identifying with an opportunity and taking responsibility for promoting it); and sponsorship (helping to gain access to necessary resources). There are successful examples of Russian organizations (Sberbank, Beeline, Gazprom Neft, etc.) creating the conditions necessary for employees to display entrepreneurial initiative and search for new solutions in the form of products and services for clients.

Sustainable development competencies in the field of entrepreneurship are the basis of an empirical study conducted by Ploum, et al. [5] among 402 potential entrepreneurs. As a result of the research, the authors present 6 key competencies required for an agent of change in the field of entrepreneurship: strategic management competence and competence in actions; diversity acceptance and interdisciplinary competence; systematic thinking, normative competence; far-sighted thinking competence; interpersonal competence [5, p.119]. Ahmad [6] uses nine clusters of entrepreneurial competencies, including strategy, commitment, concept, opportunity, organization and leadership, relationships, learning, personality, and technical competencies.

As a result of analysing the structures and indicators of entrepreneurial competencies in various works, we chose the framework "EntreComp: The Entrepreneurship Competence Framework", prepared by the research team Bacigalupo, et al. [22]. This framework fitted the key requirements with regards to the choice of materials for the development of the assessment tools for analyzing the level of development of entrepreneurial competencies.

2 Methodology

To collect empirical research data, the authors developed and tested the test method "Development of students' entrepreneurial competencies" (Pesha and Shavrovskaya). As described above, the method is based on "EntreComp: The Entrepreneurship Competence Framework" [22].

The self-assessment of competencies is carried out in the methodology based on three interrelated clusters of competencies: ideas, opportunities and resources for action. In each of these clusters, 5 competencies are evaluated. The assessment is therefore carried out for 15 competencies.

The rating scale consists of 7 levels. In "EntreComp: The Entrepreneurship Competence Framework" there are 8 indicator levels. In our version, the basic level is 1 because in the process of working with the indicators, we did not find a critical difference in levels 1 and 2 in the authors' progress model.

The scale looks like this:

1. Basic. Competencies are not expressed.

2. Intermediate 1. The establishment of independence. Willingness to take action independently and together with others.

3. Intermediate 2. The establishment of independence. Willingness to accept and share responsibility.

4. Advanced 1. The acceptance of responsibility. Work under the guidance of and in collaboration with others.

5. Advanced 2. Take responsibility for making decisions and working with other people.

6. Expert 1. Encourage innovation, transformation and growth. Take responsibility for participating in complex developments in a specific area.

7. Expert 2. Encourage innovation, transformation and growth. Make a significant contribution to the development of a specific area.

The test consisted of 49 questions, each of them with 4 possible answers. Each response option correlates with one of the levels of competence development. Each answer is awarded from 0 to 8 points. 0-2 basic level; 2,1-3,9 intermediate level; 4-5,9 advanced level; 6-8 expert level.

The testing was conducted online using resources https://app.onlinetestpad.com and https://docs.google.com/forms. The invitation to take the test was sent via messenger. Upon completing the test, each participant automatically received a profile of their personal entrepreneurial competencies, with comments on each of the 15 components.

3 Results and Discussion

3.1 Participants

In total, 436 students took part in the survey, of which 195 from Russia and 241 from the Czech Republic.

In Russia, students from three universities took part in the testing: USUE (Yekaterinburg), The Ural Federal University (Yekaterinburg), Omsk State University (Omsk). All the test participants from the Czech Republic were students of the University of Technology and Business in České Budějovice. The test was completed by full-time and part-time students of bachelor's and master's degree programmes. The distribution of the participants by gender, location/residency and field of study is presented in Table 1.

https://doi.org/10.1051/shsconf/20219002007

Table 1. Distribution of participants by gender, location/residency and field of study

Characteristics of respondents	Russia (m_1) (number of persons)	Czech Republic (m_2) (number of persons)
Gender		
Male	30	72
Female	165	169
Location/residency		
Capital	0	12
Megapolis	138	0
Big city (over 25,000 inhabitants)	0	71
Small town (up to 25,000 inhabitants)	44	81
Countryside	13	71
Field of study		
Human resource management	158	170
Economics of business	37	70
Economics and management	0	1

Source: Authors

Figure 1 shows the distribution of the test participants by age. The most represented age group in the study is young people aged 21-25 years, and the least the older age group 40-50 years (31 people). In future research, it will be interesting to compare the results obtained from the testing of these age groups, particularly in light of their different life experiences.

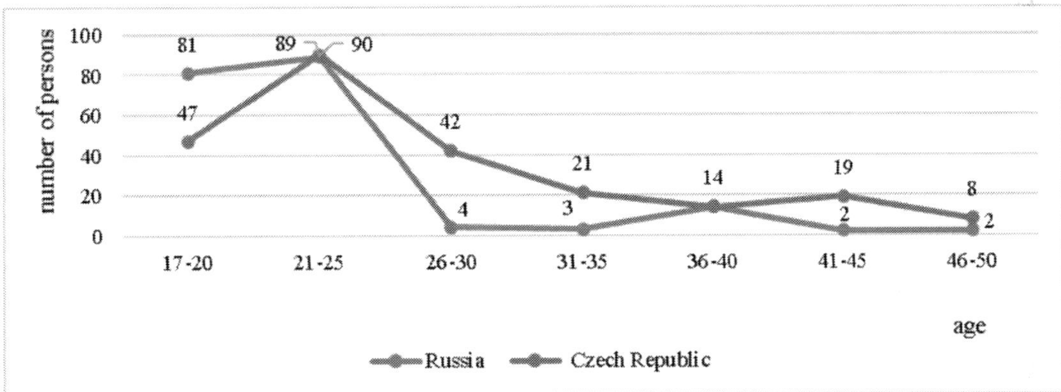

Figure 1. Age of participants (number of people)

Source: Authors.

3.2 Main results of the research

At first glance, and as expected, the overall level of development of entrepreneurial competencies is approximately the same among the participants from both countries (see Figure 2). Nevertheless, a comparative analysis of the results using t-statistics showed the presence of statistically significant differences (p-value=0.018577), whereby the value of the student's t-test was 2.36. The critical value of the student's t-test was 1.972, at a significance level $\alpha=0.05$. As a result, the hypothesis was confirmed that most of the students participating in the test do not have a lower than average level of competence development. As can be seen in Figure 2, most of the respondents have an Intermediate 2 or Advanced 1 level of entrepreneurial competence development. The Advanced 2 level of entrepreneurial competence development was found in 18 students from the Russian universities and in 15 students from the Czech Republic. In one student from Russia and two students from the Czech Republic, the development of entrepreneurial competencies was assessed at the Expert I level. The Russian student was a 22-year-old studying the master's degree programme "Human resources management", whereas the two Czech were 33 years old and 45 years old, respectively, and both studying "Business Economics". It is clear that achieving an expert level of development of sustainable development competencies in the field of entrepreneurship can be achieved at any age. No participants were identified as having a Basic or Expert 2 level of competence development. A Basic level of competence development assumes a lack of knowledge regarding the application of existing software in the field of resource management, a lack of initiative and

 https://doi.org/10.1051/shsconf/20219002007

a lack of action in relation to the implementation of business ideas. An Expert 2 level of competence development reflects a level of experience that goes beyond the standard and is more dependent on the application of entrepreneurial competencies in practice. Therefore, while studying at university, the probability of achieving this level is low.

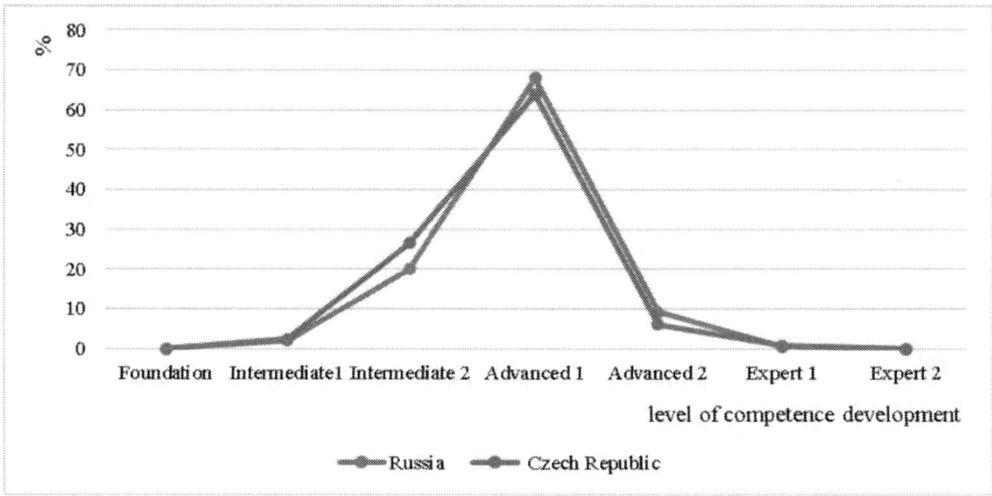

Figure 2. Level of development of entrepreneurial competencies of students in Russia and the Czech Republic (as a percentage of the total number of participants)

Source: Authors.

The main descriptive statistics for the assessment of the entrepreneurial competencies of the students in Russia and the Czech Republic are presented in Table 2. There is no mode in either sample (there are several indicators with the same frequency value). The sample has a close to normal distribution. The coefficient of variation $v \leq 30\%$ ($Vm1=13.23\%$; $Vm2=15.08\%$) means that the population is homogeneous, the variation is weak, and the data obtained can be trusted. The asymmetry is insignificant and the distribution curves are peaked.

Table 2. Main descriptive statistics for the assessment of the entrepreneurial competencies of the students in Russia and the Czech Republic

Statistics	Russia (m1)	The Czech Republic (m2)
Average	234	227
Standard error	2.22	2.129074
Median	235.5	226
Standard deviation	30.92	33.05211
Sample variance	956	1092.442
Excess	0.12	0.211963
Asymmetry	-0.105	0.076684
Interval	179	209
Minimum	140	122
Maximum	319	331
Reliability level (95.0%)	4.405927	4.194057

Source: Authors

Scope diagrams for the three key clusters are presented below, each of which includes 5 competencies (Figures 3-5). In the scope diagrams, the competency scores are arranged in pairs.

Innovative behaviour and creativity are important for creating new products, restructuring processes, and generating new types of services [23]. Innovativeness is inextricably linked to finding ways to meet the needs of innovation and decision-making [24, 25]. These competencies are combined in the cluster "Ideas and opportunities" (see Figure 3). The values for the observations of the relevant competencies range from 1.33 to 8 points, which reflects the wide range of respondents' assessments. Under "Vision", the largest interquartile range (IQR) is 2.6 points (M2). A slight asymmetry in the data is to be seen in the evaluated competencies. The closest ratings in both sample sets were achieved under "Valuing ideas". The diagram shows three outliers in terms of competence assessments: "Spotting opportunities" - 1.5 points (M2), "Creativity"- 7.2 points (M2), and "Ethical and sustainable thinking" - 7.5

points (M2). The data for the span chart was checked using t-statistics, which showed the presence of significant statistical values in the assessments of the two groups for three competencies: "Creativity", value of student's t-test=3.94. (p-value=0.000095); "Vision", value of student's t-test=4.43 (p-value=0.000012); "Ethical and sustainable thinking", value of student's t-test=3.54. The differences are statistically significant (p=0.000451).

The average score for "Ideas and opportunities" among the Russian university students was 4.6 points, and among the Czech students 4.4 points. The average scores correspond to the Advanced 1 level of competence development. The participants are therefore likely to be able to analyse problems, identify opportunities for solving them, have the skills to generate ideas and make decisions, are able to develop a strategic vision for solutions, and maintain a sufficient degree of ethical thinking, as evidenced by actions to achieve goals.

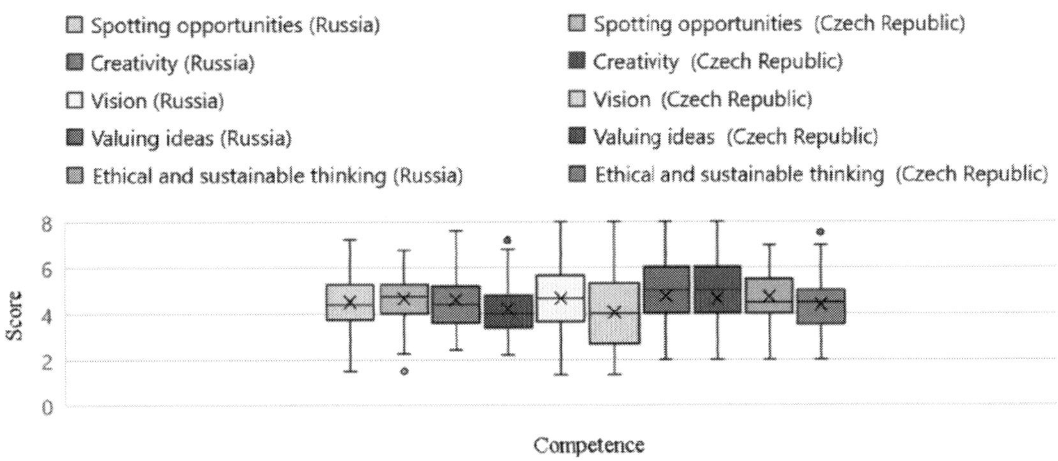

Figure 3. Box plot for competencies in the cluster "Ideas and opportunities"

Source: Authors.

It is assumed that entrepreneurial activity and the initiation thereof requires a fairly high level of investment with a high level of risk because the success of any initiative depends on entrepreneurial competencies in the field of resource management [24, 26]. For the cluster "Resources", and based on the scope diagram in Figure 4, it can be said that the results of the assessments in the two populations is fairly similar. The values for the observations of the relevant competencies range from 0 to 8 points, which reflects the participants′ choice of all the possible answers. The diagram shows two outliers: "Mobilizing resources" (M1=1.33 points) and "Mobilizing others" (M1=0.66 points), the values of which are significantly different to the average values within the cluster. The most significant IQR for the assessment of "Financial and economic literacy" in both sample sets (3 points), as well as the widest range of observations, is for the same competence, which shows that the financial and economic abilities of the respondents in both sample sets is completely different. The interquartile range of assessments for this cluster in the two sample populations shows significant differences in the competencies "Motivation and perseverance", "Mobilizing resources" and "Mobilizing others". This fact indicates that there are statistically significant differences in the assessments of the respondents in the two countries. Statistically significant differences in the indicators of the two sample sets are also confirmed by the results of the t-statistics: "Motivation and perseverance", value of student's t-test=4.40 (p-value=0.000014); "Mobilizing resources", value of student's t-test=5.78 (p-value=0.000000); "Mobilizing others", value of student's t-test=3.49 (p-value=0.000537).

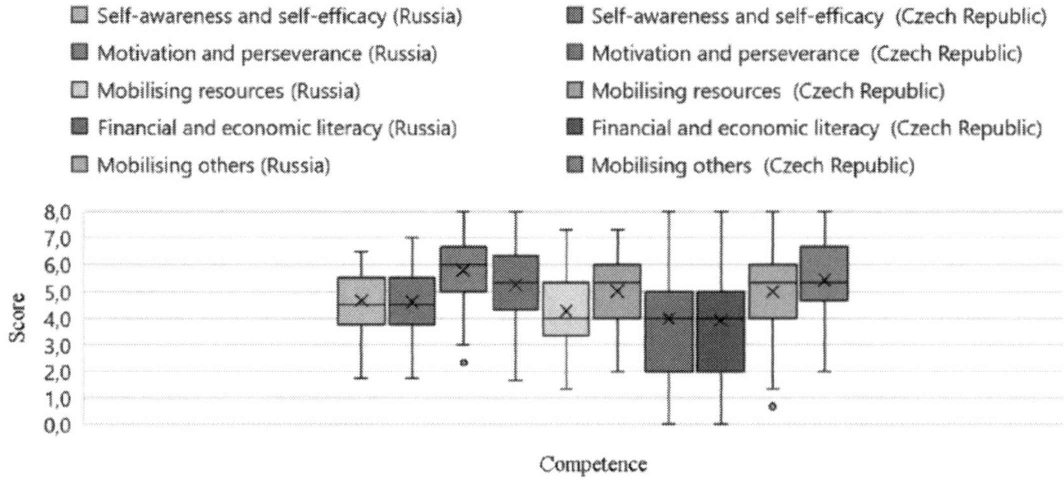

Figure 4. Box plot for competencies in the cluster "Resources"

Source: Authors.

The average assessment of the students' competencies in this cluster was M1= 4.74 points, M2= 4.83 points. The average scores correspond to the Advanced 1 level of competence development and are slightly higher than the scores for "Ideas and opportunities". Most of the students who took part in the test are probably focused on success, have the ability to solve problems, and are able to manage and mobilize resources and people to achieve relevant goals.

Learning and developing as a person is an important component of entrepreneurial competence [6,27]. The same can be said about the manifestation of initiative [26], planning and management [5], risk management [21], and interaction with people [5, 6]. For the competencies in the cluster "Into action", the assessment focuses on the students´ ability to act and apply their knowledge and skills in various situations (see Figure 5). In the span diagram, there are a large number of outliers, which means a large variety of data estimates for the students' 5 entrepreneurial competencies.

There are only statistical differences with regards to the assessment of "Learning through experience". In general, students from both countries gave similar answers to the questions in this block, with identical values accordingly. For the other 4 competencies, the values of the ratings show statistically significant differences: "Taking the initiative", value of student's t-test=2.07, (p-value=0.038762); "Planning and management", value of t-criterion=3.10 (p-value=0.002032); "Copying with uncertainty, ambiguity and risk", value of t-criterion=4.16 (p-value=0.000038); "Working with others", value of t-criterion=4.89 (p-value=0.000001). The interquartile ranges (IQR) for these four competencies also show significant differences in the two sample populations. The average values of respondents' ratings for this area of competence were: M1= 4.9 points, M2=4.7 points. Indicators of average values, medians, and modes are close in value, and the sample population has a normal distribution. The difference in respondents' ratings in both sets is large enough to suggest that most of the respondents have a single level of competence development in this area. In terms of competencies, the scope of assessments is everywhere, from the minimum to the maximum. At the same time, the level of reliability (95.0%) varies from 0.14 to 0.26.

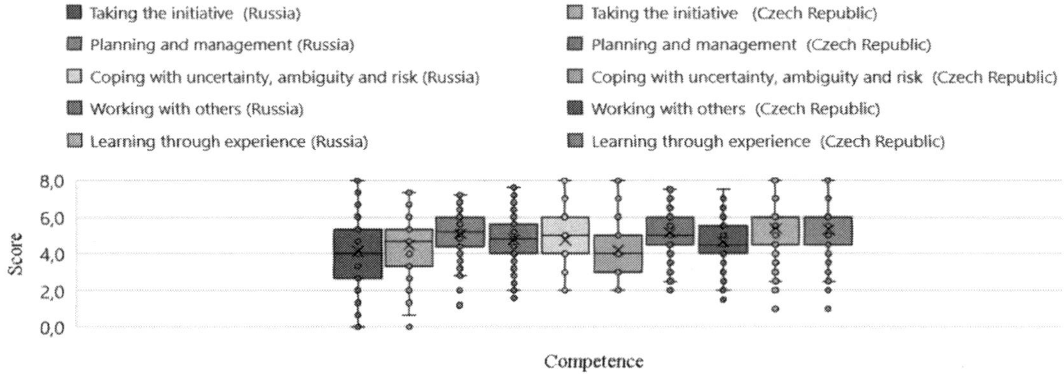

Figure 5. Box plot for competencies in the cluster "Into action"

Source: Authors.

4 Conclusion

The research presented here involved the study of different approaches to the classification of entrepreneurial competencies, as proposed by scientists around the world [5, 16, 17, 18, 20]. The analysis was carried out to determine a universal model of entrepreneurial competencies as a basis for the development of a methodology for their assessment, in particular a model with the ability to track the dynamic development of university students. We chose "EntreComp: The Entrepreneurship Competence Framework", as prepared by Bacigalupo, et al. [22]. The prepared evaluation method, as presented in this article, was tested on university students in Russia and the Czech Republic.

Based on the test results, several conclusions were drawn and tasks identified for the psychometric debugging of the test in order to increase the reliability of its results. Additional elements should be added to the methodology to avoid choosing "socially expected responses" by participants, which cause serious distortions in the results. This issue is currently being addressed by adding a "lie scale" that includes "trick questions". In addition, while working on the assessment methodology, the conclusion was drawn that it was necessary to expand the list of indicators of certain competencies, as well as to include the possibility of choosing an answer if the participant did not understand the context and/or did not have any experience of the situation presented in the question.

When testing the methodology, the task was set to conduct a comparative analysis of the level of entrepreneurial competencies of students in higher education institutions in the Russian Federation and the Czech Republic.

As a result of the study, it was found that the students who participated in the study in Russia and the Czech Republic have a fairly high ability to actively participate in society. The hypothesis was confirmed that most of the participating students do not have a lower than average level of competence development. According to the analysis, and as is graphically presented in Figure 2 and the scope diagrams, most of the respondents have an Advanced 1 level of entrepreneurial competence development. Students are ready to take the initiative and promote ideas to create new products that meet the needs of the individual, the environment, the country and society as a whole. Students showed a sufficient level of competence in the field of managing their lives, motivation and career.

At the same time, during the study, the conclusion was drawn that the hypothesis that there are no statistically significant differences in the assessment of the level of development of students' entrepreneurial competencies in both countries was partially confirmed. Specifically, the respondents in the two sample groups received similar ratings for the 5 competencies: "Spotting opportunities", "Valuing ideas", "Self-awareness and self-efficiency", "Financial and economic literacy" and "Learning through experience". The level of development of these competencies among the students of universities in Russia and the Czech Republic was similar. On average, the listed competencies of the students were at the Advanced 1 level of development, as characterized by the ability to take responsibility and work under the guidance of others. At the same time, the level of development of "Financial and economic literacy" competencies produced average development indicators corresponding to the Average 2 level, and "Learning through experience" to the Advanced 2 level.

For the 10 competencies not included in the list above, there are significant statistical differences in respondents' ratings. The highest average scores were given to the competence "Motivation and perseverance" (M1=5.79 points, M2=5.26 points). Students are aware of their needs, aspirations and desires in the short, medium and long term, and are able to identify and evaluate their strengths and weaknesses, work in situations of uncertainty, and cope with failures.

Despite the revealed statistical differences in the assessments of students' entrepreneurial competencies in Russia and the Czech Republic, it is possible to state that the results obtained are quite close. Students in both sample groups have approximately the same level of development of entrepreneurial competencies, which is clearly shown in Figure 2.

We agree with the conclusions of previous studies that higher education is one of the main platforms for the formation of the competencies required to be an entrepreneur in modern society [13,15]. We believe that the development of entrepreneurial competencies should be nurtured throughout the educational process in higher education institutions through the implementation of relevant academic programmes and disciplines. Where earlier it was put forward that entrepreneurship training should be implemented within the framework of special courses [28], now the development of a wide range of supra-professional competencies needs to be the focus of attention in every academic discipline [29, 30].

The results of the study presented in this article can form the basis of an international project aimed at developing the entrepreneurial competencies of students from two countries. In this regard, it will be interesting to analyse innovation and entrepreneurship training as part of the main bachelor's and master's degree programs in the two countries. The question of the application of technologies in the educational process and special courses for students on entrepreneurship, in particular in the study field "Human Resource Management", is also interesting from the point of view of trends in joint research between scientists from the Russian Federation and the Czech Republic.

Acknowledgment

The reported study was funded by RFBR, project number 19-29-07435

Reference

1. UNESCO, *Education for sustainable development goals: learning objectives* [online]. Available at: http://unesdoc.unesco.org/images/0024/002474/247444e.pdf (2017)
2. G. De Haan, The BLK '21'programme in Germany: a 'Gestaltungskompetenz'-based model for Education for Sustainable Development. *Environmental Education Research*. **12**(1), 19–32 (2006)
3. A. Wiek, L. Withycombe, C. L. Redman, Key competencies in sustainability: a reference framework for academic program development. *Sustainability science*. **6**(2), 203–218 (2011)
4. S. Taimur, H. Sattar, Education for Sustainable Development and Critical Thinking Competency. In W. Leal Filho, A. Azul, L. Brandli, P. Özuyar, T. Wall (eds), *Quality Education. Encyclopedia of the UN Sustainable Development Goals*. Springer, Cham (2019)
5. L. Ploum, V. Blok, T. Lans, O. Omta, Toward a validated competence framework for sustainable entrepreneurship. *Organization & environment*. **31**(2), 113–132 (2018)
6. N. H. Ahmad, *A Cross Cultural Study of Entrepreneurial Competencies and Entrepreneurial Success in SMEs in Australia and Malaysia* [online]. Available at: https://digital.library.adelaide.edu.au/dspace/bitstream/2440/48199/10/Ahmad2007_PhD.pdf
7. R. Gerards, S. van Wetten, van C. Sambeek, New ways of working and intrapreneurial behaviour: the mediating role of transformational leadership and social interaction. *Review of Managerial Science*. 1–36 (2020)
8. J. W. Hernandez Gonzalez, J. Diaz Castro, Intra Entrepreneurship Impact on Business Growth of MSMEs in the Tourism Sector in Villavicencio. *Estudios Avanzados*. **29**, 75–85 (2018)
9. G. Pinchot, Intrapreneuring: Why you don't have to leave the corporation to become an entrepreneur. *University of Illinois at Urbana-Champaign's Academy for Entrepreneurial Leadership Historical Research Reference in Entrepreneurship*. (1985)
10. M. N. M. Shariff, M. S. Basir, An attitude approach to the prediction of entrepreneurship on students at institution of higher learning in Malaysia. *International Journal of Business and Management*. **4**(4), 129–135 (2009)
11. F. A. Argandona Gomez, M. C. Persico Jimenez, A. M. Visic Matulic, J. I. Bouffanais Cuevas, Case study: A teaching methodology, in higher education, for the acquisition of integrative and entrepreneurial competencies. *Tec Empresarial*. **12**(3), 7–16 (2018)
12. M. A. González, L. M. Llanos, P.M. Rosario, The role of higher education in development of entrepreneurial competencies: Some insights from Castilla-La Mancha university in Spain. *Administrative Sciences*. **9**(1) (2019)
13. S.-M. Wang, H.-P. Yueh, and P.-Ch. Wen. How the New Type of Entrepreneurship Education Complements the Traditional One in Developing Entrepreneurial Competencies and Intention. *Frontiers in Psychology*. **10** (2019)
14. Ch. A. Julijevič, How and why to teach students entrepreneurship [online] Available at: https://cyberleninka.ru/article/n/kak-i-zachem-obuchat-studentov-predprinimatelstvu-polemicheskie-zametki/viewer (2017)
15. K. Bischoff, K. C. Volkmann, D. B. Audretsch, Stakeholder collaboration in entrepreneurship education: an analysis of the entrepreneurial ecosystems of European higher educational institutions. *Journal of Technology Transfer*. **43**, 20–46 (2018)
16. R. S. Mansfield, D. C. McClelland, L. M. Spencer, J. Santiago. *The Identification and Assessment of Competencies and Other Personal Characteristics of Entrepreneurs in Developing Countries: Final Report Project No. DAN-5314-C-00-3074-00* (1987)
17. W. U. Wei-Wen, A competency-based model for the success of an entrepreneurial start-up. *Wseas Transactions on Business and Economics*. **6**(6), 279–291 (2009)
18. H. Kaur, A. Bains. Understanding the concept of entrepreneur competency. *Journal of Business Management & Social Science*. **2**(11), 31–33 (2013)

19. B. Inyang, R. O. Enuoh, Entrepreneurial Competencies: The Missing Links to Successful Entrepreneurship in Nigeria. *International Business Research*. **2**(2) (2009)

20. B. Bird, Toward a theory of entrepreneurial competency. *Advances in Entrepreneurship, Firm Emergence and Growth*. **21**, 115–131 (2019)

21. J. C. Hayton, D. J. Kelley, A competency based framework for promoting corporate entrepreneurship'. *Human Resource Management*. 45, 407–427 (2006)

22. M. Bacigalupo, P. Kampylis, Y. Punie, G. Van den Brande, *EntreComp: The Entrepreneurship Competence Framework*. Luxembourg: Publication Office of the European Union (2016)

23. C. Santandreu-Mascarell, D. Garzon, H. Knorr, Entrepreneurial and innovative competences, are they the same? *Management Decision*. **51**(5), 1084–1095 (2013)

24. S. Mitchelmore, J. Rowley, Entrepreneurial competencies: a literature review and development agenda. *International journal of entrepreneurial Behavior & Research*. **16**(2), 92–111 (2010)

25. R. C. Becherer and J. G. Mauer, The Proactive Personality Disposition and Entrepreneurial Behavior among Small Business Presidents. *Journal of Small Business Management*. **37**(1), 28–36 (1999)

26. K. Pascoe, K. Mortimer, Identifying entrepreneurs through risk taking behaviour: illegal downloading. *Journal of Research in Marketing and Entrepreneurship*. **16**(2), 183–199 (2014)

27. A. A. Mamun, N. B. C. Nawi, S. F. F. B. Shamsudin, Examining the effects of entrepreneurial competencies on students' entrepreneurial intention. *Mediterranean Journal of Social Sciences*. **7**(2), 119–127 (2016)

28. B. Johannisson, University training for entrepreneurship: A Swedish approach. *Entrepreneurship and Regional Development*. **3**(1), 67–82 (1991)

29. C. MacDermott, L. Ortiz, Beyond the business communication course: A historical perspective of the where, why, and how of soft skills development and job readiness for business graduates. *IUP Journal of Soft Skills*. **11**(2), 7–24 (2017)

30. L. Yan, Y. Yinghong, S. M. Lui, M. Whiteside, K. Tsey, Teaching "soft skills" to university students in China: The feasibility of an Australian approach. *Educational Studies*. **45**(2), 242–258 (2019)

Development of Entrepreneurial Competencies of Employees within the Framework of internal Entrepreneurship Programs in the Region

Anastasiya Pesha[1], *Marina* Shavrovskaya[1,*], *Oksana* Borodina [2]

[1] Ural State University of economics, Department of labour economics and HR management, 8 Marta 62, 620144 Yekaterinburg, Russia
[2] Omsk State University, Department of Regional Economics and Human Resource Management, Mira Avenue, 55-A, 644077 Omsk, Russia

Abstract. The global economic environment is becoming more active and complex. Technological and social changes affect the internal processes of companies associated with the diversification of production, the introduction of new lines of products and services, the modernization of business processes. In this article, the authors present the results of a study into the ways ten large industrial companies (fuel, chemical and petrochemical, food industry, defence industry and agricultural technology) in the Omsk region develop domestic internal entrepreneurship. As a result of narrative interviews with HR directors of the selected companies, the authors formulated two approaches to the development of internal entrepreneurship. The first method is based on the development of a creative environment that stimulates the generation of initiatives to improve business. The second method involves companies implementing business projects through employee development programs. The advantages and disadvantages of both methods are subsequently highlighted. The conclusions draw attention to the problems of the development of internal entrepreneurship within large industrial companies and the need to create a special environment and infrastructure to promote the business initiatives of employees.
Keywords: entrepreneurial competencies; internal entrepreneurship; personnel; bank of ideas; innovation

1 Introduction

Many companies around the world pay great attention to attracting and developing internal entrepreneurship to maintain their competitiveness. The entrepreneurial spirit of employees determines the implementation of innovations in the company's activities and the maintenance of business sustainability in a rapidly changing economic and social reality [1, 2]. The foundations of the concept of internal entrepreneurship were laid by scientists such as Macrae (1976), Nielsen, Peters and Hisrich (1985), Vesper (1984), Pinchot (1985) and others [1, 3, 4, 5]. Three decades ago, the founders of the concept noted that internal entrepreneurship was the main driver of large-scale changes in business and the economy.

Over the past 10 years, the number of studies in the field of entrepreneurial competencies in general and internal entrepreneurship in particular has been increasing. According to WoS data, the number of publications on the subject of this paper (the search term "intrapreneurship") increased more than 4 times from 2010 to 2019 (from 14 publications in 2010 to 61 publications in 2019). Today, new concepts of internal entrepreneurship are emerging, which point to the need to study entrepreneurial competencies in terms of "potential" (opportunities for implementation in the field of entrepreneurship), and not only the "capital" (availability of special education and entrepreneurial experience) of a person [6, 7].

The hypotheses about a positive correlation between the effectiveness of internal entrepreneurship, the development of entrepreneurial competencies and business growth and economic effect [8] are derived from and supported by empirical data [7]. Companies focused on engaging employees in innovative internal corporate behaviour should create an organizational climate that promotes entrepreneurial initiatives. The internal business climate, which is characterized by a "positive correlation between management support, remuneration system and employee autonomy", seeks to tap into employee innovation and to transform initiatives into real goods and services [9]. New ways of working (NWW), which involves internal competitions for the development of a culture of initiative and innovation within a company, enables employees to gain recognition for their work and for their importance to the company. Within this context, the culture and conditions for the manifestation of initiatives within an organization have a significant impact on the development of internal entrepreneurship [6, 10]. As a result of the activities of internal entrepreneurs, new products and services are

* Corresponding author: marina_bel@list.ru

introduced, new internal corporate companies are established, new markets and business development horizons are opened [11, 12, 13].

Many large Russian companies are currently paying attention to the development of domestic internal entrepreneurship, realizing that in this area there are significant reserves for increasing productivity and labour efficiency [14, 15]. In Soviet times, companies had systems for submitting and implementing innovation proposals, which helped to improve business processes. During the years of Perestroika, the system almost disappeared. However, it is now being revived because many companies understand the key role the level of entrepreneurial competencies of their employees plays.

The management of industrial companies based on internal entrepreneurship has been studied in the scientific works of authors such as Bukhvalov, Molodchik, Popov and Pustovojt (2012) and Dondokova (2004) [16, 17]. Industrial companies, like many others, need to create a special environment and infrastructure for the development of internal entrepreneurship, which facilitates the manifestation and implementation of business initiatives.

2 Methodology

The article presents the results of research into the practice of developing internal entrepreneurship in ten large industrial companies (fuel, chemical and petrochemical, food industry, defence and agricultural engineering) in the Omsk region, which are highlighted on the website of the city administration https://www.admomsk.ru. The authors analyse existing approaches to the development of internal entrepreneurship in industrial companies, which are the main employers of graduates and the main providers of internships for students. The main goals of this work are to present and analyse the existing approaches to the development of internal entrepreneurship of companies in the region, as well as to highlight their advantages and disadvantages.

The method of collecting information was a narrative interview with HR specialists who are responsible for the development of internal entrepreneurship in their respective company, as well as for the collection, analysis and implementation of employee initiatives and/or projects. This method allowed us to collect qualitative data, which were further analysed and grouped into existing approaches to the development of internal entrepreneurship among employees.

The authors selected HR specialists in those companies where this work is carried out (for various reasons, many Omsk companies are not engaged in the development of internal entrepreneurship, despite the fact that in Soviet times they had a system for submitting innovation proposals). In addition, not all companies are allowed to disclose information about existing business processes, so only those HR specialists who could share tools for developing internal entrepreneurship participated in the interviews.

During the interview, it was important for the authors to get answers to the following questions:

1. How is the work on getting initiatives from employees organized?

2. How are employees involved and motivated to take the initiative?

3. What criteria are used to make a decision on whether an initiative will be implemented or not?

4. What advantages and disadvantages are there in the approaches to the development of internal entrepreneurship?

In addition, document analysis was used as a method. In total, half of the companies were able to share standards or presentations that describe systems for collecting initiatives from employees, their method of evaluation, and the methods used by the companies to motivate employees to submit them.

3 Results and Discussion

To study the features of the work on the development of internal entrepreneurship at industrial companies in the Omsk region, two main approaches were identified.

3.1 Development of a creative environment that stimulates the generation of initiatives

An industrial company is usually a well-established and stable structure that is difficult to quickly rebuild or break. However, by tradition, such organizations have their own systems that encourage various improvements and innovations. Some companies pay a fixed amount of money to initiators of changes in production, others build systems such as "The factory of initiatives", "The bank of initiatives", "The pipeline of initiatives", and so on. These are tools for improving the efficiency of a company by creating an environment for developing, reviewing and implementing employee initiatives.

The structure of such systems usually contains two basic elements – an administrator who receives, registers and curates ideas, and an initiative committee, which reviews, evaluates, makes a decision and summarizes the results of their implementation.

The functioning of this system is approximately the same everywhere and consists of several stages:

1. The author of the initiative formulates and submits it in a certain way in the company. Some organizations require the initiator to write down their idea and bring it to the responsible person, whereas others have special boxes for collecting them, whilst the most advanced companies may use corporate portals or groups in messenger or on social networks.

2. The administrator registers the initiative. A number of companies have developed special classifiers for this purpose. There is an instruction for classifying and maintaining a register of initiatives in one of the studied companies.

3. An initiative is sent for evaluation to experts of relevant departments. A special methodology for evaluating initiatives has been developed at one of the military-industrial companies. At this stage, the initiative can be refined or combined with other ones that are similar in content. At this stage, initiatives can either be screened out, finalized or justified.

4. The revised initiative is sent to the finance department, where the economic effect is evaluated. At this stage, the idea's passport is formed and issued.

5. At a committee meeting of experts, the initiative is evaluated, ranked and a decision subsequently made on its implementation. By consolidating the experience of various industrial companies, the authors developed a matrix of criteria for evaluating initiatives based on their usefulness and availability.

Table 1. Criteria for evaluating initiatives

Criterion	Scores		
	1	2	3
Usefulness			
Compliance with company's strategy	The initiative is useful for achieving business goals, but not relevant in the short term (1-3 years)	The initiative corresponds to the organization's strategy, but its implementation does not affect the achievement of the company's priority goals	The initiative corresponds to the organization's strategy, and its implementation affects the achievement of the company's priority goals
Expected economic effect	Less than RUB 50,000	From RUB 50,000 to RUB 500,000	More than RUB 500,000
Probability of achieving the expected result	Probability of achieving the expected result is less than 30%	Probability of achieving the expected result is from 30% to 80%	Probability of achieving the expected result is more than 80%
Impact on business processes	Implementation of the idea will affect one category of business processes	Implementation of the idea will affect two or more categories of business processes	Implementation of the idea will affect the results of the company as a whole
Influence on company's image	A significant impact on the company's image	There is an impact on the formation of a positive image of the company in the "eyes" of one of the interaction groups (the state, consumers, employees, including potential ones)	It will influence the formation of a positive image of the company in the "eyes" of more than one of the interaction groups (the state, consumers, employees, including potential ones)
Availability			
Financial resources	The initiative requires an increase in the current business plan	The initiative requires a reallocation of resources between items, but within the current business plan	The initiative is implemented within the current business plan
Human resources	It is necessary to organize the interaction of departments more than one functional block to implement the idea	The interaction of departments of one functional block to implement the idea	The work of employees of the profile division is enough
Administrative resources	The general manager plays the main role	The managing director plays the main role	The division head plays the main role
Time resources	It will take six months or more from the initiation of the idea to the result	It will take three to six months from the initiation of the idea to the result	It will take less than three months from the initiation of the idea to the result

Source: Authors.

When an Initiative (I) is evaluated according to two criteria, with scores for the two parameters usefulness and availability, it is assigned to one of four categories (see Figure 1):

- A – implementation zone (high availability and high usefulness);

- B – decision-making zone (high availability and low usefulness or low availability and high usefulness);

- C – rejection zone (low utility and low availability);

In order to motivate employees to generate various initiatives, industrial companies use different systems. The main objectives of these systems are to:

- create an information base for generating initiatives;
- create a communicative and creative environment;
- create a positive image of the initiators;
- support initiatives and feedback.

All these systems can be classified into three groups:

1. Motivation to participate, whereby ideas are encouraged.
2. Motivation through results, whereby employees receive remuneration:
a. percentage of the economic effect of the implementation results;
b. fixed amount for each initiative.

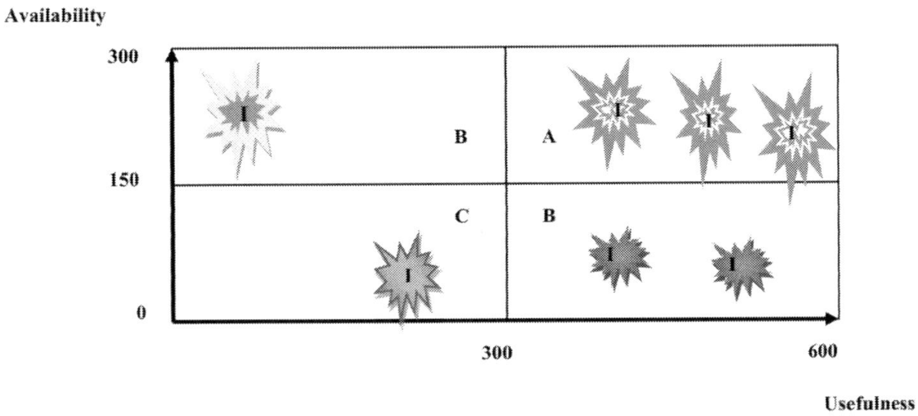

Figure 1. Evaluation of initiatives by expert committee

Source: Authors

3. Motivation through consolidation, whereby an employee is put in charge of the implementation of their own initiative, thereby creating an opportunity for career growth.

In several companies, the authors found interesting mechanisms to measure performance against the initiatives of its employees. By consolidating this information, three groups of performance indicators can be distinguished:

1. Employee engagement
   a. engagement rates (number of initiators / number of personnel)
   b. number of initiatives filed in a particular period (N)
2. Content of initiatives
   a. areas of improvement (business process groups)
   b. structure of initiatives by category (A/N, B/N, C/N)
   c. percentage of rejected initiatives
3. Performance
   a. implementation rate (number of initiatives implemented / number of initiatives submitted)
   b. coefficient of achievement of economic effect (factual savings / planned savings)
   c. costs per 1 initiative (∑costs / number of all initiatives)
   d. potential coefficient (number of initiatives without investment / number of all initiatives)
   e. total efficiency coefficient (∑effects / ∑costs)

The advantages of this approach are that:

1. all employees of the organization are involved in the process of generating ideas. Everyone has an equal opportunity to implement an initiative;
2. the environment for the development, review and implementation of initiatives of employees is formed within the company.

The disadvantages of this approach are that:

1. for large scale companies in the region, it is difficult to create an environment for the manifestation of entrepreneurial initiatives because of the large number of employees, structural divisions, difficulties in establishing communication processes, as well as the need to introduce additional jobs for organizing, planning and controlling the process;
2. the process from the inception of an idea to the assessment and implementation thereof is very long.

3.2 Implementation of business projects through personnel development programs

Without negating the value of the first approach, many industrial companies implement the initiative of internal entrepreneurship through personnel management. In this case, the promotion of business initiatives is implemented through training and development programs with a business focus. This may occur through, for example, the organization of corporate scientific and practical conferences for employees.

The main essence of this approach is that talents or initiators are assigned the task of developing and protecting a business project aimed at improving business processes or implementing business initiatives within the framework of developing corporate programs [18].

Employees can work on projects in small groups or alone. A mentor can join the project. As a rule, a certain period of time, for example a calendar year, is given to develop a business initiative. After the development stage, business projects are defended, winners are identified and rewarded, and the best ideas implemented.

This approach, in our opinion, has a number of advantages:

1. The talents within an organization are those with the potential to occupy a more responsible position. Members of the talent pool are ambitious, purposeful, innovative and ready to take steps to develop themselves and the company.

2. Project work is conducted usually in groups. When talented, high-potential employees interact, a synergy effect occurs and the output of promising business initiatives increases significantly (compared to the first approach).

3. Motivation to improve the business process and introduce initiatives for business development is a powerful tool for increasing employee engagement and loyalty, as well as the basis for creating an environment for the development of internal entrepreneurship.

The disadvantages:

1. The environment for the development of internal entrepreneurship is formed only in a few personnel (for example, in the talent pool) and participants of scientific and practical conferences. The participants of scientific and practical conferences are often talents.

2. Low risk tolerance of employees of Russian industrial companies. The fear of restrictions can become a stumbling block at the implementation stage of an employee's business initiative [19].

3.3. Discussion

For the development of internal entrepreneurship, it is necessary to create a certain environment, or even an ecosystem in a company [9, 17, 20, 21]. The results of our research confirm the data of previous research on the importance of creating a supportive environment that includes, among other things, financial mechanisms, decision-making procedures, motivational and internal communication systems, and an innovative corporate culture. The key point is that the environment should promote the birth and development of internal entrepreneurs and the implementation of business initiatives. As the cases of the analysed companies show, internal initiative competitions can foster a culture of innovation in which employees can feel accomplished and receive recognition for presenting new ideas, as confirmed by the research of other scientists [10]. The relationship between a freely accessible open workplace and intrapreneurial behaviour is mediated by transformational leadership [6]. This area also needs to be studied by analysing the correlation between management styles in companies and the level of internal entrepreneurship.

Creating an environment for internal entrepreneurship should be accompanied by modernization of the industrial company infrastructure, including, among other things, technological equipment, financial support for initiatives and assessment of their investment attractiveness, and the training and development of employees' entrepreneurial competencies. While developing entrepreneurial competencies, it is important to pay attention to the employees' ability to assume responsibility for envisioning the necessary product, market, and management strategies [1]. Developing the ability of subordinates to take responsibility in Russian companies, where the authoritarian management style often prevails, is not such an easy task. In addition, it is rather difficult to teach people business thinking, which involves calculating risks and thinking about management strategies. Educational institutions should certainly assist in this by preparing graduates with a high level of entrepreneurial competence [22].

4 Conclusion

Internal entrepreneurship is not an innovation, its appearance in the last century is due to trends and features of economic development. At present, when the competition between industrial companies is strong and there is economic turbulence and dynamic technological development, internal entrepreneurship can become a strategic resource for a company's development. In this regard, the managements of companies should pay attention to creating an innovative internal corporate environment that promotes the involvement of a large number of employees in the process of developing business initiatives.

There are a number of problems associated with risk, independence, fear, financial support, and other factors on the way to implementing internal entrepreneurship. It is the shared opinion of the authors that the main obstacle to the development of internal entrepreneurship in industrial companies is the low level of development of employees' entrepreneurial competencies and the low level of involvement of people in the business processes of a company. In this area, we plan further research, in which we will try to identify areas of development in the studied companies and come up with a concept that will enable large industrial companies to successfully implement internal entrepreneurship.

References

1. G. Pinchot, Intrapreneuring: Why you don't have to leave the corporation to become an entrepreneur. *University of Illinois at Urbana-Champaign's Academy for Entrepreneurial Leadership Historical Research Reference in Entrepreneurship* (1985)
2. S. Veenker, P. V. D. W. SijdeDuring, A. Nijhof, Organisational conditions for corporate entrepreneurship in Dutch organisations. *The Journal of Entrepreneurship*. **17**(1), 49–58 (2008)
3. N. Macrae, The coming entrepreneurial revolution: A survey. *The Economist*. **25**, 41-46 (1976)
4. R. P. Nielsen, M. P. Peters, R. D. Hisrich, Intrapreneurship strategy for internal markets-corporate, non-profit and government institution cases. *Strategic management journal*. **6**(2), 181–189 (1985)
5. K. H. Vesper, *Three faces of corporate entrepreneurship - A pilot study*. In J. A. Hornaday, F. Tarpley, Jr., J. A. Timmons, K. H. Vesper (Eds.) *Frontiers of entrepreneurship research*. Wellesley, MA. Babson College, 294–326 (1984)
6. R. Gerards, S. van Wetten, C. van Sambeek, New ways of working and intrapreneurial behaviour: the mediating role of transformational leadership and social interaction. *Review of Managerial Science*. 1-36 (2020)
7. T. Lans, M. van Galen, J. Verstegen, H. Biemans, M. Mulder, Searching for Entrepreneurs among Small Business Owner-Managers. *Proceedings of the ECER*. (2010)
8. J. W. Hernandez Gonzalez, J. Diaz Castro, Intra Entrepreneurship Impact on Business Growth of MSMEs in the Tourism Sector in Villavicencio. *Estudios Avanzados*. **29**, 75–85 (2018)
9. Z. Poduška, J. Nedeljković, D. Nonić, T. Ratknić, M. Ratknić, I. Živojinović, Intrapreneurial climate as momentum for fostering employee innovativeness in public forest companys. *Forest Policy and Economics*. **119** (2020)
10. J. R. Campos-Blázquez, P. Morcillo, L. Rubio-Andrada, Employee Innovation Using Ideation Contests: Seven-Step Process to Align Strategic Challenges with the Innovation Process. *Research-Technology Management*. **63**(5), 20–28 (2020)
11. C. Blanka, An individual-level perspective on intrapreneurship: a review and ways forward. *Review of Managerial Science*. **13**(5), 919-961 (2019)
12. S. Kingma, New ways of working (NWW): work space and cultural change in virtualizing organizations. *Culture and Organization*. **25**(5), 383–406 (2019)
13. J. C. Rigtering, G. U. Weitzel, K. K. Muehlfeld, Increasing quantity without compromising quality: How managerial framing affects intrapreneurship. *Journal of Business Venturing*. **34**(2), 224–241 (2019)
14. E. A. Bubenok, Obosnovanie vedushchej roli intraprenerstva (vnutrennego predprinimatel'stva) i brendinga v obespechenii konkurentosposobnosti kompanij [Rationale for the leading role of intrapreneurship (internal

entrepreneurship) and branding in ensuring the competitiveness of companies]. Vestnik Bryanskogo gosudarstvennogo universiteta. **3**, (2014)

15. O. E. Ustinova, Korporativnoe predprinimatelstvo kak strategicheskiy resurs innovatsionnogo razvitiya kompanii [Corporate entrepreneurship as a strategic resource for innovative development of a company]. *Voprosy innovatsionnoy ekonomiki*. **10**(2), 993–1006 (2020)
16. N. Y. Bukhvalov, A. V. Molodchik, V. L. Popov, K. S. Pustovojt, Upravlenie organizacionnymi innovaciyami na predpriyatii oboronno-promyshlennogo kompleksa [Organizational innovation management at the company of the military-industrial complex] *Kreativnaya ekonomika*. **10**, (2012)
17. S. C. Dondokova, *Upravlenie promyshlennym predpriyatiem na osnove vnutrennego predprinimatel'stva* [*Industrial company management based on internal entrepreneurship*]. Doctoral dissertation, [Vost.-Sib. gos. tekhnol. un-t] (2004)
18. T. Keil, R. G. McGrath, T. Tukiainen, Gems from the ashes: Capability creation and transformation in internal corporate venturing. *Organization Science*. **20**(3), 601–620 (2009)
19. A. Pesha, T. Kansafarova, Influence of the deficit of highly qualified personnel on the competitiveness and investment attractiveness of industrial companys. In *The 12th International Days of Statistics and Economics*. 1357–1366 (2018)
20. U. Cantner, J. A. Cunningham, E. E. Lehmann, M. Menter, Entrepreneurial ecosystems: a dynamic lifecycle model. *Small Business Economics*. 1–17 (2020)
21. P. G. Svensson, T. Q. Mahoney, Intraorganizational conditions for social innovation in sport for development and peace. *Managing Sport and Leisure*. **25**(3), 220–238 (2020)
22. A. Y. Chepurenko, How and why to teach students entrepreneurship. *Educational Studies*. **3** (2017)

Formation of Entrepreneurial Skills in Students in a Changing World

Maryna Salun[1], *Kateryna* Zaslavska[2], *Radka* Vaníčková[3,*], *Kristýna* Šindelková[3]

[1]Simon Kuznets Kharkiv National University of Economics, Economy and Law Faculty, Nauky avenu 9a, 61001 Kharkiv, Ukraine
[2]Simon Kuznets Kharkiv National University of Economics, Management and Marketing Faculty, Nauky avenue 9a, 61001 Kharkiv, Ukraine
[3]Institute of Technology and Business, Faculty of Corporate Strategy, Department of Management, Nemanická 536/7, 370 01 České Budějovice, Czech Republic

Abstract. In today's global world, a number of concepts exist for the typology of a manager's personality. The personality of a manager is evident through their daily practical actions and the degree of validity thereof. The entrepreneurial skills of students is related to whether they view entrepreneurship as a career alternative. The purpose of educating successful managers is to enable them to apply and implement the principles of managerial skills in business practice. This includes the fulfilment of managerial roles and the use of basic tools to accomplish this, the principles of coaching, the creation and effective management of work teams, and knowledge of HRM techniques such as brainstorming. The aforementioned is based on the assumption that the desired result can be effectuated by one factor, but also by a combination of several factors. Qualitative comparative analysis (QCA) was therefore applied to the demands for entrepreneurial skills on the labour market and the presence of this issue in the curricula of Ukrainian universities.
Keywords: entrepreneurial hard skills; entrepreneurial soft skills; entrepreneurial competencies; entrepreneurial education; start-ups.

1 Introduction

There are four basic typologies of a manager´s personality, namely analytical, pragmatic, extrovert, and social. The social personality of a manager is focused on wellbeing and good relationships, whereby the manager can empathize with the situation and feelings of subordinate employees. A social manager does not like to listen to criticism or unpleasant things. However, they are very loyal and tend to help other employees. The analytical personality of a manager is focused on details, recognizing and continually assessing things. They require work to be carried out carefully, not only by and for himself, but also by and for his co-workers, for which clear and strict rules apply and must be observed. In contrast, the pragmatic personality of a manager is focused on performance, whereby the focus is on finding solutions to complex and challenging situations or tasks. Pragmatic managers usually derive satisfaction from achieving success. In practice, the manager is not only strict on themselves, but also with others and knows exactly what goals they want to achieve. The extrovert personality of a manager is focused on experience, which stems from excitement, fantasy and imagination. An extrovert manager does not like to strictly follow set standards and rules. They like to present their ideas here and now, while extensive and boring interpretations lack essence and represent a waste of time for them.

Leadership skills, as well as other abilities and skills, need to be acquired through practice and hard work [1]. However, there are many different ways to be a good manager-leader [2]. Many managers use different management styles [3]. The most appropriate management style is management that is flexible, adaptive, applicable to the circumstances, as different situations require different management methods. The leader must also trust the vision in every situation. The difference between a manager and a true leader lies in the following practical experience: a manager orders, a leader leads; a manager explains, a leader inspires; a manager requires discipline, a leader mentors; a manager delegates tasks, a leader delegates authority; a manager is above a team, a leader is part of a team.

Universities are critical for developing the levels of motivation and capabilities of graduates to effectively engage in entrepreneurial activity [4]. For this, students require a set of professional knowledge, especially

* Corresponding author: vanickovaradka@gmail.com

practical skills in the field of management and decision-making, differentiated according to three basic levels of management, namely top, tactical and operational, which are crucial for fulfilling defined corporate goals. They also need to learn the basic typologies, functions and roles of managers, and management styles [5].

The concept of competence sometimes refers to competencies as a result (what is done) and sometimes as a process (how it is done) [6]. In our opinion, an integrated approach that emphasises personal attributes, delivered performance and context [7] fully describes the essence of competencies. It is defined as a set of knowledge, skills and attitudes necessary to exercise a given activity, but also the person's performance in a given context in terms of behaviours used at work and the subsequent achievements. The main constituents of competence are knowledge, skills and attitudes [8], whereby it is quite important to have synergy between them.

Entrepreneurial skills, practices and approaches are undergoing profound changes at present, which must be taken into account when shaping students' learning trajectories. For example, the dynamic development of start-ups in the field of social entrepreneurship necessitates the formation of knowledge about the development of society; such knowledge becomes vital to ensure the professional significance of entrepreneurship. At the same time, the most relevant professional skills of an entrepreneur remain verbal and written communication skills to create sustainable entrepreneurial connections and communities.

The modern Ukrainian start-up market [9] is characterized by rapid growth in the total costs of offers (on average + 40.19% annually); a doubling of the number of projects invested in and their average costs over a three year period; intensification of mergers and acquisitions (91.18% in the total structure); allocation and gradual growth of investments at the stage of supporting an innovative idea; stabilization of investment volumes at the initial stage of project formation and the stage of development of an innovative product; activation of the start-up movement at universities with the formation of project start-up teams and the involvement of consultants from the academic environment.

The current entrepreneurship training initiative, "The Entrepreneurship University", implemented by the network of YEP start-up incubators in cooperation with the Ministry of Education and Science of Ukraine, the Ministry for the Digital Transformation of Ukraine, the Ukrainian Foundation of Start-ups with the support of USAID´s "Competitive Economy of Ukraine" [10], focused on developing entrepreneurial skills in and through higher education, is a response to the modern development of flexible innovative business and the growth of self-employment in the youth segment of the labour market [11], whereby the lack of entrepreneurial skills is considered to be a barrier to student employment opportunities.

The contribution aims to supplement the current discussion on entrepreneurial skills by clarifying the current composition and content thereof, by highlighting global (professional) skills, and by predicting future skills that may affect the profile of the profession and that may need to be included in curricula.

2 Methodology

As part of the research, literature was reviewed on the issues of enhancing entrepreneurship in the student environment, the formation of the start-up movement in universities, standards of higher education in Ukraine in specializations in order to identify the entrepreneurial competencies prescribed in them, and research results from international organizations on current and future entrepreneurial skills and competencies.

Basic entrepreneurial skills were presented comprehensively, often in the form of specific disciplines that are offered on Coursera and which meet the needs of current businesses. For example:

H1: Accountancy stimulates the generation of accounting skills, the creation of accounting, financial and tax information, and reporting under state or international standards.

H2: The communication skills of an entrepreneur are developed in practice by discussing a task or problem with two or more people in written or verbal form.

H3: Financial analysis skills are allocation investment opportunities of risk or uncertainty.

H4: Management provides for the formation of the skills needed to develop a company's strategy and coordinate the efforts of its employees.

H5: Marketing as an entrepreneurial skill is the ability to create positive, trusting, long-term relationships with potential and actual customers, allowing enterprises to establish rules and regulations for doing business in the regional and global markets.

H6: Sales as an entrepreneurial skill is focused on an effective system for bringing the company's products and services to the market and conducting transactions with real customers.

Quality comparative analysis (QCA) is not primarily an evaluation technique, the original application lies in comparative political science. QCA analysis seeks to explain individual, unique cases where the investigated phenomenon occurred or did not occur in a set of mutually comparable cases. The method is suitable for the evaluation of several dozen cases. It does not evaluate the strength of the dependence between the variables, but identifies the circumstances of how the intervention works. QCA is suitable for testing the theory of change. It is often supplemented by case studies and is also a supporting tool for statistical methods. The analysis indicates the necessary and/or sufficient or irrelevant conditions that (do not) lead to the observed phenomenon. It is also possible to perform the analysis at multiple levels simultaneously, i.e. it is possible to evaluate different and seemingly "unrelated" conditions of the investigated phenomenon at the same time without applying statistical methods.

The main steps of the methodological procedure were:

- determination of the hypotheses based on the theory of change, including the identification of the dependent variable and conditioning factors, i.e. gaps;
- data collection;
- creation of a truth table of the presence/absence of individual variables;
- calculation of the parameters of the individual variables with logical minimization;
- interpretation of the obtained results.

The selection of the variables was based on the practical experience and professional skills of the authors of the contribution, which were subsequently discussed with all stakeholders. For each factor, a hypothesis, respectively an assumption was made, that argues and explains the relationships and contexts of the influence of a factor on the resulting indicator.

3 Results and Discussion

The analysis conducted showed that there are gaps in the formation of students' entrepreneurial skills in Ukrainian universities. This is primarily connected to the specific demands from businesses in a rapidly changing world. This is reflected in the Global Skills Index 2020 Report [12], which focuses on the skills that are most important for businesses and high-tech areas, that is, the skills of a modern entrepreneur as well as start-up participants.

The following entrepreneurial skills are currently considered important [13]: creative thinking; leadership; the ability to take risks; adherence to business ethics; as well as business storytelling; computer skills; critical thinking; flexibility; logical thinking; non-verbal communication; planning; working with social media; strategic planning; team building; time management; and trend setting. Oosterbeeket et al. [14], describe the ten most important determinants of successful entrepreneurship. These are: the need for achievement; need for autonomy; need for power; social orientation; self-efficacy; endurance; risk-taking propensity; market awareness; creativity; and flexibility.

The aforementioned key skills of a modern entrepreneur do not give an indication of which disciplines enable the formation of these skills and competencies at university. For example, creative thinking includes the development of emotional intelligence, the ability to understand root problems, the ability to brainstorm, and the ability to build a team. Similarly, adhering to business ethics presupposes a deep knowledge and understanding of business planning, project management, the ability to work independently, consistently, and purposefully. Entrepreneurial competencies are therefore ambiguous and include several personal characteristics, attitudes and skills of a particular person.

Entrepreneurial skills can also viewed in reverse, based on the most important elements of entrepreneurship learning [15]. In this case, a certain combination of knowledge, abilities, skills and competencies, consisting of awareness of the enterprise (start-up), the field of business (business skills in the form of accounting, business planning, etc.) and creativity, shape the effectiveness of entrepreneurial decisions and business in general. This combination of knowledge, skills and competencies makes entrepreneurship learning challenging (see Table 1), whereby the question of the best way to develop an entrepreneurial skills package within the constraints of higher education remains unresolved.

Table 1. Students' entrepreneurial skills and learning elements

Groups of competencies	Students' entrepreneurial skills	Learning elements
Common entrepreneurial competencies	To form and evaluate an idea To see problems as opportunities To identify key stakeholders To assess the needs for business development Emotional intelligence and business ethics	Business planning; Project management
Competencies connected with management	To know how to build relationships with stakeholders and manage them	Business storytelling; Business law; Team building
Competencies connected with entrepreneurship	Business plan development Business growth planning Setting standards and performance criteria Marketing analysis and sales strategies Choice of financial strategies Use of professional counselling skills	Operations management; Strategic planning; Accounting, audit, analysis of activities; Marketing; Risk management
Entrepreneurial values	Initiative Willingness to take responsibility Autonomy Intuitive decision making with limited information Ability to interact and persuade Result orientation Willingness to take risks	Participation in business games, hackathons, start-up, battles
Awareness of the entrepreneurial lifestyle	Understanding the advantages and disadvantages of being an entrepreneur versus being an employee	

Source: Authors.

It is the shared opinion of the authors that the presented way of grouping entrepreneurial skills and competencies is archaic. Modern understanding of knowledge, abilities, skills and competencies is associated with their division into hard and soft skills:

Entrepreneurial hard skills [16] consist of those competencies that enable the management of human resources, marketing, production, management and finances.

Entrepreneurial soft skills [17] consist of those competencies that enhance one of 13 dimensions: initiative; seeing and taking advantage of opportunities; perseverance; searching for information; focus on the high level of performance; commitment to the job; orientation towards efficiency; systematic planning; problem-solving; confidence; power of persuasion; influencing strategies; and assertiveness.

Of the entrepreneurial hard skills listed for non-business department students, the production element is the most significant in terms of its influences on the intention to become an entrepreneur [16, 18]. This is because this competence becomes the basis for choosing a professional role in a team with the participation of a student in business games, hackathons, start-up battles, during an internship or practice.

Of the main entrepreneurial soft skills listed for non-business department students, initiative, self-confidence and persistence are the most significant influences on entrepreneurial activity. They have a significant impact on the intention to become an entrepreneur.

Entrepreneurial intent is usually defined as the intention of an individual to start a new business venture [19]. Despite the importance of measuring intent in the entrepreneurial setting, universities are more interested in measuring learning and performance, given that they are expected to measure program learning outcomes [20].

As the entrepreneurial process is not linear, creativity and finding structure in an unstructured process are central [21]. Entrepreneurship is a complex process that lacks linearity of business thinking, which requires students to therefore master uncertain environments [22]. So, educators in entrepreneurship have the task of giving students the necessary skills with which to survive in a fast and rapidly changing environment. The development of entrepreneurial skills can be improved by providing a learning environment in which students interact with real businesses in live projects [23].

In Ukraine, the development of educational programs and curricula is carried out based on the national higher education standards [24]. Within the education system, these should act as the conduit for the development of entrepreneurial competencies in both non-business department students and students of economic and managerial specializations. Currently, 150 national standards of higher education for Bachelor's and Master's degrees in 105 specializations have been developed and implemented. An analysis thereof made it possible to identify common competencies in terms of soft skills (see Table 2).

Table 2. Students' entrepreneurial soft skills according to the Higher Education Standards of Ukraine

Level of education	Students' entrepreneurial soft skills	Learning elements
First level of higher education (Bachelor´s degree)	Ability to communicate in official and foreign languages both verbally and in writing Ability to communicate with representatives of other professional groups of different levels (with experts from other fields of knowledge/types of economic activity) Ability to generate new ideas (creativity); ability to work in a team Interpersonal skills Ability to adapt and act in a new situation Ability to identify, pose and solve problems	Native and foreign languages (profession-oriented) Participation in business games, hackathons, start-ups, battles
Second level of higher education (Master´s degree)	Ability to make informed decisions Ability to generate new ideas (creativity) Ability to develop and manage projects Ability to communicate in a foreign language The ability to motivate people and move towards a common goal	Project Management Foreign languages (profession-oriented)

Source: Authors.

It should be noted that "entrepreneurship" or "self-employment" is not found in the explanatory notes as a field of graduate employment. At the same time, consultancy activities in line with the branch of knowledge and specialization is noted present in 10% of Bachelor's degree standards and 15% of Master's degree standards.

Youth Business International (YBI), the global network of expert organizations, has developed four groups of entrepreneurial soft skills (see Table 3) that are currently the most valuable for young entrepreneurs and which need to be developed and strengthened in a changing world [25].

Table 3. YBI entrepreneurial soft skills framework

Skill group	Today's skills	Future skills
Intra- and interpersonal skills	Positive self-concept	Mindfulness (new) Self-awareness and self-realization (amplified)
	Communication skills	Business storytelling (new) Negotiating networks, crowds, and movements (amplified)
	Social and interpersonal skills	Working with networks, crowds and movements (new) Empathy and compassion (amplified) Emotional/social intelligence and influence (amplified)
Creative and problem-solve skills	Planning, organization, execution and management	Adaptive planning and management (new) Time management (amplified)
	Higher-order thinking skills	Systems thinking (new) Complex problem-solving (amplified)
	Creativity and innovation	Co-design (new) Co-creativity and innovation (amplified)
Information, opportunity, and risk management skills	Opportunity seeking, recognition and assessment	Building and managing online reputation (new)
	Information seeking and management	Complex information management (amplified)
	Coping with ambiguity, uncertainty and risk	Wellness management (new) VUCA coping skills (amplified)
Growth mindset skills	Persistence, tenacity and resilience	Grit (new)
	Flexibility and adaptability	Agility (new)
	Future orientation	Active learning skills (new) Motivation to continuously learn (amplified)

Source: [25].

It should be noted that modern Ukrainian entrepreneurs are already actively using entrepreneurial soft skills:

Mindfulness - most entrepreneurs keep accounting records of their activities, form and submit reports to the tax service and other regulatory authorities.

Working with networks, crowds and movements, and building and managing their online reputations. Even before the pandemic, most entrepreneurs built their company websites in order to create an active digital presence, attracting subscribers and delivering valuable content to them. This includes pages on the most popular social networks - Instagram, Facebook, Telegram - for interaction with their clients, for conducting market research, and for promoting or testing new business ideas.

The need for systems thinking is compensated for in the entrepreneurial environment by the growth of the consultancy activities of narrow specialists and the intensification of applied scientific research at universities.

As a result, some entrepreneurial soft skills have been strengthened as skills of today and ceased to be skills of the future. Within this context, when comparing the YBI entrepreneurial soft skills framework with that of the entrepreneurial soft skills required under the Higher Education Standards of Ukraine, it becomes very clear that the latest standards, some of which were only adopted in 2020, are hopelessly outdated and require updating in line with current trends and the challenges of a changing world.

Entrepreneurship education is a complex set of measures that affects all levels of education (including teacher training), all subjects, and all parties involved in the education process: policymakers, education administration, school administrations, teachers, students, neighbouring communities, etc. [26]. This realization makes it necessary to involve all stakeholders in the educational process.

The feasibility and necessity of teaching entrepreneurial skills to non-business department students is justified by Martin Lackéus [27]. He focuses on key entrepreneurial competencies like knowledge, skills and attitudes that affect a person's readiness and ability to become an entrepreneur:

Marketing skills – conducting market research; assessing the marketplace; marketing products and services; power of persuasion; getting people excited about your ideas; dealing with customers; communicating a vision.

Resource skills – creating a business plan; creating a financial plan; obtaining financing; securing access to resources.

Opportunity skills – recognizing and acting on business opportunities and other kinds of opportunities; product/service/concept development skills.

Interpersonal skills – leadership; motivating others; managing people; listening; resolving conflicts; socializing.

Learning skills – active learning; adapting to new situations; coping with uncertainty.

Strategic skills – setting priorities (goal setting) and focusing on goals; defining a vision; developing a strategy; identifying strategic partners.

These entrepreneurial skills can be formed among university students by focusing their activities on long-term teamwork to solve the real problems of external stakeholders. That is, setting complex educational tasks that are multidisciplinary in nature, that require social interaction for their implementation, and that make use of process focused and practice-based learning methods.

4 Conclusion

The basic competencies of a manager contribute to the successful performance of the relevant work activity. These competencies are usually assessed according to professional knowledge, such as acquired education and professional skills, which a manager acquires mainly through regular training and sufficient practice. The role of manager in an organization is also shaped by social maturity, which manifests itself through character traits. It is projected in the style of management and management work, in which the manager plays the role of a manager. The manager performs tasks in the usual way, repeatedly so. The individual style of managerial work is shaped by the personality of the manager, but also by the conditions in which the manager works, such as the team. In practice, there are ten typical characteristics a manager should have, namely vision, commitment, openness, integrity, creativity, assertiveness, generosity, modesty, justice, and a sense of humour.

Entrepreneurial skills, knowledge and mindset are crucial factors contributing to the growth of the economy and the benefit of society [28]. The focus in the educational process is to not only help students develop the skills to become fully fledged entrepreneurs but to present a universally applicable skills set.

Currently, the formation of entrepreneurial skills in students needs to focus on the ability to take risks, the ability to quickly adapt to changing environmental conditions, and the ability to interact with stakeholders, including remotely through networks, crowds and movements. These entrepreneurial soft skills should be formed

https://doi.org/10.1051/shsconf/20219002009

at the first level of higher education (Bachelor´s degree). With regards to the formation of entrepreneurial hard skills at the Bachelor´s degree level, this mainly concerns business plan development and business growth. This includes basic planning, organization, marketing, sales and finance skills. Further entrepreneurial soft skills at the Bachelor´s degree level should include business storytelling, and for entrepreneurial hard skills, adaptive planning and management. By focusing on these areas, the education system will support the integration of Bachelor´s degree graduates from non-business departments into the job market, not only as employees, but also as entrepreneurs.

For the second level of higher education (Master´s degree), analytical skills should be refocused on overcoming resource constraints and maintaining social/community values in a competitive environment. The current emphasis on the awareness of the regulatory aspects of entrepreneurial activity needs to also shift towards creativity, creative thinking, and non-standard management decision-making in order to adapt to the challenges of a changing world.

References

1. I. Krakovetskaya, A. Dalibozhko, I. Slesarenko, Forming hard and soft skills for university graduates employability enhancement. *Proceedings of the 31st International Business Information Management Association Conference, IBIMA 2018: Innovation Management and Education Excellence through Vision 2020*. 2454-2466 (2018)
2. A. Dardiri, I. Alfianto, T. Kuncoro, H. Usman, Sugiono, Entrepreneurial leadership for excellent technical and vocational education institutions. *World Transactions on Engineering and Technology Education*. **16**(4), 441–447 (2018)
3. Y. M. L. Malaikosa, N. Ulfatin, N. S. Degeng, Maisyaroh, Curriculum management and learning with creative economics to shape life skills. *International Journal of Innovation, Creativity and Change*. **12**(12), 444–456 (2020)
4. S. Tomy, E. Pardede, An entrepreneurial intention model focusing on higher education. *International Journal of Entrepreneurial Behavior & Research*. **26**(7), 1423-1447 (2020)
5. J. Turner, G. Mulholland, Enterprise education: Towards a framework for effective engagement with the learners of today. *Journal of Management Development*. **36**(6), 801–816 (2017)
6. C. Rowe, Clarifying the use of competence and competency models in recruitment, assessment and staff development. *Industrial and Commercial training*. **27**(11), 12–17 (1995)
7. A. Gonzi, Competency-based learning: a dubious past-an assured future? In D. Boud, J. Garrick, (Ed.) *Understanding learning at work*. Londres: Routledge, 180–194 (1999)
8. H. Brandao, J. Borges-Andrade, Causas e efeitos da expressão de competências no trabalho: para entendermelhor a noção de competência [Causes and effects of the expression of skills at work: to better understand the notion of competence]. *Revista de Administração Mackenzie*. **8**(3), 32–49 (2008)
9. Ukrainian Venture Capital & Private Equity Association. Access to Information, Ukrainian Venture *Capital and Private Equity Overview 2019*. [online]. Available at: http://uvca.eu/ua/news/investments-into-ukrainian-startups-in-2019-overview (2020)
10. Ministry of Education and Science of Ukraine, *Access to Information, Presentation of the Entrepreneurial University Initiative*. [online]. Available at: https://mon.gov.ua/ua/events/22-veresnya-onlajn-prezentaciya-iniciativi-pidpriyemnickij-universitet (2020)
11. L. Gulyayeva, L. Zhuk, Startap-dvizheniye v Ukraine kak innovatsionnaya model' dlya ustoychivogo razvitiya biznesa [Start-up movement in Ukraine as an innovative model for sustainable business development]. *Ikonomika na ustoychivoto razvitiye*. **2**, 21–30 (2016)
12. Coursera. *Access to Information, Global Skills Index 2020* [online]. Available at: https://www.coursera.org/gsi (2020)
13. A. Doyle, Essential Skills for Entrepreneurs. *Access to Information, The Balance Careers*. [online]. Available at: https://www.thebalancecareers.com/list-of-skills-entrepreneurs-need-2062391 (2020)
14. H. Oosterbeek, M. VanPraag, A. Ijsselstein, The impact of entrepreneurship education on entrepreneurship skills and motivation. *European Economic Review*. **54**(3), 442–454 (2010)
15. A. I. Frank, Entrepreneurship and enterprise skills: A missing element of planning education? *Planning Education & Skills*. **22**, 635–648 (2007)

16. C. M. Chou, C. H, Shen, H. C. Hsiao, S. C. Chen, A Study on Constructing Entrepreneurial competence indicators for business department students of vocational and technical colleges in Taiwan. *World Transactions on Engineering and Technology Education*. **8**(3), 316–320 (2010)

17. L. M. Spencer, *Competence at work: Models for superior performance*. New Jersey: John Wiley Sons Inc. (1993)

18. B. P. D. Riyanti, Ch. W. Sandroto, M. T. Warmiyati D.W., Soft Skill Competencies, Hard Skill Competencies, and Intention to Become Entrepreneur of Vocational Graduates. *International Research Journal of Business Studies*. **9**(2), 119–132 (2016)

19. N. Krueger, Entrepreneurial Intentions are Dead: Long Live Entrepreneurial Intentions. In A. Carsrud, M. Brännback (eds). *Understanding the Entrepreneurial Mind. International Studies in Entrepreneurship*. New York: Springer (2009)

20. N. Mozahem, R. Adlouni, Using Entrepreneurial Self-Efficacy as an Indirect Measure of Entrepreneurial Education. *The International Journal of Management Education*. In Press. (2020)

21. G. Linton, M. Klinton, University entrepreneurship education: a design thinking approach to learning. *Journal of Innovations and Entrepreneurship*. **8** (2019)

22. H. M. Neck, P. G. Greene, Entrepreneurship education: Known worlds and new Frontiers. *Journal of Small Business Management*. **49**(1), 55–70 (2011)

23. J. Chang, A. Rieple, Assessing students' entrepreneurial skills development in live projects. *Journal of Small Business and Enterprise Development*. **20**(1), 225–241 (2013)

24. Ministry of Education and Science of Ukraine, *Access to Information, Higher Education Standards*. [online]. Available at: https://mon.gov.ua/ua/osvita/visha-osvita/naukovo-metodichna-rada-ministerstva-osviti-i-nauki-ukrayini/zatverdzheni-standarti-vishoyi-osviti (2020)

25. Youth Business International (YBI). *Access to Information, Entrepreneurial soft skills for the future: a scoping study*. [online]. Available at: https://www.youthbusiness.org/file_uploads/YBI-SoftSkillsForTheFuture.pdf (2019)

26. V. Bikse, I. Riemere, The Development of Entrepreneurial Competences for Students of Mathematics and the Science Subjects: The Latvian Experience. *Procedia - Social and Behavioural Sciences*. **82**, 511–519 (2013)

27. M. Lackéus, *Entrepreneurship in education* [online]. Available at: https://www.oecd.org/cfe/leed/BGP_Entrepreneurship-in-Education.pdf (2015)

28. Y. Maaravi, B. Heller, S. Amar, H. Stav, Training techniques for entrepreneurial value creation. *Entrepreneurship Education*. **3**, 215–238 (2020)

Assessment of Conflict Resolution Training at University

Nadezhda Sivrikova[1,*], *Tatyana* Ptashko[1], *Elena* Chernikova[1], *Artem* Perebeynos[1], *Vitaly* Kharlanov[2]

[1] South Ural State Humanitarian Pedagogical University, Department of Social Work, Pedagogy and Psychology, Chelyabinsk, Russian Federation, Lenin prospect 69, Russia
[2] South Ural State University, Legal institute, Department of Theory of State and Law, Constitutional and Administrative Law, Russian Federation, Chelyabinsk, Lenin prospect 76, Russia

Abstract. The growth of conflicts in a changing world makes conflict resolution competence an important component of entrepreneurial competencies. Higher education should empower students to effectively prevent and resolve conflicts. The problem of assessing the educational process in the field of conflict resolution training and the result of this training is discussed in this article. The authors cite the results of an empirical study, the purpose of which was to determine the needs of students of the State University (Chelyabinsk, Russia) with regards to conflict management preparation. The survey relied on an analysis of the self-assessment of students' own skills and commitment to the principles of tolerance, an assessment of the educational process, and the need to increase their own conflict management competence. 152 students of the State University took part in the survey, of which 63 were male and 89 female. The Likert scale was used. The results of the study showed that students assess the level of their own conflict resolution competence as high, as do they the educational process at the university. The limitations of the study are discussed at the end of the article.
Keywords: conflict resolution competence, higher education, conflict management, conflict resolution training, entrepreneurial competencies

1 Introduction

The existence of objective contradictions in the educational process makes it inevitable that conflicts arise between the participants of the educational process [1]. Conflict management is an essential component of teaching. It is only through it that the peaceful coexistence of the subjects of the educational process can be ensured [2], the conflict resolution competence of children develop [3], and the psychological climate in schools be favourable [3]. Therefore, professionals who need to meet the modern requirements of the labour market should be ready for conflict prevention and resolution [4].

The need for conflict management training is raised by employers, students and teachers alike. Competencies related to conflict management are present in all professions as standard. Social competence, which includes the ability to manage conflicts, is at the top of the list of key competencies needed to successfully operate as a professional in any field [5]. Students in many assisting professions speak of their desire to increase their own training in conflict resolution, e.g. health workers [6, 7], staff managers [8] and teachers [2, 9, 10]. However, there is a lack of data on how existing conflict management training at universities is assessed. Therefore, the purpose of the research presented in this paper was to study how students at the State University (Chelyabinsk, Russia) assessment their conflict management training.

In scientific literature, the problems of assessing the quality of conflict management training are presented at different levels. As part of the discussion of the prospects for the development of vocational education, the ability to establish educational programs to ensure the formation of a conflict management culture is analysed [5], as is the conflict management competence [1] of future specialists. Scientists also discuss the effective conflict management techniques that a teacher should possess [4, 9, 10]. They recognize that teachers should be encouraged, as a strategic training line, to acquire opportunities to properly manage conflict and social relations [2].

Studies on the development of courses and programs to develop conflict competence indicate a low level of competence in the field of conflict management if there is no focused work in this direction [10–12]. Researchers who compared the level of conflict culture in students of different specialities came to a similar conclusion [5]. A more optimistic picture is presented in the results of student self-assessments with regards to the formation of their knowledge and skills in the field of conflict resolution [13]. These studies conclude that there is a need to improve conflict management training programs at universities, too.

* Corresponding author: bobronv@cspu.ru

The level of conflict management training is assessed through qualitative studies aimed at analysing problems and ways of resolving conflicts used by practitioners [14–16].

The assessment of the quality of student training in conflict resolution is therefore based on different methodological principles. However, it is generally not an independent focus of study. Findings on the need to improve conflict management training are based on indirect data. Therefore, it is important to study students' views on the quality of their preparation for conflict resolution.

The assessment of the conflict resolution training of future specialists is comprehensive. This is because being prepared for conflicts in itself represents a complex phenomenon. Bisembayeva believes that its structure includes: conflict knowledge, skills, mastery of diagnostic techniques, forecasting, prevention and resolution of conflicts [12]. Other researchers from Kazakhstan describe the structure of conflict competence as a set of personal, motivational, meaningful, behavioural and reflexive components [11]. Spanish researchers view this from a classical competency structure that includes knowledge, ability and skills [2].

Many researchers view conflict behaviour strategies as an important indicator of conflict resolution training. They believe that constructive behaviour in a conflict situation can contribute to achieving a higher level of development of the cooperating parties. Researchers have shown that mediators are more likely to choose pro-social conflict resolution strategies, and their colleagues without special training are more likely to choose a care strategy [17].

In addition to assessing the level of knowledge, ability and skills in the field of conflict management, the assessment of conflict resolution training includes taking into account two other criteria: compliance of the educational program with the tasks of conflict readiness preparation [1] and the need of students for conflict resolution training [2].

As a result, three criteria for assessing conflict resolution training in university are considered in this study. These are a self-assessment of the level of strength of personal skills and abilities related to conflict management, compliance of the educational program with the tasks of conflict readiness preparation, and the need for conflict resolution training.

The aim of the study was to determine the need of university students for training in conflict management.

2 Methodology

The survey of students of the Department of Pedagogics of the State University (Chelyabinsk, Russia) was conducted in 2020. In total, 152 students (63 male and 89 female) participated in the survey. The distribution of the participants according to study year is shown in Table 1.

Table 1. Characteristics of the research participants (n=152)

Year of study	Number of people (%)
2	56.9%
3	19.6%
4	17%
5	7.8%

Source: Authors.

The survey consisted of 17 questions, each with a 5-point response scale (1. Completely disagree; 2. Disagree; 3. I don't know; 4. Agree; 5. Completely agree). The developed tool enabled the evaluation of three variables: 1. Representation of conflict training elements in training; 2. Self-assessment of the skills and abilities required for effective conflict resolution; 3. Need for conflict training. The Alpha Cronbach correlation coefficient was used to determine the reliability (internal consistency) of the developed instrument. The coefficients for the different questionnaire blocks ranged from 0.938 to 0.716, which is acceptable.

A descriptive and quantitative analysis of the information collected was carried out. IBM SPSS Statistics 23 software was used for mathematical processing of the results.

3 Results and Discussions

The findings show (see Table 2) that, in general, students believe their own conflict resolution skills to be high (sample averages approach 4 points on the Likert scale).

The participants of the study assessed the level of their own skills with regards to a) recognising the meanings of messages depending on the context of the communication situation, and b) predicting the actions of people on the basis of the analysis of real situations, below that of other skills. Despite this, the students still rated their skills as more than satisfactory. Students believe that they possess the skills to perceive and interpret the personality of others, and that these skills are more developed than their other skills.

Students rate their own commitment to the principles of tolerance quite highly (see Table 3). Sample averages approach 4 points.

Students who took part in the study rate their own internal resilience below other principles. The largest number of students noted that they fully agreed that they followed the principle of non-violence (80%).

Students highly appreciate the system of conflict resolution training at the university (see Table 4). They partially (40.6%) or completely (48.4%) agree that the educational process at the university prepares students for the employment functions related to the resolution of conflicts. At the same time, only 21.3% indicated that they have no need to increase their own conflict resolution competence, with up to 74.8% stating they do.

Table 2. Results of students´ self-assessment of conflict resolution skills (n=152)

Studied parameters	Likert response option (%)					mean value	standard deviation
	1	2	3	4	5		
Adequacy of perception and interpretation of another's personality	0.6	2.6	4.5	56.8	35.5	4.24	0.72
Adequacy of evaluation of emotional state by non-verbal reactions	0.6	4.5	17.4	46.5	31.0	4.03	0.86
Recognition of the meaning of verbal messages depending on the context of the communication situation	1.3	3.9	27.7	52.3	14.8	3.76	0.80
Predicting people's actions based on real situation analysis	0.0	7.1	17.4	57.4	18.1	3.86	0.80
Reflexive hearing	0.6	6.5	20.6	40.0	32.3	3.96	0.92
Analysis and evaluation of actions of conflicting parties	0.0	3.9	16.1	42.6	37.4	4.12	0.82
Making responsible and informed decisions	0.0	5.2	12.9	47.1	34.8	4.12	0.83

Source: Authors.

Table 3. Results of students´ self-assessment of acceptance of the principles of tolerance and non-violence (n=152)

Studied parameters	Likert response option (%)					Mean value	Standard deviation
	1	2	3	4	5		
Position as equals	0.6	3.9	16.8	45.8	32.9	4.07	0.85
Refusal of violence	2.6	3.2	6.5	7.7	80.0	4.59	0.94
Submission to rules	1.9	3.2	11.0	35.5	48.4	4.24	0.92
Positive purposes	1.9	2.6	11.6	36.1	47.7	4.25	0.91
Internal stability	4.5	3.2	21.9	38.1	32.3	3.91	1.04
Personal choice	1.9	1.9	20.6	34.2	41.3	4.10	0.93

Source: Authors.

Table 4. Students´ self-assessment of system of conflict resolution training and need to increase preparedness (n=152)

Studied parameters	Likert Response Option (%)					Mean value	Standard deviation
	1	2	3	4	5		
Assessment of the conflict training system at the university	1.9	2.6	6.5	40.6	48.4	4.31	0.86
Need to increase conflict resolution competence	21.3	0.0	1.3	2.6	74.8	4.1	0.50

Source: Authors.

4 Conclusion

The results of the study make it possible to conclude that students of the State University assess the level of their own conflict resolution competence as high, as they do the elements of the educational process at the university. However, at the same time, many of them feel that there is room for improvement on this front.

It should be noted that the study presented has a number of limitations. First, the conclusions are drawn from an analysis of student self-assessments, which are subject to social desirability factors. This is especially true for assessing the level of conflict competence.

Secondly, students of different courses took part in the presented study, and the studied variables are dynamic and vary depending on the degree of mastery of the educational program.

The study also does not analyse the profession being mastered by the respondents.

Despite the aforementioned limitations, the results of the study contribute to the evaluation of conflict resolution training at the university. They can serve as a basis for changing existing or developing new educational programs.

References

1. M. G. Sergeeva, N. M. Simonova, E. Y. Pryazhnikova, Y. A. Volkova, and T. M. Rozhnova, Technology of pedagogical conflict management within an educational institution. *Revista Inclusiones.* **7**, 217–228 (2020)
2. M. C. Lopez Lopez, R. Tejero Olmedo, and A. Fernandez Herreria, School conflict and its management in initial teacher training. A case study. *Profesorado-Revista De Curriculum Y Formacion De Profesorado.* **21**, 293–314 (2017)
3. L. García-Raga, R. M. B. Bonet, and J. M. Lasagabaster, Meaning and sense of school mediation based on high school mediators' views. *Revista Espanola de Orientacion y Psicopedagogia.* **29**, 79–93 (2018)
4. S. A. Djumaeva, Technology for the Elimination of Conflicts in the Process of Higher Education by Means of Pedagogical Correction. *International Journal of Innovative Technology and Exploring Engineering.* **9**, 2670–2676 (2020)
5. O. Shcherbakova and A. Tatarinceva, Conflictological culture of personality in a professional environment (by the example of pedagogues and engineers). *Society. Integration. Education: Proceedings of the International Scientific Conference.* **1**, 524 (2016)
6. F. K. Cheng, Mediation skills for conflict resolution in nursing education. *Nurse Education in Practice.* **15** (2015)
7. S. Vandergoot, A. Sarris, N. Kirby, and H. Ward, Exploring undergraduate students' attitudes towards interprofessional learning, motivation-to-learn, and perceived impact of learning conflict resolution skills. *J. Interprof. Care.* **32**, 211–219 (2018)
8. E. A. Karpova, Development prospects of conflictological competence. *Akmeology*. 248–256 (2015)
9. Y. del C. Camacho Sojo, Mediation in conflicts: challenges in the on-going education of elementary school teachers. *Revista de Mediación.* **11** (2018)
10. N. Gunduz, H. Uzunboylu, and D. Ozcan, Developing and testing a scale to assess teachers' attitudes toward peer mediation of student disputes. *Social Behavior and Personality.* **45**, 1745–1760 (2017)
11. M. B. Alpysbayeva, Z. A. Karmanova, V. V. Bobrova, A. M. Sharzadin, and R. B. Mazhenova, Development of conflict management competency in social work undergraduate students. *Science for Education Today.* **9**, 58–80 (2019)
12. K. Bisembayeva, N. V. Sivrikova, Formation of conflictological competence of future educational psychologists. *Herald of South Ural State Humanitarian Pedagogical University bulletin.* 147–154 (2018)
13. P. Iranzo-García, M. Camarero-Figuerola, C. Barrios-Arós, J. M. a. Tierno-García, S. Gilabert-Medina, What do teachers think about school leadership competences and their pre-service training?. *REICE. Revista Iberoamericana Sobre Calidad, Eficacia y Cambio en Educacion.* **16**, 29–48 (2018)
14. E. M. Blunk, E. M. Russell, C. J. Armga, The role of teachers in peer conflict: implications for teacher reflections. *Teacher Development.* **21**, 597–608 (2017)
15. G. Ciuladiene, R. Raudeliunaite, Peer mediation as tool to resolve school based conflitcs: the opinion of an school staff (case study). *EDULEARN Proceedings: 8th International Conference on Education and New Learning Technologies.* **16**, (2016)
16. E. P. da Costa, S. Sá, Teacher Narratives on the Practice of Conflict Mediation. *Advances in Intelligent Systems and Computing.* **861**, (2019)
17. R. Pulido, T. Fajardo, L. Pleguezuelos, R. De Gregorio, and P. Educativa, School mediation in the region of Madrid: analyzing the impact of education on teachers and students at the 'Las Americas' high school in Parla. *Revista de Mediación.* **3**, 32–43 (2010)

Current Managerial Competencies in Strategic Management and Decision-Making in the Czech Republic

Jarmila Straková[1,*], *Jan* Váchal[1], *Filip* Čaniga[1]

[1]Institute of Technology and Business, Department of Management, Okružní 517/10, 37001 České Budějovice, Czech Republic

Abstract. Despite the increasingly strong paradigm of management, which is especially visible in the corporate management sector, there have been changes in the perception of strategic management and decision-making. An important indicator of this process is the level or the application of managerial competencies in practice. Extensive research, which was started by the Department of Management of the Institute of Technology and Business in České Budějovice in 2016, and is still going on, has shown the lack of literacy of managers in this field, especially in the case of older managers. From the perspective of objective scientific knowledge, it can be stated that in scientific and research terms, there is a significant deficit in the corporate sector; the transfer of new knowledge in practice is stagnant or non-existent. This makes the situation all the more urgent, as managers who possess knowledge of this issue consider their knowledge to be obsolete and only partly applicable in terms of their needs for business management. The basic prerequisite to remedy this unfavourable situation consists in knowing the actual situation of individual companies of various sizes and the sectors of the national economy. The submitted contribution presents partial results in terms of the knowledge and application of internal and external strategic analyses by companies operating in the Czech Republic. The research included a set of 456 companies from the entire Czech Republic (109 micro-companies, 140 small companies, 124 medium-sized companies, and 83 large companies). The results revealed a critical state in terms of managerial competencies and skills in companies in the Czech Republic concerning the knowledge and application of strategic situation analysis methods. More than 50% of managers appeared to have only minimum knowledge of this issue.
Keywords: managerial competencies; strategic management; strategic decision-making; strategic analyses

1 Introduction

The issue of strategic management and decision-making has been perceived as important by the corporate sector since 1970s. In the last few decades, the international business environment has changed dramatically (development of transport infrastructure, improvements in production processes, reduction of trade barriers, more efficient business communications, increasing number of people with international business experience, etc.). Currently, strategic management is an integral part of each business´ management regardless of its size or specialization.

With the growing dynamics and development of the corporate environment, there are growing demands being placed on the managements of business entities, as well as on the managers´ skills and abilities. The existing principles, rules, and the ways managers act and make decisions are failing; new approaches and methods to solve problems are being sought. The voices of business owners and managers calling for uniqueness and originality in close relation to a competitive business strategy, or its creation and implementation, are being heard more and more often. A successful business strategy guarantees the competitiveness, efficiency and long-term sustainability of a company. It should be stated that the scientific and research platform, which includes universities, does not fully ensure the needs of corporate practice.

The ideas and especially the practical instructions of Michael Porter [1-4], the man behind the modern concept of corporate competitiveness, i.e. competitive advantage and competitive strategy, are not known in Czech corporate practice. The principles declared in Porter´s publications, e.g. the statement that the essence of competition is not to beat the competition but to provide the customer with unique, original value in the form of a unique product [5], remain unknown to Czech managers, with exception to top businesses or businesses under foreign ownership. Achieving and sustaining a competitive position is conditioned by the strategic management and decision making on the part of business owners and top managers, who are confronted with the reality of a competitive market every day, and who have to solve issues in accordance with the interests of their businesses, knowledge and skills. Ignorance of the above principles for creating business strategy, or their neglect in business practice, means a clear handicap for creating and running a competitive business, which may play a very negative role in ensuring the stability and development of Czech companies, especially in times of crisis.

* Corresponding author: strakova@mail.vstecb.cz

As strategic management is developing as an independent scientific discipline, the requirements being placed on managers at various levels, from the operational to the highest, are increasing. Strategists should be active players, not only consultants; they should not focus on the long-term future only; they should be those who influence, not give commands [6-7]. The basic managerial competencies may include knowledge of suitable analytical tools or strategic methods of situation analysis in order to know the internal resources of a company and to analyse the mezzo-environment, or the nearest corporate environment, and the macro environment [8-9]. The survey conducted by the authors of this contribution showed that the issue of situation analysis from the perspective of knowledge and application of its methods when formulating business strategy remains a new and unknown issue for many managers [10]. A similar opinion is expressed by authors such as Kislingerová and Nový [11] and Mallya [12]. It is a fact that in strategic management and planning, many companies use a high degree of intuition, practical knowledge and skills of successful managers; in many cases, certain models of strategic management are taken over from foreign companies without deeper theoretical knowledge and implementation skills. The results achieved in companies managed in this way do not correspond with the original assumptions of top management.

As a result of the aforementioned paradigm, the decision-making of managers is becoming more and more difficult and demanding with regards to preparatory work, and especially analytical work. Managers, especially managers at the top level, make decisions in a highly uncertain environment, with a high degree of uncertainty and business risk. Only a very small part of management has the ability to predict the future development of a company [13]. According to Keřkovský and Vykypěl [14], one of the prerequisites for successful strategic decisions of managers is strategic situation analysis or the evaluation of business environment factors, that is, the requirements and changes in the behaviour of customers, competitors, suppliers, as well as the development of macroeconomic factors including the characteristics of internal corporate resources [15]. Jakubíková [16], Dedouchová [17], Mallya [12], Šulák and Vacík [18], Frynas and Mellahi [19], Kislingerová and Nový [11], Veber et al. [20], Gomes [21], Nigel and Campbell [22], Ginter, Duncan and Swayn [23] have a similar opinion on adequate knowledge of strategic management and decision-making tools. According to Jakubíková [16], strategic situation analysis is a complex approach capturing the important factors affecting the activities of companies in context and whose results are a basis for creating strategies. Despite the considerable reservations of managers in the corporate sector concerning the methods of situation analysis, usually due to their large number and often missing methodology, the knowledge of mainly the methods that are, according to M. Porter, necessary for the creation of business strategy (analysis of the internal structure of the sector, value chain), and, as the authors of the contribution believe, the analysis of internal corporate resources, are essential. A similar opinion is expressed by other authors [24-25], who state that within strategic management, it is recommended to use some analyses taken from situation analysis, especially the Porter´s Five Forces model, value chain, or strategic map of the sector; other analyses should be rather used as a supporting tool for developing a corporate strategy. This assumption is based on the fact that when choosing suitable analyses, it is necessary to consider their structure and contents, as some of them do not have a universal character and cannot therefore be used in a unified way [26-28].

The authors of this contribution believe that knowledge of suitable situation analysis methods is one of the basic managerial competencies that managers, especially top managers, should possess [10, 29-30]. This opinion is supported by the partial results obtained by means of a research survey. As documented in the presented results of the survey conducted among Czech managers, knowledge of strategic methods for analysing the internal and external environment are greatly lacking. The task of academia is to solve this situation mainly through the accelerated transfer of new knowledge to the corporate sector.

2 Methodology

The research was carried out on a sample of 456 companies from the whole Czech Republic: 109 micro-companies (up to 10 employees), 140 small companies (up to 50 employees), 124 medium-sized companies (up to 250 employees) and 83 large companies (with more than 250 employees). Besides the classification by size, the companies were classified by sector (sector differentiation of the CR) into companies operating in the manufacturing and industry sector (201) and in the services sector (255). The test sample of companies was created in cooperation with the branch of the Czech Statistical Office in České Budějovice, as well as the Czech Statistical Office in Prague in order to ensure its representativeness (representation of all size categories and relevant sectors). The objective of the contribution is to inform about the current state of managerial knowledge and skills in the area of strategic situation analysis in the corporate sector of the Czech Republic.

For data collection, a special team of interviewers was set up. The interviewers were trained to be able to ensure the collection of relevant and objective data; they were provided with information material for top managements, whereby the importance of the issue being addressed was emphasized and the request to address a top manager of the company involved in the research was formulated. The form of the research was a broadly focused questionnaire survey, which included basic issues concerning corporate strategic management and decision-making processes. The form of the questions enabled their mathematical and statistical processing. The return rate of the questionnaire survey was high (about 85%). The participating companies were mostly from the Region of South Bohemia; the test sample, however, included companies from the whole the Czech Republic.

The responses to the questionnaire survey and selected statistical analyses are used to assess the importance of the world´s best-known methods of strategic situation analysis (both internal and external) and to determine the dependence of business strategy on the individual analysed methods according to the size and sector categorization of the participating companies. Pearson´s chi-squared test was used to determine the dependence / independence of the selected categorical variables (methods of strategic situation analysis).

Formula 1. Pearson´s chi-squared test of independence

$$x^2 = \sum_{i=1}^{r}\sum_{j=1}^{c}\frac{(n_{ij}-e_{ij})^2}{e_{ij}} \quad (1)$$

where:

$\chi 2$ chi-square

Xi empirical frequencies (actual)

Npi theoretical frequencies (expected)

The tests consider two hypotheses, the zero hypothesis (H_0) and the alternative hypothesis (H_A). H_0 - there is no dependence between the selected variables. H_A – there is dependence between the selected variables. The evaluation criterion for the confirmation or rejection of the hypotheses in p-value, which achieves the values from the interval of <0.1>. In case the resulting p-value < α (at the determined significance level of $\alpha = 0.05$), H_A is confirmed.

3 Results and Discussion

Tables 1 and 2 show the dependence of business strategy creation and formulation on the use of internal and external strategic methods according to company size and the sector they operate in. In the case of internal analyses (Table 1), strong dependence was proved in the "Analysis of value chain" from the perspective of both tested sectors (manufacturing and industry sector and services sector). This result confirms the previously declared importance of this method for the formulation of sector differentiation of companies, since the tested sectors are completely different in their nature, which indicates the significantly differentiated importance of the individual value-creating processes.

No dependence was proved in the case of other methods, neither in relation to company size nor sector classification.

Table 1. Dependence of business strategy creation on internal strategic analyses according to company size and sector

	Company specialization	Company size
Internal strategic analyses	p-value	p-value
SWOT analysis	0.8921	0.4663
Portfolio analysis	0.3586	0.8461
Analysis of internal company resources	0.3162	0.4976
Key success factors	0.9874	0.1927
BCG matrix	0.871	0.4078
Analysis of value chain	**0.008428**	0.1927
SPACE analysis	0.913	0.3004
Analysis of company exposure	0.6613	0.9451
Model GE	0.6139	0.7167

Source: Authors.

https://doi.org/10.1051/shsconf/20219002011

Table 2. Dependence of business strategy formulation on external strategic analyses according to company size and sector

External strategic analyses	Company specialization	Company size
	p-value	p-value
Analysis of competitors	0.6514	0.1242
Analysis of industry internal structure – Porter´s Five Forces	0.7713	0.671
STEP analysis	0.8112	0.3615
Strategic maps	0.198	0.437
Scenarios (predictions of external corporate environment development)	0.429	0.6221
Analysis of industry attractiveness	0.1391	0.8971
Analysis of economic characteristics of industry	0.935	0.7879
Analysis of interest groups	0.8808	**0.04944**
Analysis of field lifecycle	0.5083	0.4595
Analysis of industry drivers	0.4083	**0.01219**
Strategic clock	0.8145	0.3031

Source: Authors.

In the case of external strategic analyses (see Table 2), dependence was confirmed in terms of size classification, specifically in the case of "Analyses of interest groups" and "Analyses of industry drivers". In addition to the analysis of interest groups, which undoubtedly has an influence on the creation of business strategy, the authors expected dependence to be proved in the case of the analysis of industry internal structure, which is, along with the analysis of value chain, considered to be the basic analytical apparatus for the creation of successful business strategy.

The graphs in Figures 1 – 8 show the results of managers´ opinions on the application of external and internal strategic analyses according to company size and sector. The results show that none of the strategic situation analysis methods is used by more than 50% of the companies examined (from all tested categories). This could be considered a critical situation in terms of managerial skills and knowledge concerning strategic situation analysis in Czech companies. The results are a warning signal in terms of the ongoing innovation process. With this level of knowledge and degree of application of strategic situation analysis in Czech companies, a positive course can hardly be expected, especially its influence on the intensification and profitability of production. The findings also indicate possible problems in the implementation of digitization in the management of Czech companies.

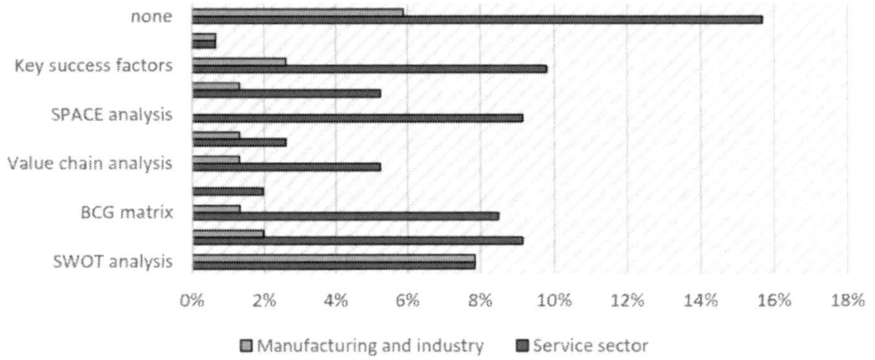

Figure 1. Use of internal strategic analysis by owners of micro-companies

Source: Authors.

https://doi.org/10.1051/shsconf/20219002011

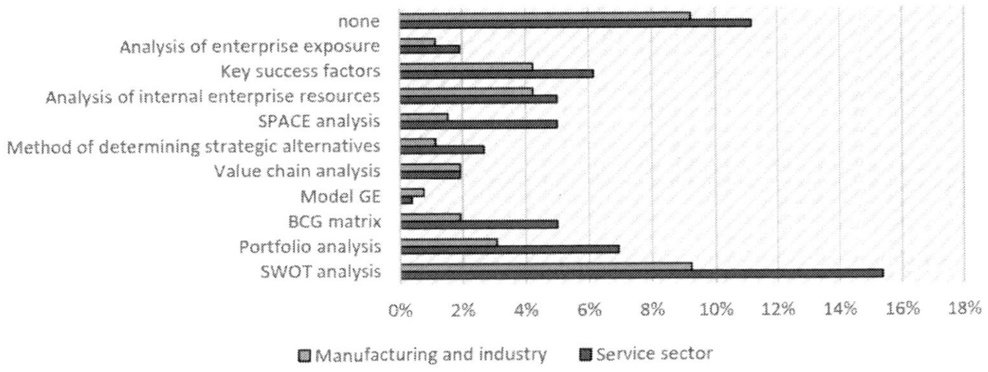

Figure 2. Use of internal strategic analyses by owners of small companies

Source: Authors.

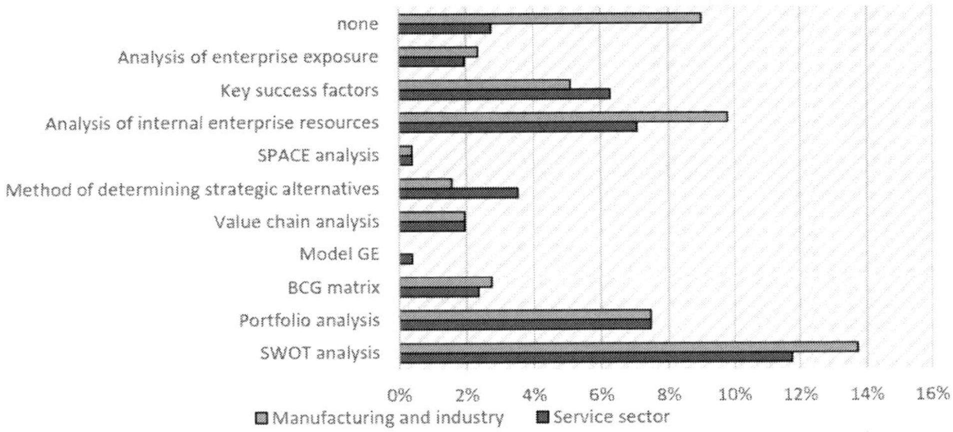

Figure 3. Use of internal strategic analyses by managers of medium-sized companies

Source: Authors.

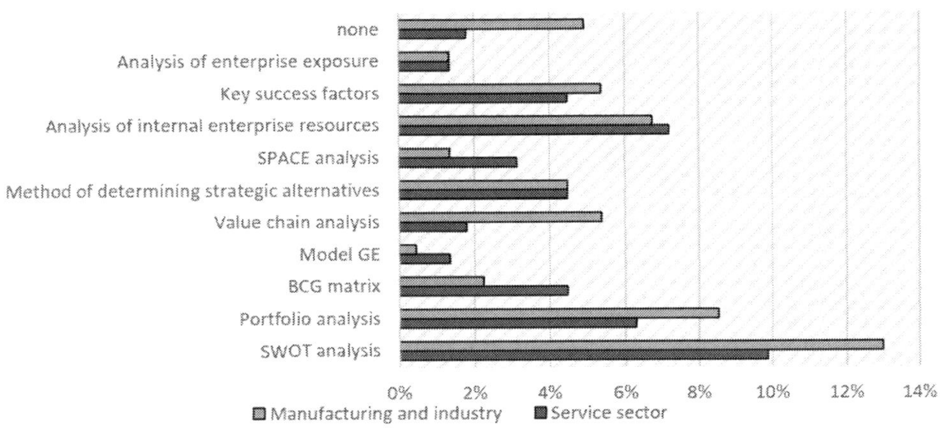

Figure 4. Use of internal strategic analyses by managers of large companies

Source: Authors.

https://doi.org/10.1051/shsconf/20219002011

As mentioned above, Figures 1 – 4 show the low level of knowledge and use of internal strategic analyses. In the case of micro-companies, this is evident in the services sector, where the "Analysis of key success factors" prevails, while in manufacturing companies, this is the knowledge and use of the world´s best known "SWOT analysis", "Analysis of internal corporate resources", and "Portfolio analysis". However, the results achieved are highly alarming from the point of view of managerial competencies.

In the case of large companies, the results are slightly more favourable. In this category, where the knowledge and competencies should be most evident, a changing trend in the creation of corporate strategy and the weakening influence of strategic situation analysis, as described by several Czech and foreign authors, can be observed.

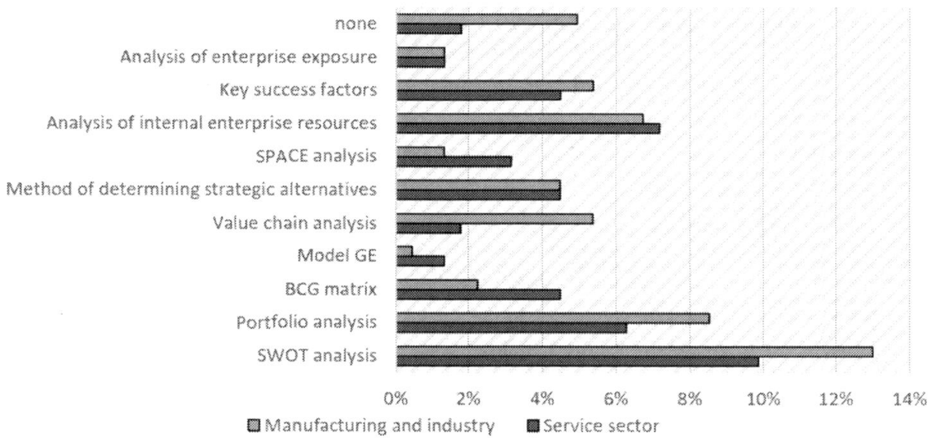

Figure 5. Use of external strategic analyses by owners of micro-companies

Source: Authors.

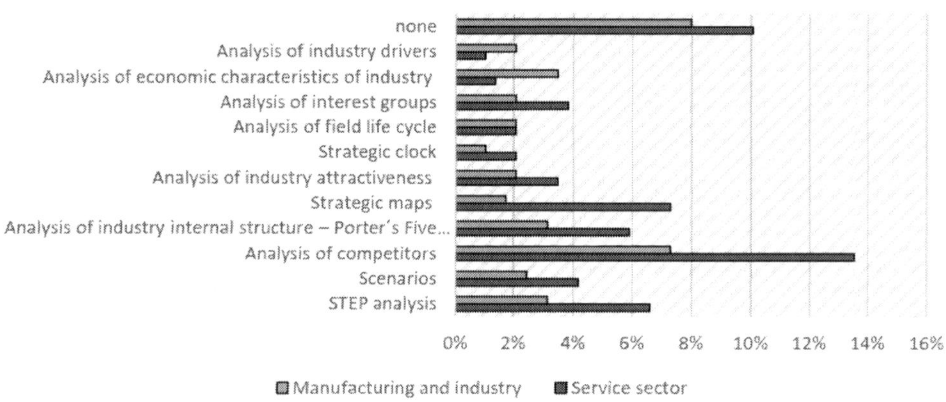

Figure 6. Use of external strategic analyses by owners of small companies

Source: Authors.

https://doi.org/10.1051/shsconf/20219002011

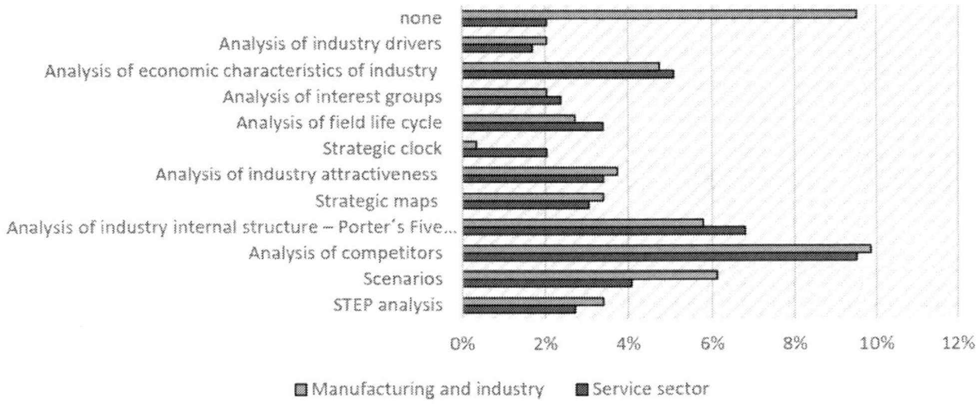

Figure 7. Use of external strategic analyses by managers of medium-sized companies

Source: Authors.

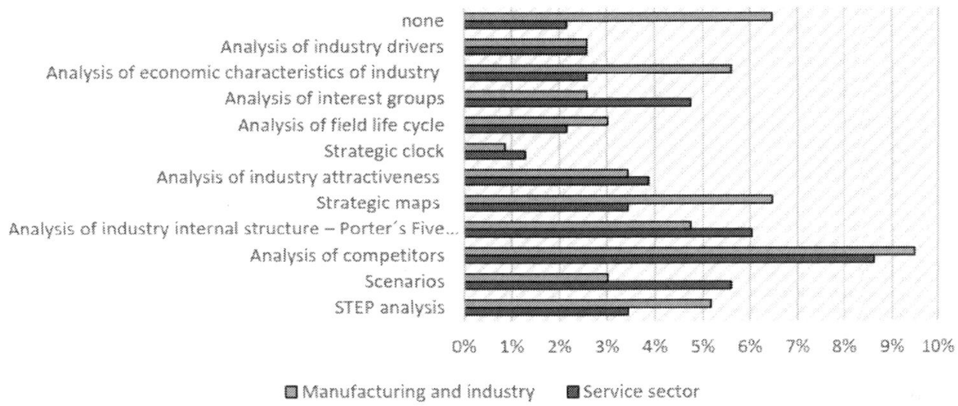

Figure 8. Use of external strategic analyses by managers of large companies

Source: Authors.

Figures 6 – 7 show the predominance of using "Analysis of competitors". The owners of micro- and small companies appear to have deeper knowledge of external strategic methods than the owners of manufacturing companies.

In the case of medium-sized and large companies, the knowledge and use of "Analysis of competitors", "Strategic maps", and "Analysis of industry internal structure" prevail. Nevertheless, even with regards to these size categories, it can be stated that the results achieved are worrying from the perspective of the managerial competencies of top level managers. Even in the case of external methods, there is an evident declining trend in the importance of strategic situation analysis in business management and the creation of corporate strategy.

In conclusion, for all the aforementioned outputs it can be stated that, in accordance with the research conducted, the methods of strategic situational analysis in the corporate sector are gradually losing importance in terms of the mezzo- and macro-environments. This confirmed the assumption of the need to find a new tool for generating corporate strategy that includes both changes in a company´s internal environment and in its immediate surroundings. It is the shared opinion of the authors of this contribution that one of the most important new findings is the identified importance of the value chain for the profitability of companies. The importance of this statement is amplified by the high degree of economic growth at the time the research survey was conducted. This leads to the discussion in terms of the analysis and evaluation of value flows in a company as a basis for its strategic development, competitive advantage, or stability, that is, a generator of a successful business strategy.

4 Conclusion

From the outputs of the research concerning managerial knowledge and competencies in the area of strategic management and decision-making, the following conclusions can be drawn:

- In terms of strategic situation analysis, the critical state of the managerial competencies and knowledge of Czech managers was proved. More than 50% of managers have only a minimum or zero knowledge of the issue.
- The survey showed that the knowledge of Czech managers concerning strategic methods for the analysis of the internal and external environment are greatly lacking and that the task of academia is to solve this crisis by accelerating the transfer of new findings to the corporate sector.
- In the case of internal strategic analyses, a strong dependence was identified between business strategy and the "Analysis of value chain" according to sector differentiation, which confirms the importance of this method in the creation of business strategy.
- In the case of external analyses, a high degree of dependence was identified between business strategy and two methods, namely "Analysis of industry drivers" and "Analysis of interest groups" according to company size differentiation. Surprisingly, the dependence of business strategy on the "Analysis of industry internal structure", which in combination with "Analysis of value chain" represents an important analytical tool for creating successful business strategy, was not confirmed.
- In terms of the knowledge and use of internal strategic analyses, the world´s best known and most commonly used SWOT analysis prevails in SMEs.
- In the case of knowledge and the use of external strategic analyses, for all size categories, the "Analysis of competitors" prevails; there are slight differences for the "Analysis of industry internal structure", "Strategic maps", and "Analysis of industry attractiveness".
- In the case of large and (partially) medium-sized companies, a declining trend in the use of strategic situation analysis in the creation of business strategy can be observed.

Despite the partially positive results for large companies, the situation can be considered as an appeal to both the corporate sector and research platforms to change this situation. On the side of companies, there needs to be greater motivation to implement new findings in the field of strategic management and decision-making in corporate practice. However, this must correspond with the offers, or transfer of newly proposed and verified methods and tools for business strategy from the side of research organizations, including universities. A positive change in the given situation may be the fact that more and more requirements and demands are being made, in particular by small and medium-sized companies, for new, innovative forms of management and tools that correspond to the ever-intensifying competitive environment in the local, regional, Czech and international markets.

Despite the wide range of companies in the sample, which included two important sectors (manufacturing industry and services), the authors of this contribution are aware of the limitations of the outputs obtained. The main limitation is considered to be the fact that the research was conducted in a period of strong economic growth, and many of the managers´ views were affected by this fact. The research will therefore be repeated as the economic cycle changes. Another important factor the authors are aware of is the identified significant difference between the professional competence of younger managers compared to the managers of the older generation, who have a high degree of professional, practical skills and experience, but at the same time are less flexible, creative, and more sensitive to changes in the external environment.

References

1. M. Miyamoto, Application of competitive forces in the business intelligence of Japanese SMEs. International *Journal of Management Science and Engineering Management*. **10**(4), 273–287 (2015)
2. H. Ryu, J. Lee, B. Choi, Alignment Between Service Innovation Strategy and Business Strategy and Its Effect on Firm Performance: An Empirical Investigation. *IEEE Transactions on Engineering Management*. **62**(1), 100–113 (2015)
3. R. Kaplinsky, Globalisation and Unequalisation: What Can Be Learned from Value Chain Analysis? *The Journal of Development Studies*. **37**(2), 117–146 (2000)
4. W. Williams, D. Lewis, Strategic management tools and public sector management - The challenge of context specificity. *Public management review*. **10**(5), 653–671 (2008)
5. J. Magretta, *Michael Porter jasně a srozumitelně* [*Michael Porter clearly and intelligibly*]. Praha: Management Press (2012)
6. S. Clegg, Ch. Carter, M. Kornberger, J. Schweitzer, *Strategy: Theory and Practice*. London: SAGE Publications Ltd. (2011)
7. B. D. Keillor, *Understanding the Global Market: Navigating the International Business Environment*. Santa Barbara: ABC–CLIO, LLC. (2013)

8. P. Knott, A typology of strategy tool applications. *Management Decision.* **44**(8), 1090–1105. (2006)
9. F. A. Frost, The use of strategic tools by small and medium-sized companies: an Australasian study. *Strategic Change.* **12**(1), 49–62 (2003)
10. J. Straková, P. Pártlová, J. Dobrovič and J. Váchal. Situational analysis and its role in the process of strategic business management. *Polish Journal of Management Studies.* **18**(1), 353–364 (2018)
11. E. Kislingerová, I. Nový et al., *Chování podniku v globalizujícím se prostředí* [*Business behavior in a globalizing environment*]. Praha: C.H. Beck (2005)
12. T. Mallya, *Základy strategického řízení a rozhodování* [*Basics of strategic management and decision making*]. Praha: Grada (2007)
13. Z. Virglerová, L. Homolka, L. Smrčka, K. Lazányi, T. Kliestik, Key determinants of the quality of business environment of SMEs in the Czech Republic. *E+M Ekonomie a Management.* **20**(2), 87–101 (2017)
14. M. Keřkovský, O. Vykypěl, *Strategické řízení. Teorie pro praxi* [*Strategic management. Theory for practice*]. Praha: C. H. Beck (2006)
15. R. Murray-Webster and Office of Government Commerce. *Management of risk: guidance for practitioners*. London: TSO (2010)
16. J. Jakubíková, *Strategický management* [*Strategic management*]. Praha: Grada (2008)
17. M. Dedouchová, *Strategie podniku* [*Company strategy*]. Praha: C. H. Beck (2001)
18. M. Šulák, E. Vacík, *Strategické řízení v podnicích a projektech* [*Strategic management in companies and projects*]. Praha: VŠFS (2005)
19. J. G. Frynas, K. Mellahi. *Global strategic management*. New York: Oxford University Press Inc. (2011)
20. J. Veber et al. *Management: základy, prosperita, globalizace* [Management: basics, prosperity, globalization]. Praha: Management Press (2004)
21. E. Gomes, *Mergers, acquisitions, and strategic alliances: understanding the process*. New York: Palgrave Macmillan (2011)
22. E. Nigel, D. Campbell, *Strategic management for travel and tourism*. Oxford: Butterworth-Heinemann (2002)
23. P. M. Ginter, W. J. Duncan, L. E. Swayne, *Strategic management of health care organizations*. 7th ed. San Francisco, CA: Jossey-Bass (2013)
24. P. Wetherly, D. Otter, *The Business Environment: Themes and Issues in a Globalizing World*. Oxford: Oxford University Press (2014)
25. G. B. Dagnino, D. R. King, J. Tienari, Strategic management of dynamic growth. *Long Range Planning.* **50**(4), 427–430 (2017)
26. E. Gomes, Y. Weber, Ch. Brown, S. Y. Tarba. *Mergers, acquisitions, and strategic alliances: understanding the process.* London: Palgrave Macmillan (2011)
27. H. M. Abdulhussien, S. Hamza. Strategic Management Accounting Techniques in Romanian Companies: An Empirical Study. *Studies in Business and Economics.* **7**(2), 126–140 (2012)
28. P. Williams and E. Naumann. Customer satisfaction and business performance: a firm-level analysis. *Journal of Services Marketing.* **25**(1), 20–32 (2011)
29. J. Straková. Strategic analysis methods and their influence on stability and development of small and medium-sized companies in the Czech Republic. *Studia Commercialia Bratislavensia.* **10**(38), 196–214 (2017)
30. J. Straková, P. Pártlová, J. Váchal, J. Dobrovič and I. Rajiani, Use of the Value Chain in the Process of Generating a Sustainable Business Strategy on the Example of Manufacturing and Industrial Companies in the Czech Republic. *Sustainability.* **12**(4), (2020)

AUTHOR INDEX